Heinrich Walter

Vegetation of the Earth
and Ecological Systems
of the Geo-biosphere

Third, Revised and Enlarged Edition

Translated from the Fifth, Revised German Edition
by Owen Muise

With 161 Figures

Springer-Verlag
Berlin Heidelberg New York Tokyo

Professor Dr. Heinrich Walter
Egilolfstraße 33
70559 Stuttgart 70, FRG

Translated from the German edition „Vegetation und Klimazonen",
published by Eugen Ulmer, Stuttgart, 1984.

First edition   1973
2nd printing   1975
3rd printing   1977
4th printing   1978
Second edition   1979
2nd printing   1983
Third edition   1985
2nd printing   1994

ISBN 3-540-13748-3 Springer-Verlag Berlin Heidelberg New York Tokyo
ISBN 0-387-13748-3 Springer-Verlag New York Heidelberg Berlin Tokyo

Library of Congress cataloging in Publication Data. Walter, Heinrich, 1898-. Vegetation of the earth and
ecological systems of the geo-biosphere. (Heidelberg science library). Translation of: Vegetationszonen
und Klima. Bibliography: p. Includes index. 1. Botany – Ecology. 2. Vegetation and climate. 3. Phytogeo-
graphy. 4. Plant physiology. 5. Life zones. I. Title. II. Series. QK901.W2613. 1985. 581.5'42. 84-14143.

Media conversion: fotosatz & design, Berchtesgaden.
Printing and bookbinding: Beltz Offsetdruck, Hemsbach/Bergstr.

2131/3130-54321 – Printed on acid-free paper

Foreword to the First English Edition

Ecology is current, exciting, relevant, and offers guides to action and even some hope of harmony and order, as well as a congenial environment for mankind in an overpopulating world. Plant ecology is basic to general, animal systems, paleo-, and human ecology. Plants are the primary producers. They dominate the flow and cycling of energy, water, and mineral nutrients within ecosystems. The structure of the vegetation determines much of the character of the landscapes in which other organisms live and prosper, including men and women.

Plants are immediately at hand for study. They are evident, mobile only in certain stages, familiar, easily identified, and related to a rich literature on the properties of various kinds of plants. If we know why plants grow where they do, we know a good deal about why organisms other than plants live where *they* do.

Plant ecologists need a general botanical background. Professor Walter has that background. His multivolume textbook series, "Introduction to Plant Science," includes books on general botany, systematics, ecology in a strict sense (2nd ed., 1960), plant geography (2nd ed., with H. Straka, 1970), and on vegetation. Ellenberg covers the last topic with books on principles of vegetation organization (1956) and the vegetation of central Europe (1963). These texts are in German, published by Ulmer (Stuttgart), and the reader can be referred to reviews of these (Ecology 38(4):666–68, 1957; 43(2):346, 1962; 47(1):167–68, 1966; and J. Ecology 55(1):234–45, 1967).

It is not enough for the plant ecologist to be well-grounded in principles. Our so-called ecological principles need continual reexamination, questioning, and testing. Principles must be drawn from, and applied to the specific ecological relationships of plants in particular ecosystems. Concrete ecosystems are the testing ground for principles as well as their source. So the ecologist needs an overview of the earth's plants and its vegetation. What are the possibilities and actualities of plant growth and vegetation organization?

Professor Walter's two volumes in German entitled "Vegetation of the Earth Considered Ecophysiologically" (1964 and 1968) are an admirable summary of much of this basic plant ecology. They are separated by a language barrier from many persons who need to use them. They have been justly praised by A. Löve (Ecology 50(6):1105–6, 1969) and by Grubb (J. Ecology 58(1):315–16, 1970.[1] They record Walter's very extensive firsthand knowledge of much of the earth's vegetation. They add the progress made in our ecophys-

---

1 English edition of Vol. 1 cf. J. Ecology 60(3):940–41 (1972)

iological understanding of how and why plants grow where they do throughout the world since Schimper's founding work, "Plant Geography on a Physiological Basis" (1898, 1903 in English, 2nd German edition by von Faber in 1935). Walter's books are indispensable to every practicing plant ecologist, as Löve says. Their 1593 pages are a rich feast. They also stand apart, separated by their awesome scholarship, bulk, cost, and rich detail. Students who lack a good geographical schooling or have only a rudimenary plant taxonomic background may find them difficult to digest. Further, Walter's books are traditionally scientific, international in intent and scope, but this tradition contrasts with a widespread general retreat into parochialism which sometimes seems to be a corollary of the potentially wide intellectual scope of ecology. Or perhaps we all just have too much to do.

In 1970, Professor Walter published a small volume on the ecology of zonal kinds of vegetation, in relation to climate, viewed causally, and covering all the continents. The English edition of that book you have in your hand. The German edition was reviewed in Ecology (52(5):949, 1971). The book is short but replete with facts. It places these facts in a consistent frame of reference. It suggests where more factual data are needed. It is also neat, precise, very readable, an excellent summary of two larger books, and current.

In recent years ecophysiology has attracted much interest from very skilled botanists. Technical progress in instrumentation has made possible some accurate measurements of photosynthesis and transpiration in the field, rapid chemical analyses of soils and plants, the sensing of elements of the heat balances of organisms, measurements of water potential in both soils and plants, computer modeling of the photosynthetic and respiratory processes in stands of vegetation, summarization of masses of data which would have formerly overwhelmed the investigator, etc. No synthesis of the principles of ecophysiology has appeared. But Walter's short book does something else that is very valuable. It provides a framework of ecological descriptions of kinds of zonal vegetation into which ecophysiological data must fit.

Plant ecophysiology can be used variously. Plant success, adaptations to environment whether ecotypic or plastic, should be definable in ecophysiological terms. For others the goal may be understanding the physiology of vegetation, those organizations of individual plants which many botanists ignore. Progress has been made in studying the physiology of crop stands. Walter records in this book new data on biomasses which summarize the results of the physiological processes in various kinds of vegetation and express their structure.

Under the International Biological Program many kinds of ecosystems are being studied. In the U.S. these include grasslands, deserts, deciduous, coniferous, and tropical forests, and tundra. A project on "Origin and Structure of Ecosystems" says, "The fundamental biological question that this program is asking is whether two very similar physical environments acting on phylogenetically dissimilar organisms in different parts of the world will produce structurally and functionally similar ecosystems. If the answer is no, there cannot be any predictive science of ecology. In fact, knowledge acquired from studying a given ecosystem cannot be applied to an analogous ecosystem, unless similar physical environment indeed means similar ecosystem" (U.S. National Committee for the International Biological Program. National Academy of Sciences, Washington, D.C., Report 4:46). The answer to the question is obviously no in general, but the conclusion suggested does not follow. True, there are many similarities between the plants and vegetations in different parts of the world that have similar climates, soil parent materials, topographies, fire histories, and plant successional and soil developmental histories even though their biotas differ, and Walter mentions examples repeatedly. But he also mentions exceptions. Ecology is not a simple matter of adaptations to environment. It must consider that biotic diversity over the earth is an ecological factor, and Walter outlines some of the classical conclusions on floristic diversity in the first pages of his text. The evidence is already clear that the partial effects of differing genetic inputs into environmentally closely similar ecosystems persist ecologically in the form of specific structures and functions. The C-4 photosynthetic pathway is a splendid, so far almost ecologically meaningless, example.

Further, structure and function should be considered separately. They have no necessary and invariant connection. Premature correlation, such as Schimper's bog xeromorphosis, is a mistake. Walter's generalization of "peinomorphosis" represents real progress. While the apparent congruences of structure and function are striking, study of them in detail and the evident exceptions have led to solution of many ecological puzzles.

Approaching the problem from another direction, ecological principles are applied widely in such kinds of land management as agriculture, forestry, range and pasture management, pollution control, park and wilderness management, and natural area maintenance. Ecological principles evidently exist and do allow predictions. Walter's little book is a good corrective to the kind of hurried, mistaken generalization that was quoted.

Of course evolution is a dimension of ecology. Unfortunately the concept of adaptation is only used by evolutionists and not often measured. In fact, there is often a shifting from form to function and vice versa in discussions of adaptation, a substitution of two unknowns for one. Walter does not make this mistake either. He does make clear how much more research needs to be done to describe plant and vegetation structure and function, their correlations, and their relationships to environment. And his book records a fine start.

Professor Walter is a plant ecologist of very wide and long experience. He and his wife are knowledgeable, enthusiastic, indefatigible, and helpful field companions. His investigations, and those of his students, have suggested, checked, and documented his ecological ideas derived from extensive travel and residence in all the continents. His teaching has clarified their presentation.

Many have questioned whether plant ecology in its century of development since Haeckel coined the parent term in 1866 has developed any principles. This book provides an affirmative answer.

University of California, Davis   JACK MAJOR
February 8, 1972

Preface to the Third English Edition

When this paperback book was first published over a decade ago, it was a brief summary of the author's two-volume work "Vegetation of the Earth". Further evaluation of the results of ecological research from the past more than 60 years has increasingly led to the tendency to place ecological considerations in the foreground. Relevant recent publications which may be referred to are: Walter: "The Vegetation of Eastern Europe, North and Central Asia", 452 pp., Stuttgart, 1974 (an evaluation of the voluminous Russian literature); Walter, Harnickell, Mueller-Dombois: "Climate Diagram Maps of the Individual Continents and the Ecological Climate Systems of the Earth", Stuttgart, 1975; and Walter: "The Ecological Systems of the Continents – Geo-biosphere", 132 pp., Stuttgart, 1976 (principles of the systems with examples).

The increasing importance of ecological principles is reflected noticeably in the subsequent German editions of this book in the years 1972, 1975, and 1979.

In the meantime, this book has been translated into five languages. A Portuguese translation is in preparation in Brasil, and the Iranian translation was terminated by the revolution there. The English translation was published in 1973 by Springer-Verlag (New York) with the title "Vegetation of the Earth" and was reprinted in 1975, 1977 and 1978. In 1979, the second English edition, based on the third German edition, was published.

This translation of the fifth German edition is likely to be the last. This edition has been revised by the 85-year-old author, and is, in its basic contents, a condensed version of the new three-volume work "Ecology of the Earth" by H. Walter and W.-W. Breckle, of which the first and second volumes have already been published (Stuttgart, 1983), and the third is in preparation. This relationship is indicated by the subtitle of the 5th edition.

In contrast to the technically orientated, analytical biological research prevailing today, ecology strives for a synthesis; i. e., the illustration of interrelationships. The results obtained in analytical research are the individual building blocks from which, figuratively speaking, a construction or mosaic may be formed. For this reason, a great degree of knowledge is required in ecology. The best specialized knowledge in a limited discipline is insufficient. It has, therefore, been a goal of mine in the course of this long lifetime, not to pursue one single discipline in detail, but rather to gather experience and scientific knowledge on research expeditions in all regions of the earth with my life companion and co-worker, Dr. Erna Walter, where we personally investigated each floral kingdom and climatic zone, in order to obtain comparative material on a global scale (see Figure). Ecology cannot be learned and understood from

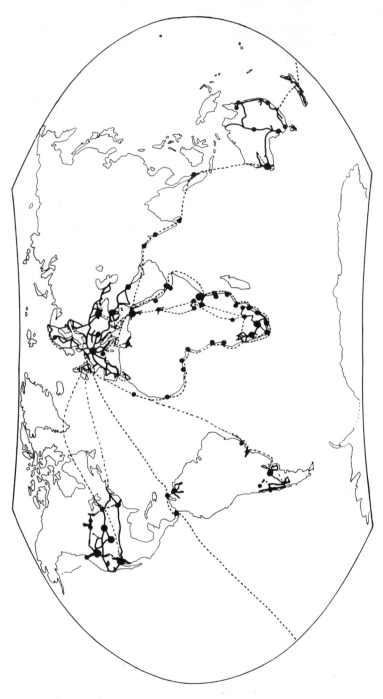

The author's research journeys, —— by car or train, ----- by sea or air; large dots indicate longer visits to research institutes. The Russian literature has been consulted for information concerning the inaccesible regions of Asia

books or in the library, but rather according to the following principle: "The ecologist's laboratory is God's nature and the whole earth is his field."

The following must be said for this 5th edition: Besides the usual additions and revisions, the supplement of the 4th edition has been included in the text. The entire section on Zonobiome II has been rewritten due to recent publications on the tropical savanna and park regions. Of special interest is the brief description by B. Frenzel on the substanial extent of the thermokarst formation in eastern Siberia, with which he became acquainted on an excursion to Yakutsk in the summer of 1982. He has kindly allowed me to include his report and the respective descriptions in this edition (see pp. 269–273). The "Conclusion" is also new, and is to be considered a type of personal testament for young ecologists. Its contents are the result of my lifelong experiences and have been published in more detail in "Testimony of an Ecologist" (Walter 1985, 4th ed.).

I would like to thank the publishers and Dr. M. Rahmann for their help in preparing the book for printing and for proofreading.

Plant names which appear in Latin in the text may be found in English in the Subject Index.

University of Stuttgart-Hohenheim   HEINRICH WALTER
Easter, 1985

Contents

Introduction:
The Classification of Ecological Systems

1. The Aims of Ecology

Ecology is a biological science of highest dignity and, just as life itself (as far as we currently know), it is limited to our own planet. Life, on the whole, is bound in a cycle—synthesis, based on the binding of solar energy, and decomposition, based on the disintegration of this bound energy. The smallest autonomous unit of life is the cell, the structure and function of which are investigated by the sciences of molecular biology and biochemistry. The single-celled microorganisms are primarily objects of the science of microbiology. The unit of life at the next-highest level of organization is the organism, with its multicellular tissues and organs. We distinguish between plant and animal organisms, which are morphologically, anatomically and functionally of quite different natures. The first group is the subject of the science of phytology (botany), and the latter is assigned to zoology. The green plants are autotrophic and anabolic, while the colorless and the animal organisms are heterotrophic and catabolic. The highest units of life are the communities of plant and animal organisms which, together with abiotic environmental factors (climate and soil), form ecosystems. These are characterized by a continous cycling of material and flow of energy. Ecology in the widest sense of the word is the science of these ecosystems, from the very smallest to the global level (the "biosphere"). This book is intendend to serve as a brief comprehensive introduction to this global ecological system.

The biosphere comprises the natural world in which man has been placed, and which, thanks to his mental capacities, he is able to regard objectively, thus raising himself above it. On the one hand, he is a child of this outer, apparent world, and dependent upon nature, but on the other hand, through the world within, he has access to the godly.

Only an awareness of both sides of his nature enables man to develop into a wise and harmonious being with the hope of divine fulfilment upon death.

It is not the sole calling of man to use nature to his own ends. He also bears the responsibility for maintaining the earth's ecological equilibrium, of tending and preserving it to the best of his ability.

If he is to do this and to avoid exploiting the environment in a way which in the long run jeopardizes his own existence, he has to recognize the laws of nature and act upon them.

We shall limit our observations to the conditions in natural ecosystems, since it would be beyond the scope of this book to embark upon a consideration of secondary, man-made ecosystems.

2. Classification of the Geo-biosphere into Zonobiomes

The biosphere is the thin layer of the earth's surface to which the phenomena connected with living matter are confined. On land, this comprises the lowest layer of the atmosphere permanently inhabited by living organisms and into which plants extend, as well as the root-containing portion of the lithosphere, which we term the soil. Living organisms are also found in all bodies of water, to the very depths of the oceans. In a watery medium, however, cycling of material is achieved by means other than those on land, and the organisms (plankton) are so different that aquatic ecosystems have to be dealt with separately. The biosphere is therefore subdivided into (a) the geo-biosphere comprising terrestrial ecosystems, and (b) the hydro-biosphere, comprising aquatic ecosystems, which is the field of hydrobiologists (oceanographers and limnologists).

Our studies are confined to the geo-biosphere (Walter 1976), which constitutes the habitat of man and is, therefore, of special interest. The prevailing climate, being the primary independent factor in the environment, can be used as a basis for further subdivision of the geo-biosphere since the formation of soil and type of vegetation are dependent upon it (see p. 3), and it has not yet been substantially influenced by man. Furthermore, it may be measured exactly and in all locations with the increasingly dense network of meteorological stations (for the principles of classification, see Walter 1976).

Climate is determined by planetary air currents in the atmosphere, and meteorologists distinguish seven genetical climate belts: (1) the equatorial rain zone, (2) the summer-rain zone on the margins of the tropics, (3) the subtropical dry regions, (4) the subtropical winter-rain region, (5) the temperate zone with year-round precipitation, (6) the subpolar zone, and (7) the polar zone.

It is mainly climate within the geo-biosphere that interests the ecologist, and this can be depicted by means of ecological climate diagrams (see p. 22). A further subdivision of the very large temperate zone of the meteorologists has proved to be useful, whereas the subpolar and polar zones have been combined to give a single arctic zone. This leaves us with nine climate zones, ecologically designated as *zonobiomes* (ZB), a biome being a large and climatically uniform environment within the geo-biosphere. The term "humid" is used to describe a climate with much rainfall, and "arid" to describe a dry climate with little rainfall. Where both terms are used, the first refers to summer and the second to winter conditions. The nine zonobiomes are as follows:

ZB I   Equatorial with diurnal climate, humid
ZB II   Tropical with summer rains, humido-arid
ZB III   Subtropical-arid (desert climate), arid
ZB IV   Winter rain and summer drought, arido-humid
ZB V   Warm-temperate (maritime), humid
ZB VI   Typical temperate with a short period of frost (nemoral)
ZB VII   Arid-temperate with a cold winter (continental)
ZB VIII   Cold-temperate (boreal)
ZB IX   Arctic (including antarctic), polar.

Each zonobiome is clearly defined by a particular type of climate diagram (p. 23), although, with a few exceptions, the zonobiomes largely correspond to soil type and zonal vegetation. This is shown by the following survey:

ZB	Zonal soil type	Zonal vegetation
I	Equatorial brown clays (ferralitic soils, latosols)	Evergreen tropical rain forest
II	Red clays or red earths (savanna soils)	Tropical deciduous forests or savannas
III	Sierozems	Subtropical desert vegetation
IV	Mediterranean brown earths	Sclerophyllous woody plants
V	Yellow or red podsdol soils	Temperate evergreen forests
VI	Forest brown earths and gray forest soils	Nemoral broadleaf-deciduous forests (bare in winter)
VII	Chernozems to sierozems	Steppe to desert with cold winters
VIII	Podzols (raw humus–bleached earths)	Boreal coniferous forests (taiga)
IX	Tundra humus soils with solifluction	Tundra vegetation (treeless)

Although the zonobiomes take this order on both sides of the equator, a certain degree of asymmetry results from the fact that in the Southern Hemisphere there is less land and the climate is more maritime and cooler. The temperature equator is about 10° to the north of the geographical equator.

Zonobiomes VI and VII are poorly developed in the Southern Hemisphere, ZB VIII is completely absent, and ZB IX is represented solely by the subantarctic islands and the southernmost tip of South America, if we ignore the icy Antarctic (which is almost entirely devoid of vegetation).

The larger zonobiomes are further subdivided into subzonobiomes (sZB) on the basis of climatic deviations. This will be indicated in each case.

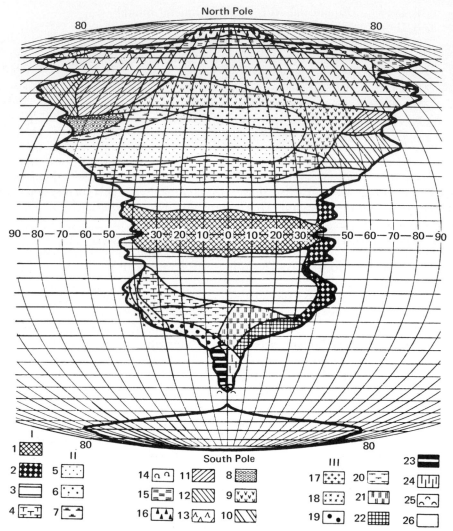

Fig. 1. "Average continent," showing the asymmetry in vegetational zones in the Northern and Southern Hemispheres. I. *Tropical zones:* (1) Equatorial rain forest; (2) tropical rain forest with trade wind, orographic rain; (3) tropical-deciduous forest (and moist savannas); (4) tropical thornbush (and dry savannas). II. *Extratropical zones of the Northern Hemisphere:* (5) hot desert; (6) cold inland desert; (7) semidesert or steppe; (8) sclerophyllous woodland with winter rain; (9) steppe with cold winters; (10) warmtemperate forest; (11) deciduous forest; (12) oceanic forest; (13) boreal coniferous forest; (14) subarctic birch forest; (15) tundra; (16) cold desert. III. *Extratropical zones of the Southern Hemisphere:* (17) coastal desert; (18) fog desert; (19) sclerophyllous woodland with winter rain; (20) semidesert; (21) subtropical grassland; (22) warm-temperate rain forest; (23) cold-temperate forest; (24) semidesert with cushion plants, or steppes; (25) subantarctic tussock grassland; (26) inland ice of the Antarctic. (Modified from C. Troll; Walter 1964/68)

3. Zonoecotones

Climate zones and hence the zonobiomes are not sharply defined but are linked by broad transitional zones known as *zonoecotones* (ZE). An ecotone is an area of ecological tension over which one type of vegetation is gradually replaced by another, e.g., deciduous forests by steppes. In zonoecotones, both types of vegetation occur side by side under the same general climatic conditions and are in a state of keen competition. Which of the two types of vegetation is successful depends upon the microclimatic conditions resulting from local relief or soil texture, so that there is a diffuse mixture of the two kinds of vegetation or a mosaiclike pattern of the two. In crossing a zonoecotone, at first one kind of vegetation is better represented, then the two are more or less equally successful, and finally, the second type begins to take over, with the first becoming more sparse. When the latter eventually disappears, the next zonobiome has been reached.

Zonoecotones are designated according to the zonobiomes they link, i.e., ZE I/II, ZE II/III, and so on. Three-cornered zonoecotones are also found at the convergence of three zonobiomes.

The geographic distribution of the individual zonobiomes and zonoecotones is illustrated in the world map at the end of the book.

4. Orobiomes

Owing to the presence of mountains, the geo-biosphere can be divided vertically as well as horizontally, and thus has to be considered threedimensionally. Mountains differ climatically from the climate of the zone from which they rise and must be considered separately. Such mountainous environments are termed *orobiomes* (OB) and can be vertically subdivided into altitudinal belts. A characteristic of all orobiomes is that the mean annual temperature decreases with altitude. In the Euro–north Asiatic region, for each additional 100 m in altitude, the temperature drop is approximately the same as the mean annual decrease in temperature recorded over a distance of 100 km from south to north. This is why mountainous altitudinal belts are about 100 times narrower than the vegetational zones on the plains. It would nevertheless be wrong to assume that altitudinal belts are merely small-scale repetitions of these vegetational zones. In Europe and North America, certain similarities immediately catch the eye, but differences are invariably present. Apart from the drop in temperature, which means a reduction in the vegetational period with increasing altitude, the mountainous climate is different from that of the plains. For example, with increasing altitude, the length of day and the position of the sun do not change, while the length of day increases from north to south in summer, and the sun's position at noontime becomes lower. Direct irradiation increases with altitude, and diffuse irradiation decreases; whereas in moving north on the plains daylength changes, and the change in radiation is exactly the opposite. Precipitation often increases in the mountains very rapidly, with increasing altitude whereas in the arctic regions precipitation is lower.

Every mountain within a zonobiome is an ecological entity with a typical sequence of altitudinal belts, generally termed colline, montane, alpine, and nival, but exhibiting considerable differences according to the zone in which they occur. For example, the successions of altitudinal belts in the mountains of Zonobiomes I, IV, and VI have little in common. For this reason, the orobiomes are distinguished according to the zonobiome to which they belong, e.g., Orobiome I, Orobiome II, and so on.

In addition, a distinction is made between uni- , inter- , and multizonal orobiomes (mountains), depending upon whether they fall within one zonobiome, form a boundary between two zonobiomes, or extend over two or more zonobiomes, as is the case with the Urals (IX–VII) and the Andes (from I–IX). The Alps, the Caucasus, and the Himalayas are examples of interzonal mountain ranges. These mountains usually form sharp climatic boundaries, and the altitudinal belts on the northern slopes differ from those on the southern side. In dealing with multizonal mountains, it is advisable to consider the constituent zones individually, each with its own particular sequence of altitudinal belts. The Andes are multizonal as well as interzonal (western and eastern slopes differ). The sequence of altitudinal belts may also be different in the innermost parts of the mountain valleys, with little precipitation and continental conditions (intramountainous sequence of altitudinal belts).

5. Pedobiomes

Exceptional situations within a zonobiome are also presented by areas with extreme types of soil and azonal vegetation. These will be termed pedobiomes, i.e., environments associated with a certain type of soil. Soils have been altered significantly by man only in regions where soil erosion, i.e., the loss of the upper levels or all of the soil has occurred, or where the soil has been cultivated or built upon.

The unmitigated effect of the overall climate is seen only in *euclimatopes* (Russian "plakor"), which are flat areas where the soil is neither too light nor too heavy, so that precipitation does not run off but penetrates the soil and is retained as interstitial water. Since the water does not seep into the groundwater quickly, it is at the full disposal of the vegetation. This is not the case with extreme chalky soils, however, and they constitute biotopes (see p. 33) that are warmer and drier than the general climate. Again, the soil may contain harmful substances such as salts ($NaCl$, Na_2SO_4) or may be extremely poor in nutrients so that the vegetation differs from that typical of the zonobiome. The vegetation of the pedobiomes is influenced to a greater extent by the soil than by the climate, and thus the same vegetational forms may occur on similar in a number of zones. This vegetation is termed *azonal vegetation*. Pedobiomes are designated according to soil type: lithobiomes (stony soil), psammobiomes (sandy soils), halobiomes (salty soils), helobiomes (moor or swamp soils), hydrobiomes (soils covered with water), peinobiomes (soils poor of deficient in nutrients; from peina, Greek = hunger, lack), and amphibiomes (temporarily wet soils), etc.

Pedobiomes often occupy vast areas, as for example the lithobiome of the basalt-covered areas of Idaho (United States), the psammobiome of the southern Namib, of the Karakum desert in Central Asia (350,000 km^2), the helobiome of the Sudd swamp region of the Nile (150,000 km^2), the bog region of western Siberia (over 1 million km^2; Fig. 154), and so on, although in other cases they may cover only small areas. The ecology of these areas has to be given special consideration when dealing with the relevant zonobiome.

6. Biomes

The word biome on its own (without a prefix) is used for the fundamental unit of which larger ecological systems are made up. A biome as an environment is a uniform area belonging to a zonobiome, orobiome, or pedobiome. The central European deciduous forest, for example, is a biome of ZB VI, Kilimanjaro is a biome of Orobiome I, the salt desert of Utah (United States) is a biome of the pedohalobiome of ZB VII, and so on. In the following global survey, it is in general not possible to consider ecological units smaller than biomes.

7. The Nature and Structure of Ecosystems

Before going on to deal with smaller ecological units, we shall take a closer look at a small-scale ecosystem, using a homogeneous deciduous forest stand of ZB VI as an example.

Continuous energy flow and cycling of nutrients take place in a plant community of this kind. The plants, together with the animal organisms and the inorganic environment, form an *ecosystem*. This ecosystem, however, *is not a closed system,* since there is an inflow of external energy from solar radiation and of matter in the form of precipitation or from gaseous exchange, dust deposits, and so on. At the same time, energy is lost in the form of heat, and matter is lost as a result of gaseous exchange or loss in drainage water. If the ecosystem comprises a definite, limited, and homogeneous community, such as a forest stand or a moor, then it is a *biogeocenosis,* which will be abbreviated as *biogeocene*. This is an entity consisting of the plants and animals, the soil (permeated by roots), and the air layer into which the plants extend. The total dry plant matter of a biogeocene is its *phytomass* and that of its dry animal matter its *zoomass*. The two together form the *biomass*. According to their role in the biogeocene, three groups of organisms may be distinguished:

1. *Producers* autotrophic, or green plants, which, in the process called photosynthesis, are able to store the energy of sunlight as chemical energy by building up organic compounds from CO_2 and H_2O. In the process, however, they extract minerals and water from the soil.

8 Introduction

2. *Consumers*–herbivorous animal organisms which (as phytophages) use plants as food and convert a small part of this matter into animal substance and predators which eat the herbivores.
3. *Decomposers*–to a large extent present in soil (saprophages, bacteria, fungi), they break down or mineralize plant and animal remains into CO_2 and H_2O, thus completing the cycling of nutrients.

In an ecosystem, the organic substance produced annually by photosynthesis is termed *gross production*. What remains after subtraction of the amount respired by plants is known as *net*, or *primary*, *production*. The substance produced by animal organisms is termed *secondary production* and is much smaller than primary production. Only a few percent of the primary production is eaten by consumers, whereas the larger portion enters the soil and is completely broken down by the decomposers with the formation of CO_2, H_2O, and mineral salts. The dead organic mass (litter) has been broken down prior to this by *saprophages* (lower animal forms). Loss of CO_2 from the soil is known as *soil respiration*. This is part of the *short cycle* which plays the major quantitative role in nutrient recycling (Fig. 2). In addition, there is a *long cycle,* involving consumers such as herbivores, phytophages, carnivores and other predatory organisms, as well as omnivores (animals which consume both plant and animal matter). Plant parasites can also be included among the consumers. Excrement and dead animal remains are returned to the soil, where animal organisms (coprophages and necrophages) prepare them for further breakdown by microorganisms.

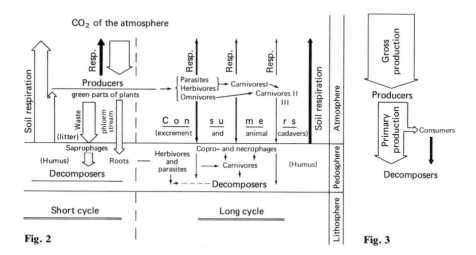

Fig. 2. Schematic representation of the short and long cycles in a deciduous-forest biogeocene, the material cycle

Fig. 3. Energy flow (thickness of arrows indicates approximate relative quantitative contribution)

The long cycle, although of little quantitative importance, plays a very considerable role in regulating the equilibrium of the ecosystem as a whole. The consumers can thus be regarded as *regulators*. As soon as a particular plant species in an ecosystem multiplies excessively, the number of animals feeding upon it also increases, which in turn lowers the plant population, with a resultant reduction in the numbers of the phytophagic species. A situation of this kind, as exists, for example, between phytophages and the zoophages which prey upon them, is described as a *cyclic oscillation in population density*. Using cybernetic terms, it can be said that the ecosystem is kept in a steady state by means of *regulatory circuits with feedback*. Although population densities fluctuate slightly, they remain within certain limits. Such fluctuations can also be elicited by the changing weather conditions from year to year, whereby sometimes one plant species is favored and sometimes another.

Overlapping regulatory circuits form *nutritional chains*. The long cycle is made up of a succession of such chains, which ensure that, despite fluctuations, the ecosystem will have a high degree of average stability. Extermination of predators or employment of other far-reaching measures disrupts the nutritional chain, so that the entire ecosystem is thrown out of balance and may even collapse (Gigon 1974).

Since phytophages and predators are often strictly specialized (i.e., feed on one particular species), one of the most important tasks ahead of zooecologists is the elucidation of the various nutritional chains down to the smallest detail.

Cycling of material is paralleled by the *energy flow*. Photosynthesis (carried out by producers) transforms solar energy into chemical energy, which is then used by consumers and decomposers. The chemical energy is continually being lost as heat in respiration or the fermentation processes of microorganisms, until it is entirely exhausted. The energy flow is depicted in Fig. 3 (right-hand side).

The following values are given by Duvigneaud (1974) for an oak forest in the Belgian mountains with a hazel shrub layer (Querceto-Coryletum) and a sparse herbaceous layer. The leaf area index for the trees was 3.87, and for shrubs, 1.83 (total, 5.70).

Phytomass (t/ha)	
Aboveground	
Leaves of trees	3.5
Twigs and branches	58.3
Trunks	180.2
Shrub layer	18.1
Herbaceous layer	0.7
Total	260.8
Subterranean	55.4
Total phytomass	316.2

Primary production per year (t/ha/yr)
Aboveground

Total litter	6.2
Loss due to feeding	0.5
Tree increase	5.9
Shrub increase	2.1
Herbaceous increase	0.6
Total	15.3
Subterranean	2.3
Total production	17.6

Dead organic substance in the soil amounted to 122 t/ha. The phytomass of the forest communities is large because of accumulation of dead wood in the trunks (150 t/ha). But even without this, the phytomass is more than 1000 times the zoomass. The following values have been given for the vertebrate zoomass in European forests: reptiles, 1.7 kg/ha; birds, 1.3 kg/ ha; and mammals (mainly small), 7.4 kg/ha: The invertebrate zoomass is considerably higher, especially the underground (up to 14 kg/ha dry weight, 90% of which are dipteran larvae). In a deciduous North American forest with *Lirioden-dron* and a total phytomass of 214 t/ha (aboveground 173 t/ha) Reichle et al. found a total invertebrate zoomass of 157 kg dry weight/ha (quoted from Duvigneaud 1974, p. 98):

			(kg/ha)
Aboveground	–	phytophage arthropods	2.43
		predatory arthropods	0.61
In litter	–	large invertebrates	8.42
		small invertebrates	3.42
In soil	–	earthworms *(Octalasium)*	140.00
		small invertebrates	2.20
		total	157.08

Zoomass can rise drastically during caterpillar epidemics *(Lymantria dispar)* in oak forests: the dry mass of 2–4 million caterpillars per hectare gives a dry weight of about 75–150 kg/ha, whereby 1–2 t of dry leafweight/ha is destroyed and 500–1000 kg/ha of excrement is produced. The entire ecosystem is thus thrown out of balance; however, this only holds true for even-aged monoculture forests. In the mixed oak forests of Eastern Europe, it has been observed that, due to the improved light conditions, the amount of wood growth of ash and lime increased after the oaks had been eaten bare by caterpillars, resulting in an overcompensation. In the 4 years after such a caterpillar epidemic, an increased wood growth of 10% was registered.

A similar delayed compensation was found even in a mixed-aged pine monoculture after infestation with *Dendrolismus pini*, resulting in the successive growth of the formerly surpressed and less infested trees. Wood production decreased to 76% in the second year and to 56% in the third year. In the 4th and 5th years it increased to 150% and 194% respectively (see Walter 1983). A moderate grazing of grasslands also stimulates the vegetative growth of the grasses to such an extent that the entire annual production (including the quantity of grass consumed) increases (see Walter 1983, pp. 4–43).

The primary production is of especial significance for the ecosystem. Production analyses indicate that its extent is less dependent on the intensity of photosynthesis than on the assimilation budget of the producers (Walter 1960), i.e., on the way in which the photo-assimilates are utilized during the course of the vegetation period. They are either utilized productively if new assimilating leaves are constantly being grown, or unproductively if the growth of woody organs occurs, the effects of which will not be noticeable until several years later.

If, for example, single-seed fruits of beech (*Fagus sylvatica*) and sunflower (*Helianthus annus*) are planted in good soil in Central Europe under identical conditions, the beech seedling produces only 1.5 g (dry weight) in the first year, whereas the sunflower produces approximately 600 g, even though the climate is not optimal for this species. The sunflower constantly produces new large assimilating leaves, while the beech seedling grows only two or three small leaves in order to use the photoassimilates for the formation of a woody stem. Although the intensity of photosynthesis in the sunflower is double that of the beech, this cannot explain the 400-fold difference in productivity. Perennial plant communities are characterized by an especially high level of primary production.

Perennials, like annual plants, mainly grow assimilating leaves during the entire vegetation period; blossoms and fruits do not appear until the end of this time. Since they possess a relatively large amount of reserves stored from the previous year, they are able to grow leafed sprouts within a short time, whereas seedlings of annuals require a long lag period before the leaf surface is able to attain its maximum size. This is why summer grain requires 10 weeks for the production of the first quarter of the dry yield, 2 weeks for the second quarter and only 1 week for the last half.

Perennials, however, are able to take advantage of the entire vegetation period very productively. This explains the large amount of phytomass aboveground and the substantial underground reserves stored in the autumn for the following year. A precise production analysis is available for a perennial stand in a river valley in Japan (Iwaki et al. 1966, see Walter 1979). It deals with a pure stand of the adventive goldenrod *(Solidago altissima)*. The primary production was determined through monthly measurements of the above- and underground phytomass. The following values were obtained:

Growth aboveground between April and October	$1\,201$ g/m^2
Growth of rhizomes and roots during the same period	294 g/m^2
Plant parts which had died during the vegetation period	283 g/m^2
Total production	$1\,778$ g/m^2

This amounts to an annual net production of approximately 18 t/ha, and corresponds to that of a western European mixed oak forest (see p. 10). It is somewhat below the production of a 50-year-old evergreen *Castanopsis cuspidata* forest in the warm temperate climate of Japan with an aboveground annual primary production averaging 18.3 t/ha. According to more recent investigations by Morzov and Belaya (see Walter 1981), the production of the natural high-growth perennials in the permanently wet and nutrient rich soils of the river valleys of Kamchatka and Sakhalin is even higher.

Kamchatka lies within the subarctic zone with low-growth *Betula ermannii* forests. The vegetation period lasts 90−110 days (the average frost-free period is only 64 days). The average temperatures are 3.5 °C in May, 10.6 °C in June, 14.3 °C in July, 13.3 °C in August and 7.2 °C in September. The perennials attain a height of 3.5 m, whereby *Filipendula camtschatica*, *Senecio cannabifolium* and *Heracleum dulce* are the dominant species (Figs. 4−6). According to Hulten (1932), bears sleep in this vegetation during the day and emerge at night to feed on the salmon in the river.

The standing phytomass attains a maximum of 31 t/ha (10 t/ha of which are underground). Since a portion of the sprout mass dies during the vegetation period, the primary production is higher than the maximum herbacious phytomass. In spite of the short vegetation period, it is estimated to be more than 16−20 t/ha annually. The perennials are used in the form of silage for animal feed.

Higher values were found on southern Sakhalin, which is much further south (approximately 45 °N), and has a warmer climate with mixed deciduous

Fig. 4. The upper Kamchatka River valley. A narrow bear path can be seen directly along the river bank; behind it is the perennial growth with *Filipendula camtschatica* and the gallery forest with *Salix sachalinensis* with dead branches. On the hill is a forest with *Betula ermannii*. (Photo E. Hulten)

Fig. 5. High perennial stand in the lower Bolschaya River valley in front of the *Salix* gallery forest. *Filipendula camtschatica* is up to 3.5 m high (compare to the man in the center of the photograph). In the *foreground* is a trodden-down stand of *Anthriscus* and on the *left* are leaves of *Urtica platyphylla*. (Photo E. Hulten)

Fig. 6. Huge individual specimen of the perennial, *Angelica ursina*. (Photo E. Hulten)

and coniferous species. The frost free period is 145−155 days and the average temperature in the warmest month is 18 °C. There, the perennials grow to 4.5 m in height and their composition is similar to those in Kamchatka, although they are more homogenous since certain species are locally dominant. In stands where *Filipendula* is the dominant species, a leaf surface index of 13−14 is typical. Where *Polygonum sachalinese* is dominant, an index of 18−21 is possible if the perennial stand receives additional light, such as may be reflected from the river. This probably accounts for the enormous amount of aboveground phytomass produced annually, which attains 30 t/ha where *Polygonium* is the dominant species (total phytomass: 70 t/ha). This primary production may, therefore, supercede the record value for a smaller area with 38 t/ha.

Although values are not yet known, the primary production of high *Papyrus* stands in the tropics may be even higher. Because of the higher temperatures at night, it must be kept in mind that losses due to respiration in the tropics are very high, so that in spite of the high gross production, the net production is reduced substantially.

Luxuriant stands of perennials are also found in the subalpine level of the western Caucasus Mountains (Walter 1974). Stands of *Alnus viridis*, which are not quite so high, are also found in the subalpine regions of the Alps. They assimilate atmospheric nitrogen, thereby improving the soil. Exact production values are not available, however, in either case.

High perennial grasses in wet nutrient-rich areas also produce a large amount of phytomass annually. In 2.3-m-high stands of reeds (*Phragmites*) on the lower Amudarya, for example, a phytomass of 35 t/ha has been registered aboveground.

8. Special Material Cycles in Terrestrial Ecosystems and the Role Played by Fire

The scheme of the deciduous-forest ecosystem shown in Fig. 2 is far from being universal. Some of the possible variations are discussed below:

1. The Role of Mycorrhizae. Most forest trees, the Ericaceae, and some other plants form mycorrhizae with fungi. This has the function of extending the root system and facilitating the uptake of mineral nutrients from soils containing much humus. The mycorrhizal fungus can also supply the host plants with organic material, as is illustrated by the saprophytic orchids *(Neottia, Corallorhiza,* etc.), the Pyrolaceae *(Monotropa,* etc.), and other families. That the mycorrhizal fungi of forest trees and the Ericaceae provide their hosts with organic substances has not yet been demonstrated; however, this might well be the case in stands growing on extremely poor, sandy soils with a raw humus layer, and Went and Stark (1968) assume that it is possible for the trees of the tropical rain forests. If this is true, the short cycle would be shortened still further since litter would not necessarily have to be mineralized.

2. The Role of Fire. It is a well-established fact that fire can often replace decomposers and bring about very rapid mineralization of the accumulated litter. Natural fires caused by lightning occurred even in the Carboniferous forests and are characteristic of all grasslands with a period of drought, woodlands of the winter-rain regions, and all conifer regions, even without the helping hand of man. Fire is, in fact, a necessity for the vegetation if the decomposers are unable to break down all of the dead litter. After fires had successfully been prevented in Grand Teton National Park, a bark beetle catastrophe resulted because the animals were able to multiply to a much greater extent in the accumulating dead wood of the *Pinus* forests. Now that the natural fires are once more allowed to take their course, the equilibrium of the ecosystem has been restored. Steppes or prairies within national parks that are totally protected against fire degenerate owing to accumulation of litter that would otherwise be periodically mineralized. In some Australian heath regions, cycling of material comes to a stop if the dead organic plant remains are not burned at least once every 50 years. Without the action of fire, all mineral nutrients are stored in the large woody fruits of *Banksia,* as well as in the hard, dead leaves of the grass trees *(Xanthorrhoea),* and cycling stops (see p. 177). After a fire, this cycling is set going once more by the components of the ash. The situation is similar in the large *Protea* stands around Cape Town (South Africa).

Thus fire is often an important natural factor in maintaining the equilibrium of an ecosystem. For the years 1961–70, exact statistics are available concerning the forest and grassland fires caused by lightning in the United States. Lightning was responsible for 34,976 fires (37% of all fires) in the Pacific states, 51,703 (57%) in the Rocky Mountains, 13,733 (2%) in the southeastern states, but only 1167 (1%) in the humid Northwest (Taylor, 1973).

3. The Role of Wind. A remarkable case of an ecosystem entirely lacking in producers has been discovered in the dune region of the Namib fog desert (pp. 142–147). The organic matter which is a prerequisite for the cycling of material is blown from the neighboring regions by the wind and accumulates on the lee slopes of the dunes. It serves as food for saprophages (tenebrionids), which are devoured by small predators (reptiles, etc.), which in turn serve as food for larger predators. In this manner, a diverse fauna, with very strange adaptations to life in moving sand, has developed, even in the absence of vegetation.

9. Smaller Units of Ecological Systems: Biogeocenes and Synusiae

Having split up the land surface (geo-biosphere) of the earth into large units (biomes), a further subdivision can be undertaken according to the respective level of information available. In cases where no exact information is available, such subdivisions cannot be carried out.

These smaller ecological units are best defined on the basis of vegetational units. In a geographical area with uniform features corresponding to a biome, even the slightest differences in water and soil conditions influence the vegetation and thus the ecosystems themselves. Direct measurements of seasonally variable environmental factors of this kind are not feasible, and their combined effect cannot be estimated. Nevertheless, we can justifiably assume that the *natural* vegetation, which is in a state of equilibrium with its environment, provides us with an integrated expression of the effect of such environmental factors. Even the smallest change in one of them can bring about a qualitative, or at least quantitative, change in the vegetation.

In view of the fact that the influence of man is now almost everywhere detectable to a greater or lesser degree, the effects of natural and anthropogenic factors have to be carefully distinguished from one another, and as far as the latter are concerned, the role of man in the past has also to be taken into consideration. In the case of forest societies, this is still detectable centuries later (the effects of clear-cutting, mode of rejuvenation, grazing, utilization of litter, etc.). It is often held that the herbaceous layer of the forest gives a good idea of the natural situation, but the herbaceous layer depends itself to a very large extent upon the composition and nature of the tree layer (amount of shade, degree of competition from roots, kind of leaf litter) and is less deeply rooted than the tree layer, so that only the upper horizons of the soil are of any significance to it. Any change wrought on the tree layer by man has its effect on the herbaceous layer too. Even the removal of hollow tree stumps decaying on the ground represents severe interference with the ecosystem.

In densely populated regions, it can be assumed that there are no ecosystems that have not in some way been influenced by man.

The basic unit of the smaller ecosystems is the *biogeocene*. It corresponds to a plant community with the rank of an association, which is the basic unit of phytocenology. Opinions differ as to the correct definition of a plant association. Whereas some advocate a definition based mainly upon the dominant species, others place more weight upon the characteristic species. Some would like to set wider limits to an association, but others would draw the confines closer. However, we cannot dwell upon this much-debated question.

Within an ecosystem, cycling of material, energy flow, phytomass, and production are determined by the dominants, which, in forests, are tree species. Characteristic species, even if represented by only a few specimens, are useful in identifying the community, but they have not the slightest influence on the ecosystem. For this reason, it should be required that each type of ecosystem be defined with respect to the dominants. The limits of a plant community can only be laid down in the field after thorough study of the history of the individual stands and careful examination of the entire area, including habitat conditions and nature of the soil profile down to the root limits. Only actual stands can be investigated ecologically, not the abstract associations used by plant sociologists.

Although the biogeocene is the fundamental unit of the ecosystem, it is not the smallest. A number of synusiae can be differentiated within it. *Synusiae* are functional communities of species of similar life-form and ecological requirements and are thus identical with the ecological groups of Ellenberg.

However, synusiae should not be confused with ecosystems: they are merely parts of ecosystems, since they have no material cycle of their own but contribute to that of the ecosystem of which they form a part. Although production of the individual synusiae is only a fraction of the total production of the ecosystem, it is important because turnover in synusiae is often more rapid than that of the ecosystem as a whole.

Typical examples of synusiae are the various groups of species with a similar rhythm of development and similar ecological requirements, such as the spring geophytes (*Corydalis, Anemone, Ficaria,* etc.) which exploit the light phase on the floor of broadleaf forests before the trees come into leaf. Other examples are provided by herbaceous plants that survive the shady phase of summer or plants with evergreen leaves. Synusiae of lower plant forms are formed by lichens on tree trunks or mosses at the base of trees.

Between the biomes on the one hand and the biogeocenes, on the other, is a wide gap which has to be filled by units of intermediary rank. These units we propose to call biogeocene complexes. They often correspond to a particular kind of landscape, have a common origin, or are connected with one another by dynamic processes. As an example, we can cite a biogeocene sequence on a slope with lateral material transport (catena) or a natural succession of biogeocenes in a river valley or a basin with no outlet. A biogeocene complex can be a temporal succession of biogeocenes (like a secondary succession) or a spatial succession, in which the biogeocenes form an ecological series as a result of gradual alterations in a habitat factor (sinking groundwater table, increasing depth of soil, etc.). The area taken up by a biogeocene complex can vary considerably. The different types have as yet been given no ecological names of their own, and we shall employ the general term biogeocene complex. There is nothing to be gained by embarking upon a theoretical discussion concerning further subdivision of these complexes before more concrete examples have been described.

10. Diagrammatic Representation of the Hierarchy of Ecological Units

Bearing in mind what has been said above, we are now in a position to construct a scheme depicting the hierarchy of larger and smaller ecological units.

The following scheme forms the basis of our classification of the geobiosphere:

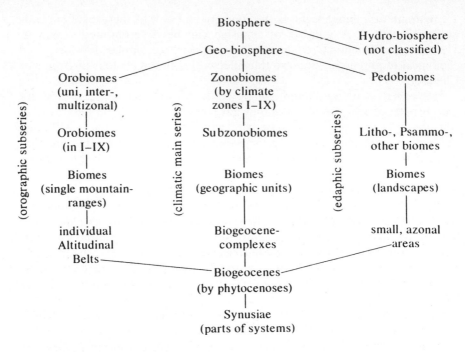

Ecological units are real. Just as a physician is only able to examine and treat living individuals and not types of human beings, the ecologist can only carry out his investigations on real ecosystems and not on abstract models thereof. The latter are merely mathematical games designed to incorporate data obtained and involving certain assumptions. Such models can never completely describe real situations but, if based upon sufficient data, can help toward a better understanding of ecosystems.

General Section

1. The Historical Factor

The geo-biosphere as we know it today is intimately linked with the history of the earth and is the result of a long process of development in the plant and animal worlds. As a result, an ecologist can never afford to neglect the historical factor.

All life began in water, and the earliest fossils of terrestrial plants originate from the transitional period between the Silurian and the Devonian. Since cormophytes do not require NaCl, the chief salt component of seawater, and since NaCl is toxic for all plants except for the halophytes, it seems that the ancestors of terrestrial plants were freshwater algae, probably living in coastal lagoons of the wet-tropical climate region. (Halophytic angiosperms are recent secondary adaptations to salty soils.) The conquest of land was made possible by the presence of large cell vacuoles, together constituting the vacuome, which provided an internal watery medium for the protoplasm; plants then developed an external cuticle which provided protection from desiccation. Stomata made possible the uptake of CO_2 for photosynthesis, and root and transpod systems compensated for transpiration losses (Walter 1967, Walter and Stadelmann 1968).

The continents were not always the same shape and previously occupied different positions with respect to the poles and the equator (Wegener's theory of continental drift). Plate tectonics and convection currents in the earth's mantle are held to be responsible for the movements of the land masses. Owing to the increasing isolation of the continents which took place in the waning Mesozoic, the flora of the earth, and the angiosperms in particular, developed along different routes, leading to the differentiation of six floristic realms (Fig. 7).

Of the phylogenetically relatively old group of conifers, the family Podocarpaceae, especially the genus *Araucaria* occur only in the Southern Hemisphere, whereas the large family Pinaceae and nearly all Taxodiaceae occur in the Northern Hemisphere. The Cupressaceae, however, are scattered over all continents.

The distribution of flowering plants, or angiosperms, the youngest branch of the plant kingdom, is much more sharply differentiated. The oldest families of this plant group are known to have existed in the Lower Cretaceous, but their main development took place in the Tertiary period, after the land masses had already split up into the different continents. In the

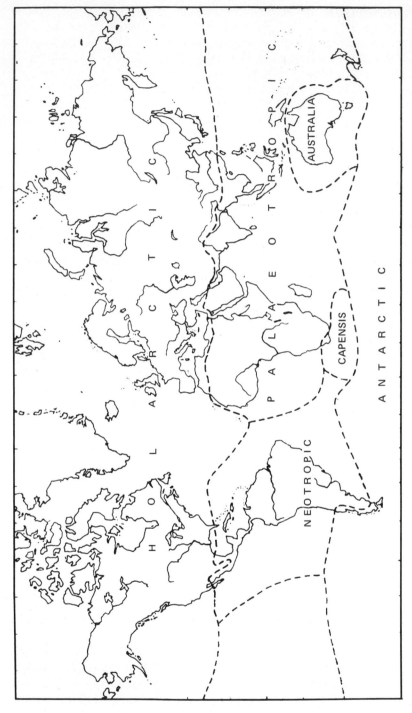

Fig. 7. The floristic realms of the earth. In New Zealand and Tasmania, antarctic, as well as paleotropic and australian, elements occur (Diels and Good; modified from Walter and Straka 1970)

Northern Hemisphere, however, this held true to a limited extent only, since it was not until the Pleistocene epoch that North America and Greenland finally separated from Eurasia. As a result, floristic differences in this area are so small that these continents can be considered as one floristic realm, the *Holarctic*. Much larger differences exist, however, between the tropical floras of the Old and New World, so that two floristic realms must be distinguished, the *Paleotropic* and the *Neotropic,* respectively. Floristically, the southern-most parts of South America and Africa, and Australia in its extreme isolation, have still less in common. For this reason, three floristic realms have been distinguished: the *Antarctic,* which comprises the southern tip of South America and the subantarctic islands, the *Australian,* which is geographically identical with the continent of Australia; and the *Capensic,* the smallest floristic realm, but one especially rich in species, in the outermost southwest corner of Africa (Fig. 7).

These six realms are not sharply delineated, and elements from one can be found far inside the next. In New Zealand, both paleotropic-melanesic elements and antarctic elements are to be found, often in mosaiclike distribution. Thus the allocation of these islands to the one or the other floristic realm is a question of informed judgment.

The animal regions of the zoologist correspond, on the whole, to the floristic realms; only the Capensic has no typical fauna.

Plant species are the building blocks of the plant communities that together constitute the vegetation of the different regions. Even if the species are not the same, extreme environmental conditions can lead to similar life-forms; these are termed *convergences,* but are, nevertheless, rather exceptional. A well-known example is provided by the stem-succulent plants, which in the arid areas of the Americas belong to the family Cactaceae and in Africa to the genus *Euphorbia.* In climatically similar arid regions of Australia, on the other hand, there are no stem-succulent plants whatsoever, although Australia is especially rich in other kinds of convergences which have not developed on other continents. In the temperate climate of New Zealand, there are none of the deciduous forests that are widespread in the Holarctic realm. Since the genetic stock of the individual floristic realms is limited, the same life-form could not necessarily develop everywhere. This is particularly marked in the Australian realm. The vegetation of this floristic realm is physiognomically very different from that of other continents, and even its mammalian fauna is unique. The Pleistocene, with its many ice ages, left a very distinct mark, especially on the Northern Hemisphere. The flora of Europe was reduced significantly and many genera became extinct, although they are still found in North America and Eastern Asia today. The ice ages affected the Sahara by bringing rain, or pluvial periods. The tropics, on the other hand, were subject to dry periods.

Obviously, the historical factor has to be taken into consideration in dealing with zonobiomes that extend over several floristic realms. A very good example is provided by ZB IV with winter rain, since it covers parts of the Holarctic, Neotropic, Australian, and Capensic realms. For the sake of convenience, it can be subdivided into five historically determined *biome groups* (Mediterranean, Californian, Central Chilean, Australian, and Capensic), which differ more with regard to flora than climate. Isolation of islands has also led to a considerable degree of endemism, i.e., the occurrence of species found nowhere else. The following figures indicate the percentage of the total flora accounted for by endemic species: Hawaii, 97.6%; New Zealand, 72%; Fiji lslands, 70%; Juan Fernández, 68%; Madagascar, 66%; Galapagos Islands (in the dry belt), 64% (only 8−27% in the humid montane belt and 12% in the coastal region); New Caledonia, 60%; Canary Islands, 50−55%; islands near mainland, 0–12%.

The further an island is situated from the mainland and the longer it has been isolated, the larger the proportion of endemic species, although oceanic currents also play a certain role.

2. Climate and Its Representation (Climate Diagrams, Homoclimes, and Climate-Diagram Maps)

Our classification of the geo-biosphere into zonobiomes is based upon the climate, by which we mean the average weather over the course of the year. Weather is defined as the result of the combined action of various meteorological factors at a given moment. But how this integration is to be arrived at is nowhere described by climatologists, although it is of great interest to the ecologist. Climate formulas and indices are of no help, and the only solution lies in a graphical representation of climate, clearly showing its seasonal course. This kind of information is provided by the *ecological climate diagram*. It is important that it contains only the most essential data from the ecologist's point of view (temperature and hydrature throughout the year), and that it be comprehensible at a glance. The Climate Diagram World Atlas of H. Walter and H. Lieth (1967) contains more than 8000 climate diagrams from stations all over the world. Fig. 8 shows how such a diagram can be constructed on millimeter paper, and Fig. 9 gives an exact explanation of the diagrams. Shown in this figure are examples of the nine zonobiomes: Yangambi on the central Congo for ZB I (humid-equatorial, diurnal climate), Salisbury in Rhodesia for ZB II (tropical summer-rain climate), Cairo on the

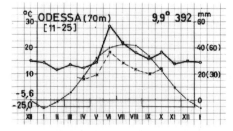

Fig. 8. Construction of a climate diagram on millimeter paper. Horizontal axis: months (each 1 cm). Vertical axis: left, mean monthly temperature (2 cm = 10 °C); right, monthly precipitation (2 cm = 20 mm of precipitation; for supplementary curve, 2 cm = 30 mm). Temperature curve, thin line; precipitation curve, thick line; supplementary precipitation curve (only in steppe diagrams), broken line. For further explanation, see text and Fig. 9. Below the O-line the frost-periods (see Fig. 9)

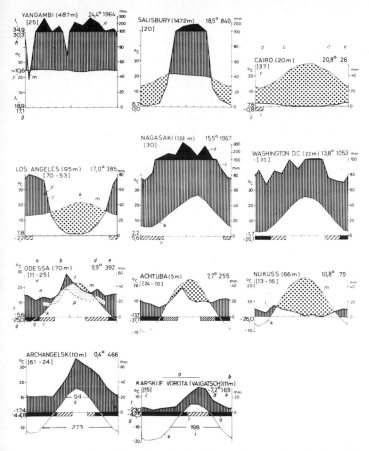

Fig. 9. Key to the climate diagrams. Abscissa: Northern Hemisphere, January to December; Southern Hemisphere, July to June (warm season in centre of diagram). Ordinate: 1 division = 10 °C or 20 mm of precipitation (figures are normaliy omitted, as in Fig. 10).

The letters and numbers on the diagrams indicate the following (see especially Odessa): (**a**) station; (**b**) height above sea level; (**c**) number of years of observation (where two figures are given, the first indicates temperature and the second precipitation); (**d**) mean annual temperature (in degrees Centigrade); (**e**) mean annual precipitation (in millimeters); (**f**) mean daily temperature minimum of the coldest month; (**g**) absolute minimum temperature (lowest recorded); (**h**) mean daily temperature maximum of warmest month; (**i**) absolute maximum temperature (highest recorded); (**j**) mean daily temperature fluctuation (h,i, and j are only indicated for tropical stations with a diurnal climate (see especially Yangambi); (**k**) curve of mean monthly temperature (1 division = 10 °C); (**l**) curve of mean monthly precipitation (1 division = 20 mm, i.e., 10 °C = 20 mm); (**m**) period of relative drought (dotted) for the climate region concerned; (**n**) corresponding relatively humid season (vertical shading); (**o**) mean monthly precipitation > 100 mm (scale reduced to 1/10, since diagrams with a monthly precipitation are often cumbersome) (black areas, perhumid season); (see Yangambi or Salisbury); (**p**) supplementary precipitation curve, reduced to 10 °C = 30 mm, [horizontally shaded area above, relatively dry period (only for steppe stations)]; (**q**) months with a mean daily minimum below 0 °C black, = cold season; (r) months with absolute minimum below 0 °C (diagonally shaded) i.e., with either late or early frosts; (**s**) number of days with mean temperature above + 10 °C (see Archangelsk), (**t**) number of days with mean temperature above − 10 °C (see Archangelsk and Karskije Vorota; s and t are only given for stations with a cold climate).

Not all of the above are available for every station. Where data are missing, the relevant places in the diagrams are left empty (see, for example, f, q and r for Nukuss) lower Nile for ZB III (subtropical desert climate), Los Angeles in Southern California for ZB IV (winter-rain climate), Nagasaki in Japan for ZB V (warm-temperate climate), Washington, D.C., in eastern North America for ZB VI (temperate climate with a short cold season), Odessa on the Black Sea for ZB VII (temperate semiarid steppe climate with a long dry season and slight drought), Achtuba on the lower Volga for ZB VIIa (temparate arid semi-desert climate with long drought), Nukuss in Middle Asia (rIII) (extreme arid desert climate with cold winters), Archangelsk (USSR) in the boreal taiga zone for ZB VIII (cold-temperate climate with a very long winter), Karskije Vorota on the island of Vaygach (USSR) for ZB IX (arctic tundra climate with a July mean temperature below +10 °C).

Figure 10 shows the climate diagrams of Orobiome I with a diurnal type of climate (páramo) and of the remaining Orobiomes II–IX. Wostok, in

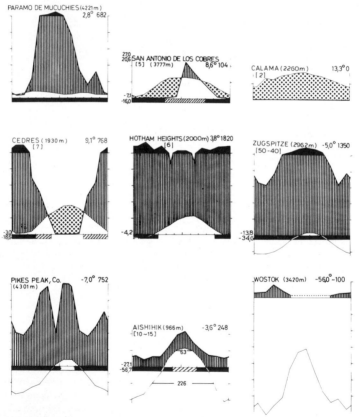

Fig. 10. Examples from mountain stations in the various orobiomes: OB I, Páramos de Mucuchies in Venezuela; OB II, Antonio de Los Cobres in the Feruvian puna, OB III, Calama in the North Chilean desert puna; OB IV, Cedres in Lebanon; OB V, Hotham Heights in the Snowy Mountains of Australia; OB VI, Zugspitze in the northern Alps; OB VII, Pikes Peak in the Rocky Mountains above the Great Plains of North America; OB VIII, Aishihik in southern Alaska; OB IX, Wostok on the ice cap of the Antarctic

Orobiome IX, with a mean annual temperature of $-56\,°C$, is the coldest station of the earth.

Climate diagrams show not only temperature and precipitation values but also the duration and intensity of relatively humid and relatively arid seasons, the duration and severity of a cold winter, and the possibility of late or early frosts. With this information, it is possible for us to judge the climate from an ecological standpoint. The aridity or humidity of the different seasons can also be read off the diagrams by using the scale $10\,°C = 20\,mm$ of precipitation. A potential evaporation curve thus takes the place of the temperature curve, and by comparing it with the precipitation curve, some idea of the water balance can be obtained. Height of the dotted area (drought) indicates the intensity of the drought, and width is proportional to its duration. Humidity is indicated in the same manner.

The ratio of $10\,°C = 20\,mm$ of rain was found by Gaussen (1954) to agree well with the actual weather conditions of the Mediterranean region. We have been able to confirm this for the climate diagrams from all climate zones. In diagrams for steppes and prairies, a scale of $10\,°C = 30\,mm$ is recommended, so that dry periods that do not attain the severity of a drought are shown.

The arid (drought) season shown in the climate diagram is only arid relative to the humid season of the particular type of climate under consideration. This is because the potential evaporation curve and the temperature curve which is used in its place are not identical, but run only more or less parallel with one another. The more arid the climate, the larger the quantitative deviation of the temperature curve below the potential evaporation curve. In absolute terms, this means that the more arid the climate as a whole, the drier the arid season in the climate diagram. An arid season on the climate diagram for a station in the steppe region, for example, is not so extreme as one for a Mediterranean station or for the Sahara. This is ecologically very fitting since the drier the climate in which plants live, the more resistant they become to dryness. For the species growing in a tropical rain forest, a month with less than 100 mm of rain is already relatively dry, and the "xerophytes" of central Europe would be "hygrophytes" in a desert region. In discussing vegetational regions, the appropriate diagram will be given in each case in order to avoid the use of long tables (Walter et al. 1975).

Climate diagrams are particularly useful for tracing homoclimes, i.e., stations with very similar or almost identical climates. This is a tedious process if lengthy climate tables have to be consulted; however, with the aid of the Climate Diagram World Atlas (Walter and Lieth 1967), a given climate diagram can be easily compared with regions where a homoclime is suspected. Figure 11 shows the homoclimes of Karachi (Pakistan), from ZB III (transition toward ZB II), and Bombay (typical of ZB II) in other parts of the world. Information about homoclimes is essential before introducing plants into regions where they have not so far been cultivated. Figure 12 shows the positions of the homoclimes of many stations in India.

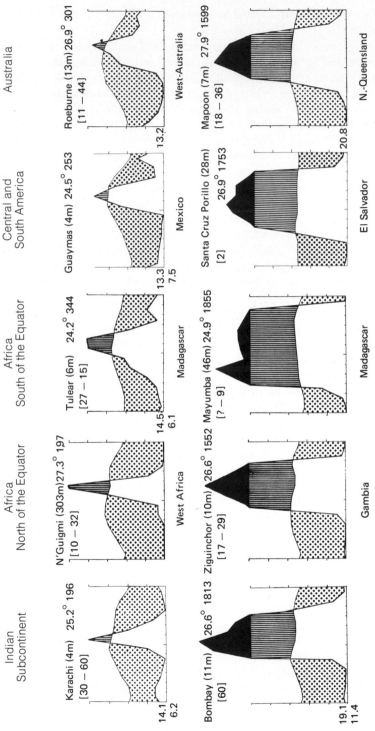

Fig. 11. Homoclimes of the stations Karachi (Pakistan) and Bombay (India) on other continents. (From Walter and Lieth 1967)

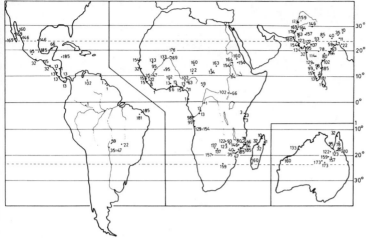

Fig. 12. Homoclimes of the subcontinent of anterior India in other parts of the earth are indicated by corresponding numbers (climate diagrams 133 and 134 are almost identical)

A climate-diagram map reveals at a glance the types of climate within a whole continent or large sections thereof. Such a map can be constructed by sticking the climate diagrams from the Climate Diagram World Atlas onto the appropriate spot on a large wall map. It becomes still clearer if the areas corresponding to drought are colored red and those corresponding the wet periods colored blue. Large black-and-white climate-diagram maps of every continent have been published elsewhere (Walter, Harnickell, and Mueller-Dombois 1975).

Because of limited space, only one small-scale map of Africa showing a few climate diagrams can be reproduced here (Fig. 13). The World Atlas contains over 1000 diagrams for Africa alone.

3. Environment and Competition

In a floristically uniform area, the structure of the vegetation is determined by the environment, and primarily by climate and soil. The climate exerts a direct influence on the vegetation as well as an indirect influence (via the soil). The interrelationships are shown in the following scheme:

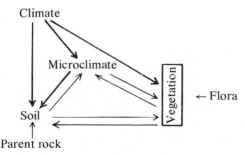

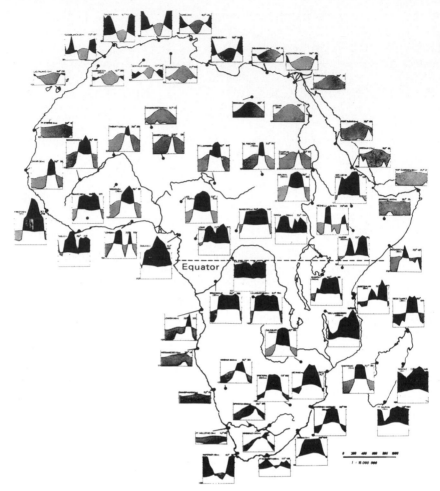

Fig. 13. Example of a climate-diagram map showing 66 stations in Africa. Zonobiomes from north to south: IV-III-II-I-II-III-IV. In the east, however, the climate to the north of the equator is drier (monsoon) and that to the south more humid (southeast trade winds) than is indicated by this sequence

   The nature of the soil and the vegetation is determined by climate, although the composition of the soil depends to a certain extent upon type of parent rock, and the flora exerts an influence on the vegetation. Soil and vegetation are so closely interrelated that they can be considered almost as one entity. Moreover, both affect to some degree the climatic conditions in the air layer nearest the ground; i.e., both influence the microclimate. The environment of plants is definable as the sum of the factors acting upon them, the physicochemical factors (without competition) being termed the *habitat* or *ecotope* and the place on which they grow the *biotope*. Habitat factors are often listed as climatic, orographic, and edaphic (soil), although such distinctions are of little ecological significance. Factors determining growth and development of plants are better considered under the following five categories: (1) heat, or temperature, conditions, (2) water, or hydrature, conditions, (3) light intensity and length of day, (4) various chemical factors (nutrients or poisons), and (5) mechanical factors (fire, grazing, gnawing, trampling by animals, wind).

Favorable thermic conditions may be provided by the regional climate or may be due, for example, to a sheltered biotope on a southern slope. Similarly, the required soil moisture may result from favorable atmospheric precipitation, limited evaporation on a northern slope, or soil structure and the proximity of groundwater. What is important is that the plants suffer no lack of water.

The widespread assumption that the distribution of plant species is directly dependent upon the physical conditions prevailing in the habitat is incorrect; these conditions are of importance *only indirectly–insofar as they influence the competitive power of the various species.* Only at the absolute distribution limit, in arid or icy deserts, or on the edge of the salt desert, are the physical environmental factors (usually one particular, extreme factor) of direct importance. Apart from such exceptions, plant species are capable of existing far beyond their natural distributional areas if they are protected from competition by other species. The northeastern limit of distribution of the European beech *(Fagus sylvatica)*, for example, runs through the Vistula region of Poland, although beech is found growing far to the north and southeast of this limit in the botanical gardens in Helsinki and Kiev. *The natural limit of distribution of a particular species is reached when, as a result of changing physical environmental factors, its ability to compete, or its competitive power, is so much reduced that it can be ousted by other species.* The range also depends upon the presence of competitors or of a certain fauna. For beech, these are hornbeam *(Carpinus betulus)* on the eastern boundary, oak *(Quercus robur)* to the north, and spruce *(Picea abies)* in the mountainous regions. That the northeastern beech limit takes a course similar to the January isotherm for $-2 °C$, that the northern limits of oak distribution follow the line indicating 4 months of the year above $+10 °C$, and that the northernmost spruce boundary coincides with the July isotherm of $+10 °C$ does not necessarily indicate a direct causal relationship. It can, at most, be concluded that, in the case of beech, the increasingly cold winter toward the east and, for oak and spruce, the shorter summer to the north, radically reduce the competitive power of these species.

The conditions under which a species occurs most abundantly in nature may be termed the *ecological optimum,* and the conditions under which it thrives best in the laboratory (phytotron or growth chamber) or in individual cultures may be termed the *physiological optimum.* These optima, are rarely identical, however, as may be seen from Fig. 14.

The distribution of a species, therefore, is not an absolute guide to its physiological requirements. For example, the fact that, in western Europe, the Scotch pine *(Pinus sylvestris)* is found under natural conditions only on

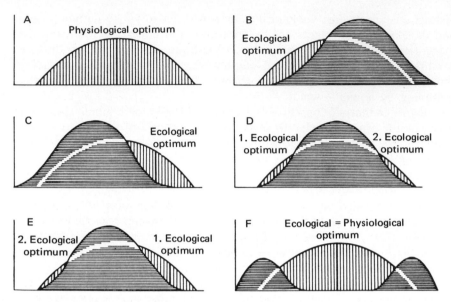

Fig. 14. Growth curves (vertical shading) of one species (A) without or (B–F) under pressure due to competition (horizontal shading). Ordinate: growth intensity and production of organic matter; abscissa: variable habitat factors. (From Walter 1960)

dry or calcareous slopes or on acid, boggy soil is due to its having been supplanted from a more suitable habitat by stronger competitors. Therefore, knowledge acquired in the phytotron about the physiological needs of a species forms an insufficient basis for either predicting or explaining its distribution in nature. Whether or not a species colonizes a habitat which is physiologically suitable usually depends, apart from the historical factor, on the nature of its competitors.

Competition is generally said to take place when the growth or development of one species is unfavorably influenced by the presence of another, without the occurrence of parasitism. It is nearly always in action wherever several species occur close together and no form of dependence exists, and it can be recognized when a particular plant develops more luxuriantly in isolation than in a plant community. Inhibition in the course of competition results mainly from the cutting off of light by surface organs or lack of water or nutrients due to root competition. Whether, apart from these factors, an important role in the competitive struggle can be attributed to certain specific substances excreted by the plants themselves (allelopathy) under natural conditions has not yet been satisfactorily determined. There are only a few examples for which this may hold true.

A distinction must be made between *intraspecific competition,* occuring between individuals of the same species, and *interspecific competition,* taking place between individuals of different species. The former type of competition eliminates the weaker individuals and helps to preserve the species. In interspecific competition, a species can achieve dominance and supplant others or, in a mixed population, a state of equilibrium may be established based on the competitive power of the individual partners. In mountainous regions, for example, it can be seen that, at the beech-spruce boundary, beech is absolutely dominant on southern slopes and spruce is dominant on the northern slopes, whereas on eastern and western slopes, the two are fairly well balanced and a mixed population is formed. This can also happen if, as seems to be the case in tropical rain forests, the seedlings of a certain species develop better beneath other species than beneath individuals of the same species.

The competitive power of a species is a highly complex phenomenon and *is only definable for a given set of environmental conditions*. It varies considerably according to stage of development, being at its weakest in seedlings and young plants and increasing with age (particularly in the case of trees). The complete morphological and physiological properties of a species are of significance in this respect. Biennial species are competitively more powerful than annuals owing to the fact that they commence their second year of growth with the large reserves accumulated during the first year. For the same reason, perennial herbs are superior to biennials from the second year onward. Woody species win out against perennial herbs if the former have not been suppressed during their early years and have succeeded in producing ligneous axial organs to raise them above the herbage layer.

As a result of competition, similar combinations of plant species occur repeatedly on similar habitats within a limited area and are termed *plant communities* (phytocenoses). Examples in central Europe are beech woods on calcareous soil together with their herbaceous flora, floodplain forests, various types of bogs and fens, and so on.

In a stable plant community, the different species are in a state of ecological equilibrium with each other and with the environment. Together with the animal organisms, they constitute a biogeocene. Apart from the influence of animals, the following factors are of importance: (1) *competition* between species (interspecific competition), (2) the *dependence* of one species (e.g., shade species) upon the presence of others, and (3) the occurrence of *complementary* species, fitted to one another either spatially or temporally so that every ecological niche is filled.

As a result of the third factor, invading species can gain no foothold in a stable community, whereas these species are much more successful if the state of equilibrium is disturbed. For this reason, long-range transport of seeds plays a role in the distribution of plants only in areas that have not been colonized, e.g., recent volcanic islands devoid of vegetation.

The equilibrium of a plant community is dynamic rather than static, in that some individuals die off while others germinate and grow. At the same time, the individual species are continually exchanging places. Quantitatively, too, the species composition exhibits certain deviations, since external conditions vary from year to year, rainy years follow upon dry ones, and so on. Consequently, sometimes one species will be favored, sometimes another. If the conditions in the habitat alter continuously in one direction, e.g., if the groundwater table rises slowly over many years, then the combination of species also changes, some species disappearing and others infiltrating from outside until finally a new plant community arises. Such a series of events is termed a *succession*. If the alterations in the habitat arise from natural causes and originate on parent rock, then it is called a primary succession; this is usually a very slow process. More often, a rapidly progressing so-called secondary succession results from human interference (draining of water meadows, deforestation, abandonment of fallow land, letting meadows grow up, etc.) or is caused by catastrophes such as hurricanes, wind, and fire. If human interference continues over a long period of time and is of a uniform nature, then an anthropogenic equilibrium develops. The plant communities arising are termed cultural formations if they are intensively utilized and semicultural formations if more extensively utilized; they constitute the vegetation in areas densely populated by man.

4. Ecotypes and the Law of Change of Biotope and Relative Constancy of Habitat. Extrazonal Vegetation

Many plant species or phytocenoses (plant communities) are of very wide distribution and, as can be seen if the areas inhabited by them are studied on the map, can apparently flourish under very different climatic conditions. There are two possible reasons for this:

1. The Existence of Ecotypes. The species as a taxonomic unit is often highly differentiated ecophysiologically, e.g., with respect to cold- or drought-resistance or climate rhythm. Thus the pine species *Pinus sylvestris* is found from Lapland to Spain and as far east as Mongolia, with only slight taxonomic variations. The Spanish race, however, cannot grow in Lapland, because it is too sensitive to cold, and the Lapland race requires too long a winter resting period to be able to survive in Spain. It is therefore very important to know the provenance (origin) of the seed employed in reforestation programs. Most taxonomically uniform species are made up of many such ecotypes, or they may exhibit a quantitative gradation in ecophysiological properties, in which case they are termed ecoclines.

2. Change in Biotope. The second possible reason underlying a widespread distribution is the property exhibited by a species or a phytocenosis of changing its biotope if it extends into a different climatic region. If the climate at the northern limit of the plant's range becomes colder, for example, the species will no longer be found on the plains but on locally warmer southern slopes. In other words, a change of biotope compensates for the change in climate, so that the habitat or environmental conditions remain relatively constant. Examples of this are seen everywhere. Toward the southern extremities of its range, a species increasingly seeks out northern slopes or deep and moist canyons or moves up into the mountains. If the climate becomes wetter, the plants choose the drier chalky or sandy soils, but take to the heavy, wet soils or those with a high groundwater table if the climate becomes drier. It has to be remembered that, in the Southern Hemisphere, the warmer slopes face north and, on the equator, the eastern and western slopes are warmer. In arid regions, however, it is the sandy type of soil that provides the best water supply for the vegetation (p. 116).

This ecological law can be formulated as follows: *If within the range of distribution of a plant species or phytocenosis, the climate in any way changes, a change of biotope compensates as far as possible for the change in climate. In other words, the habitat or environmental conditions remain relatively constant.*

If the vegetation of the euclimatopes is termed *zonal vegetation*, the term *extrazonal vegetation* can be applied after a change in biotope. The vegetation is no longer determined by the overall climate but is dependent upon local climatic conditions. For example, where forests bordering the rivers extend far into an arid region as gallery forests, they are extrazonal vegetation.

A severe disadvantage of maps showing the ranges of distribution of species is that zonal and extrazonal distribution cannot be distinguished. This gives a false impression of the climatic requirements of the plants, the more so since the borders of the areas of distribution often coincide with climate lines.

The law of change of biotope also has to be allowed for in determining the altitudinal belts in mountainous regions. In this case, it is manifest in the "special niche vegetation" (i.e., "extrabelt" vegetation) often occurring in locally favorable biotopes several hundred meters above or below the species' "proper" altitudinal belt. Conformity to the law can also be seen in the way in which the limits of the various altitudinal belts vary according to their exposure. Much more extreme "special niches" occur where intense irradiation and the runoff of cold air permit the growth of small stands of trees above the timberline and within the alpine belt (p. 69). In West Pamir, isolated trees have been found even as high as 4000 m above sea level and small shrubs have been found at 5000 m in gorges swept by wind, which prevents the accumulation of cold air. In contrast, in some parts of the eastern Alps where the cold air is trapped (dolines), forest vegetation has even ceased at 1270 m. The lowest temperature ($-51\,°C$) in Central Europe has been measured in one such region, Lunz am See, in Lower Austria.

Soil factors can also be involved in extrazonal vegetation. On hardly weathering dolomite (poor acid soil) in the eastern Alps, fragments of alpine vegetation can be encountered in the middle of the beech belt. The paths of avalanches and landslides also provide "special niches" since competition from trees has been eliminated, and the dwarf *Pinus montana* (krummholz) of the subalpine belt can grow in the lower levels of the forest belt.

In special biotopes of this kind, it is not unusual to meet relics of species that had a greater range of distribution under previous climatic conditions. Nevertheless, before conclusively stating that a species is a relict it is essential to have historical evidence.

5. Poikilohydric and Homeohydric Plants and Halophytes

The most important habitat and environmental factors involved in a bioclassification of the geo-biosphere are heat and water relations. Light is never a minimum limiting factor even at the poles, since the long polar nights coincide with the winter resting period of the plants.

Temperatures decrease fairly steadily from the equator toward the poles. In determining the boundary between the tropics and outer tropics, the occurrence of frost is especially important. However, the water factor plays a considerably greater differentiating role. Precipitation is very unevenly distributed (Fig. 15), the mean annual values varying from 10,000 mm (Assam) to almost zero (in the most extreme deserts). For comparison, Fig. 16 illustrates the vegetational zones, for which both precipitation and temperature are especially important. Not only is the average amount of precipitation of significance for the vegetation but local differences in water supply of the various biotopes also have a strongly differentiating effect on the plant cover. Ecologically, water plays a very special role (insufficiently emphasized by physiologists) in the life of a plant and their adaptations, a far greater one, in fact, than in animals, which are free to move.

With respect to temperature, we distinguish between animal organisms that are cold-blooded, or poikilothermic, and those that are warm-blooded, or homeothermic. The body temperature of the former, and thus the temperature of their protoplasm, is dependent upon the external temperature and varies with it, whereas the body temperature of the latter is more or less constant, and the animals are thus largely independent of the environmental temperature. For warm-blooded animals, therefore, it is meaningless to measure the external temperature and relate it to the course of physiological processes in the protoplasm. *Plants are invariably poikilothermic.* The temperature of the surrounding air thus provides a clue to the temperature of the protoplasm. Certain minor deviations of a purely physical origin are induced by strong radiation and must be allowed for in exact ecophysiological investigations. However, it is usually considered sufficient to record the air temperature.

With respect to water relations, plants behave in as complicated a manner as do animals with respect to temperature. *A distinction must therefore be made between poikilohydric and homeohydric plants.*

Protoplasm is physiologically active only when its water content is high, that is to say, when it is in a hydrated or swollen state. If the cells dry out, either the protoplasm falls into a latent condition and exhibits no detectable signs of life or the plant dies. According to the thermodynamics of imbibants, the degree of hydration depends upon the *relative activity of water* (a), where $a = p/p_0$ (p = actual water vapour tension, p_0 = saturation tension at the same temperature), that is, is equivalent to the relative water-vapor tension. Expressed in percent, this is called *hydrature,* analogous to temperature, the warmth activity (Walter and Kreeb 1970). The rel. water activity of pure water is 1 or has a hydrature of 100% hy; a solution with 0.8 rel. water activity

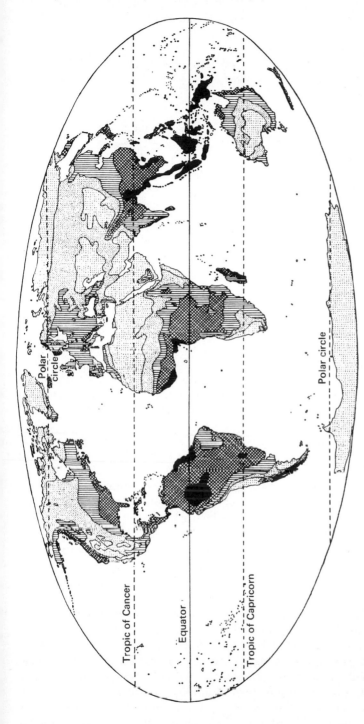

Fig. 15. Annual rainfall map. Thinly dotted, below 250 mm; closely dotted, 250–500 mm; vertical shading, 500–1000 mm; cross-shading, 1000–2000 mm; black, more than 2000 mm. Precipitation in mountainous regions was not taken into consideration. Comparison with Fig. 16 shows that temperature is also very important for the vegetation (this is reflected in the more pronounced arrangement of the zones parallel to the lines of latitude in Fig. 16). (From Walter 1950)

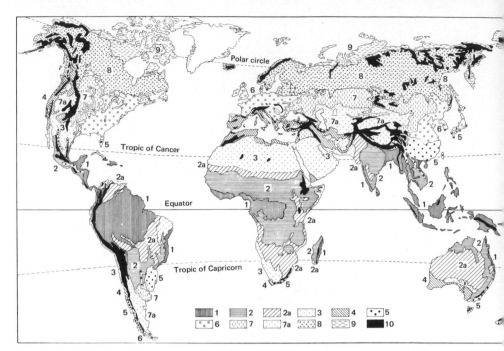

Fig. 16. Vegetational zones (much simplified, without edaphically or anthropogenically influenced vegetational regions). I. *Tropical and Subtropical Zones:* (1) Evergreen rain forests of the lowlands and mountainsides (cloud forests); (2) semievergreen and deciduous forests; (2a) dry woodlands, natural savannas or grassland; (3) hot semideserts and deserts, poleward up to latitude of 35° (see also 7a). II. *Temperate and Arctic Zones:* (4) sclerophyllous woodlands with winter rain; (5) moist warm temperature woodlands; (6) deciduous (nemoral) forests; (7) steppes of the temperate zone; (7a) semideserts and deserts with cold winters; (8) boreal coniferous zone; (9) tundra; (10) mountains. (From Walter 1964/68)

has 80% hy. In the air the hydrature is equivalent to air humidity in percent. Since the vital processes are to a large extent dependent upon the degree of hydration of the protoplasm, it is essential to know the hydrature of the protoplasm (or the activity of the water).

In *poikilohydric organisms* such as the lower terrestrial plants (bacteria, algae, fungi, and lichens), the hydrature depends entirely upon the humidity of the surrounding air. If the plants are in contact with water or if the surrounding air is saturated with water vapor, then the protoplasm of such species is hydrated and active to an almost maximum degree. In dry air, however, dehydration takes place, and the protoplasm passes into a resting condition without dying. The cells of such organisms have no vacuoles, or very small ones, so that the change in volume of the cell contents during desiccation is small and the protoplasmic structure remains undamaged. The lower limit of hydrature (atmospheric humidity) at which growth is still demonstrable is very high for most bacteria (98–94%), is varied for unicellular algae and moulds, and in only a few organisms sinks to 70%, *a value which corresponds to the extreme lower level of hydrature compatible with any sign of life.* The productivity of these poikilohydric organisms is small, and they contribute little to the entire terrestrial phytoma. For this reason, they have, until now, been afforded little attention, although they are often of much more widespread distribution, particularly in deserts, than is generally supposed. Poikilohydric organisms were probably of widespread occurrence on periodically wet areas before the conquest of land by higher plants, just as they are today on occasionally flooded clay in the desert (takyrs) which cannot be colonized by higher plants due to lack of root space. Since, however, fossilized remains of the lower plants are exceptionally rare in the older geological formations, we have no fossil evidence to support this assumption.

The terrestrial *homeohydric plants* are of much greater importance and include all the cormophytes, which developed originally from green algae. Their cells have a large central vacuole (see p. 19), and since the surrounding protoplasm is in contact with the vacuolar cell sap, its hydrature is thus not so immediately dependent upon water conditions outside the cell. The entire vacuolar cell sap, termed here the *vacuome,* constitutes an internal watery milieu for the higher plant, as already mentioned. It is this "internal milieu" which, in the course of phylogenetic development, has made possible a transition from an aquatic to a terrestrial way of life as well as a steadily improving adaptation to arid conditions. As long as terrestrial plants are able to keep the cell-sap concentration of the vacuome low, the protoplasm remains well hydrated. In other words, the hydrature of the protoplasm remains high, independent of the moisture in the surrounding air. The more water in the soil, and the better the supply to the plant via roots and transport system, the greater the degree of independence from the air moisture. Since the vascular system is only imperfectly developed in mosses, they are therefore generally confined to very damp habitats. The ferns, too, with a somewhat inefficient transport system consisting solely of tracheids, avoid dry habitats. Those mosses and the few ferns and *Selaginella* spp. that have penetrated into desert regions have had to change to a *secondarily poikilohydric way of life* in order to survive dehydration during times of drought. They achieved this ability to withstand desiccation, not normally possessed by plants with strongly vacuolated cells, by a diminution of cell size with reduction of the vacuoles, which solidify as a result of even a small water loss, thereby avoiding deformation and damaging of the protoplasm by dessicaton.

With respect to water economy, the most complete adaptation to a terrestrial way of life has been achieved by the angiosperms, which have even penetrated into extreme desert regions. Measurement of their cell-sap concentration shows that, without too greatly checking the gaseous exchange necessary for photosynthesis, they are able to keep the concentration of the cell sap low and thus maintain the hydrature of the cytoplasm at a high level. A rise in the concentration of the cell sap (decrease in osmotic potential) and a resultant dehydration of the cytoplasm is not an appropriate adaptation for desert plants, although this statement is still to be found in textbooks; rather, it is an indication of a disrupted water balance and represents a threat to survival.

The measurement of such external factors as precipitation, humidity, and water content of the soil offers little information concerning the water activity in the protoplasm (hydrature and degree of hydration), just as the measurement of external temperatures offers little information in studying warm-blooded (homeothermic) animals.

Determination of the cell-sap concentration (or osmotic potential), which is directly related to the relative water-vapor tension (hydrature), is the only means of ascertaining whether the plant has been affected by alterations in external conditions, especially by drought. On the other hand, in order to obtain information about water flow from roots to transpiration organs, measurement of the water potential is indispensable, although it gives no indication as to the state of hydrature of the protoplasm, upon which all vital phenomena in the cell depend.

In describing the habitats of the various vegetational zones, we shall, therefore, in addition to giving the usual data concerning external factors, indicate, where possible, the cell-sap concentration and its variations. This gives an idea as to the hydrature of the protoplasm and is of particular interest in discussing arid districts, where the water factor is of paramount importance. The values taken from earlier studies have been left in atmospheres, but the more recent data are given in bars since the difference between the two units (1 atm = 1.013 bar) falls within the limits of error. For ZB III, the adaptation to drought will be considered from the cybernetic point of view, and the osmotic-potential conditions will be dealt with in more detail (pp. 135–142).

These explanations are necessary because most textbooks, in dealing with water economy, make no distinction between poikilohydric and homeohydric plants.

Halophytes (salt plants) present a special problem. Salty soils of the seacoasts, and especially in deserts, are probably the most recently colonized regions. Terrestrial plants in these areas not only have to cope with the water problem, but also have to adapt to the physiological effect of the salt. The concentration of the cell sap in the vacuoles cannot be lower than that of the soil solution (and in salty soils, the latter is usually very high), otherwise a drastic drop in hydrature of the plasma would occur if other osmotically active substances such as sugar were formed in the cell sap. The problem is solved in the following way.

Salt is taken up from the soil into the cells until equilibrium with the soil solution is attained. Uptake of these electrolytes leads to additional hydration, rather than dehydration, of the plasma, which is why the organs of halophytes are hypertrophic and succulent.

However, since salts are toxic in large concentrations, halophytes have to be salt resistant, which is only possible up to certain limits. Thus soils with a very high content of salt remain bare of vegetation.

Special Section

Before we go on to discuss the individual zonobiomes, their distributions on the five continents are presented in Figs. 17–22 (pp. 40–45). In addition, smaller distinctions within the zonobiomes are referred to as well. It can be seen that the Zonobiomes VI–IX are poorly represented in the Southern Hemisphere with its cool oceanic climate.

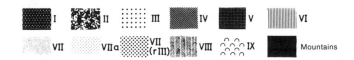

Figs. 17–22. Ecological classification of the continents. (From Walter et al. 1976). Roman numerals I–IX indicate the zonobiome (ZB). White spaces between shaded areas are zonoecotones. Further distinctions within the individual zonobiomes are indicated as follows:

    a – relatively arid for that particular ZB
    h – relatively humid for that particular ZB
  oc – climate with "oceanic" or maritime tendency in extratropical regions
  co – climate with "continental" tendency
  fr – frequent "frost" in tropical regions, at higher altitudes
  wr – prevailing winter rain, in ZB in which this is anomalous
  sr – prevailing summer rain, in a ZB in which this is anomalous
 swr – two rainy seasons (or occasional rain at any season)
  ep – "episodic" rain, in extreme deserts
 nm – nonmeasureable pricipitation from dew of fog in the deserts
(rIII) – "rain as sparse as in ZB III," e.g., ZB I(rIII) = equatorial desert
 (tI) – "temperature curve as in ZB I," e.g., ZB II(tI) = diurnal climate

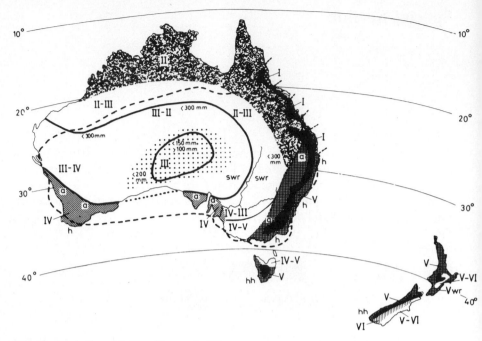

Fig. 17. Australia, with Zonobiomes I–IV

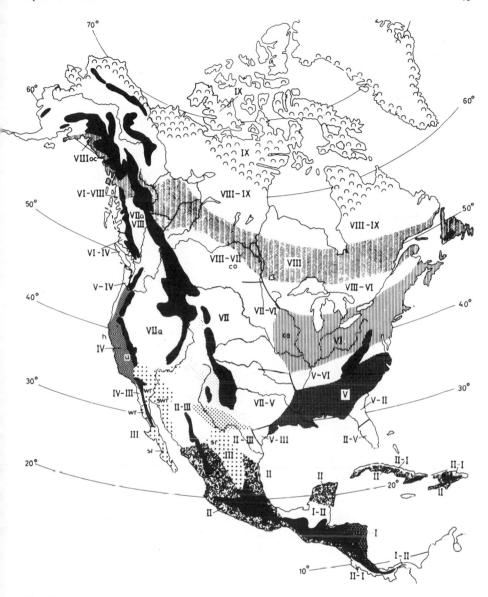

Fig. 18. North and Central America, with Zonobiomes I−IX

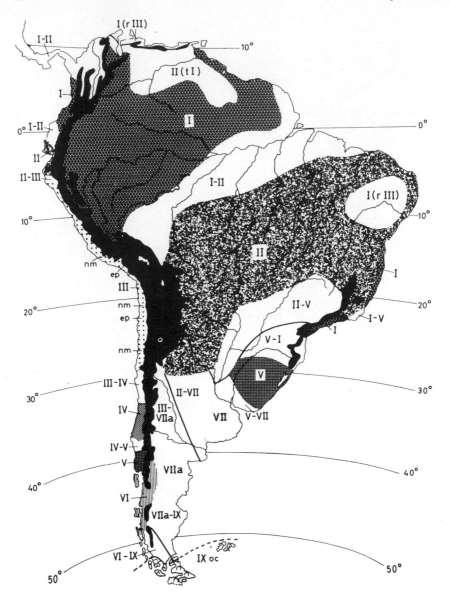

Fig. 19. South America, with Zonobiomes I–VII and IX

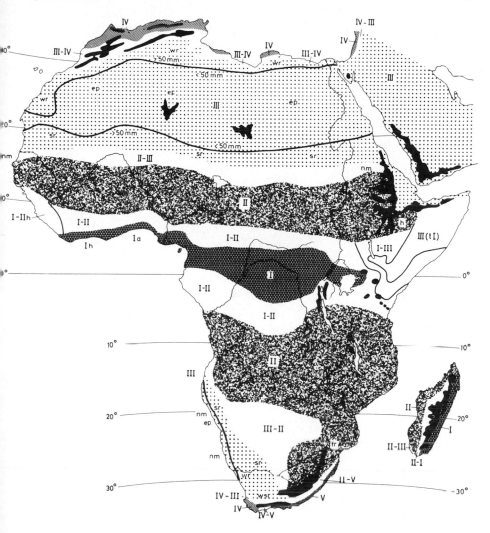

Fig. 20. Africa, with Zonobiomes I−V

44

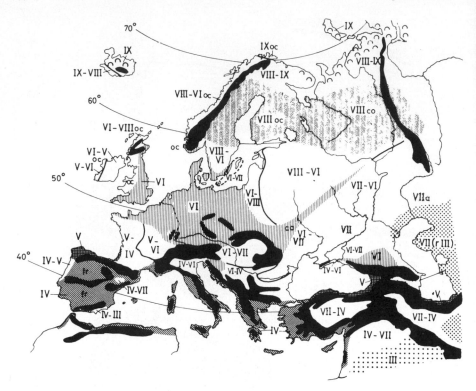

Fig. 21. Europe (plus anterior Asia), with Zonobiomes IV−IX. Owing to the influence of the Gulf Stream, the zonobiomes run more from north to south in western Europe, whereas in eastern Europe they take the normal course from west to east. From north to south: Zonobiome IX (tundra zone) with Zonoecotone VIII−IX (forest tundra); Zonobiome VIII (boreal coniferous zone); Zonoecotone VI−VIII with Zonobiome VI, both of which thin out toward the east (mixed-forest and deciduous-forest zone); Zonobiome VII (steppe zone). Zonobiomes IX, VIII, and VII continue eastward into Asia (Fig. 22). Southern Europe belongs to ZB IV (Mediterranean sclerophyllous region), offshoots of which are still detectable in Iran and Afghanistan. Zonobiome III is lacking altogether in Europe; only Zonoecotone IV−III occupies a small desertlike area in the southeast of Spain, which is the driest part of Europe. In central Europe, zonation is greatly disrupted by the Alps and other mountains. The situation in the mountainous Balkan peninsula is also complicated

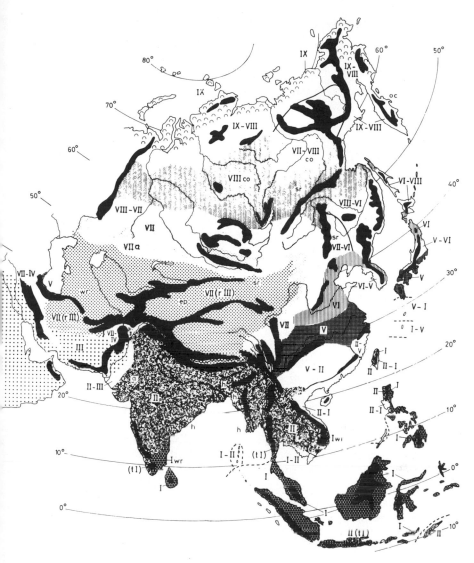

Fig. 22. Asia, with zonobiomes I−IX (for anterior Asia, see Fig. 21)

I Zonobiome of the Equatorial Humid Diurnal Climate with Evergreen Tropical Rain Forest

1. Typical Climate

In this wettest of all vegetational zones a month with less than 100 mm rain is considered to be relatively dry. Only in Malaya and Indonesia are there large areas in which the monthly rainfall does not fall below 100 mm, whereas in the Amazon basin there is merely a small area on the Rio Negro. A short dry season invariably occurs in India and is observable in most years in the Congo basin (Fig. 23).

Bogor (Buitenzorg) in Java has an extreme rain-forest climate with mean monthly temperatures varying only between 24.3 °C in February and 25.3 °C in October. The annual rainfall is 4370 mm, with 450 mm falling in the rainiest month and 230 mm in the driest. Although the daily temperature variations are negligible on cloudy days (about 2 °C), they can be as much as 9 °C on sunny days, the humidity of the air varying accordingly (Fig. 25). The rain usually falls at midday in the form of short downpours, after which the sun shines again. When the sun is at zenith, radiation is extremely strong, which means that the leaves directly exposed to its rays heat up several degrees (up to 10 °) above the already high air temperature. Consequently, even in water-vapor saturated air, large saturation deficits occur at the surface of the leaves (Fig. 24). (Overheating of as much as 10 ° to 15 °C has been recorded from unshaded *Coffea* leaves on clear days in Kenya.) Even in the rainy season, there are clear days in Bogor (Buitenzorg) when the humidity of the air sinks to almost 50% and the temperature rises to 30 °C (Fig. 25). As a

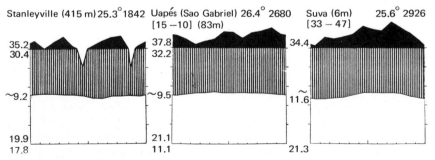

Fig. 23. Climate diagrams of stations in tropical rain-forest regions: the Congo, the Amazon basin, New Guinea. (From Walter and Lieth 1967)

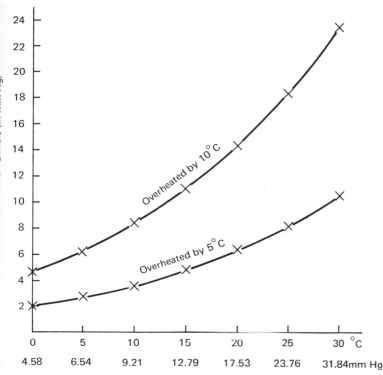

Fig. 24. Curves showing the saturation deficit at leaf surfaces if over-heating reaches 5 ° or 10 °C relative to the air temperature in water-vapor saturated air (lower line of figures = saturation pressure). Figures are millimeters of mercury

result, the saturation deficit of the over-heated leaves rises to about 40 mm Hg. Even in the wettest tropics, therefore, the leaves are at times exposed to extreme dryness for hours on end. To human beings, with a body temperature independent of the temperature of the environment, the air feels continuously oppressive and damp.

Investigators who have spent years in the tropical forests point out that, even in the perhumid region of Borneo, weeks without rainfall are observed time and again. The trees are thus exposed to recurrent dry periods, although the monthly means taken over many years give no indication of this. It is not surprising, therefore, that the leaves are adapted to a high degree to resist transpiration losses. They are equipped with a thick cuticle, are leathery but not xeromorphic (for example, the rubber tree *Ficus elastica, Philodendron),* and can radically reduce transpiration by closing their stomata, thus preserving a high degree of hydrature of the protoplasm. The cell-sap concentration is usually only about 10–15 bar. Many of these species can tolerate the dry air of heated apartments and are commonly found in Europe as indoor plants. The situation of species growing in forest shade, however, is vastly different. *The microclimate prevailing in the interior of a rain forest is*

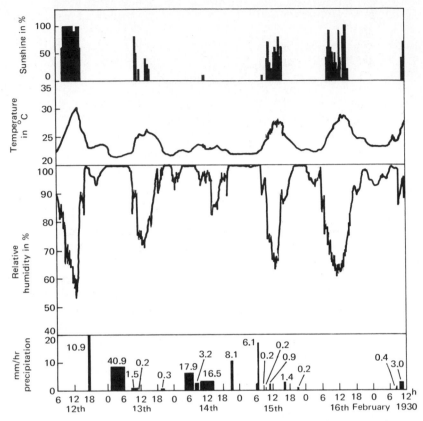

Fig. 25. Daily course of weather factors in Bogor (Java) during the rainy season (compare sunny February 12, when the air humidity sank to almost 50% with overcast February 14). The figures for rain indicate absolute quantities in millimeters. (According to Stocker; from Walter 1960)

much more equable—especially on the ground itself, where no direct sunlight falls. Variations in temperature are almost nonexistent at this point, and the air is constantly saturated. Owing to the high air humidity, even the slightest atmospheric cooling at night leads regularly to the formation of dew on the treetops, which serves to moisten the leaves of the lower layers as it drips down. Light conditions, too, are of great importance to forest plants. The irregular contours of the tree canopy and the strongly reflecting, leathery leaves ensure that light penetrates deep into the interior of the forest, although its intensity at the forest floor is very small and, depending on the structure of the forest, is measured as 0.5 to 1% of full daylight (as in temperate deciduous forests) or even as little as 0.1%.

2. Soils and Pedobiomes

Ignoring recent volcanic soils and alluvia, the soils in rain forests are usually very old, often reaching back as far as the Tertiary period. Weathering effects penetrate many meters down in silicious rock; the basic ions and silicic acid are washed out, leaving the sesquioxides (Al_2O_3, Fe_2O_3), so that a so-called laterization takes place and reddish-brown loam (ferrallitic soils or latosols) is formed, with no visible stratification into horizons. Litter decays very rapidly, and wood is destroyed by termites, which, because they live in subterranean colonies, are not obvious in the rain forest. The setting up of an experimental plot in the Congo was hampered by the presence of termite nests in 25% of the cleared ground. As a rule, the reddish-brown soil lies immediately below a thin litter layer. In consequence of the high rainfall, flat areas readily become water-logged and even swampy, so that the truly typical soil is only to be found on slightly elevated ground or on slopes.

The soils are extremely poor in nutrients and are acid (pH 4.5 to 5.5), which would, at first sight, appear to be contradicted by the luxuriance of the vegetation. In fact, *almost the entire nutrient reserves required by the forest are contained in the aboveground phytomass.* Each year, a part of this phytomass dies off and is rapidly mineralized, and the nutritional elements thereby released are immediately taken up again by the roots. Despite the high rainfall, there is no loss of nutrients due to leaching; the water in the streams has the electrical conductivity of distilled water and is at the most colored slightly brown by humus colloids. According to Went and Stark (1968), mineralization of litter does not necessarily have to occur in order to provide nutrients for the plants. On very poor sandy soils in a rain forest near Manaus on the Amazon, these authors found that feeding rootlets of the trees at a depth of only 2 to 15 cm possess a mycorrhiza by means of which they are directly connected with the litter layer through the hyphae of the fungi. The trees can thus exploit the fungi to obtain their nutrients in organic form directly from the litter, in just the same way as saprophytic flowering plants. In this manner, leaching of nutrients from the soil by rain is hindered. The quantity of leaves falling daily amounts to 4.5–12.6 g dry mass/m^2.

With such a rapid cyling of nutrients, the rain forest can grow for thousands of years on the same site, but as soon as it is deforested, and the wood burned, an intensive leaching of the suddenly mineralized nutritional reserves occurs. Only a small portion is adsorbed by the soil colloids and can be utilized for a few years by cultivated plants. If cultivation is discontinued, a secondary forest develops; these forests, however, never attain the luxuriant habit of the original forest. If this is once more cleared for temporary cultivation, fresh loss of nutrients takes place due to leaching until, after a series of such exploitations, the soil is capable only of supporting *Pteridium* or *Gleichenia* spp. After the burning of such areas, grasses like alang-alang *(Imperata cylindrica)* or other species gain a foothold.

The tropical rain forest on poor soils is inhospitable to settlers and is usually avoided by man; very primitive tribes often seek refuge in its depths.

A striking contrast is presented by the former areas of virgin forest on young, nutrient-rich volcanic soils which are today densely settled, cultivated land (Java, Central America, etc.).

The large nutritional reserves contained in the phytomass of the virgin forest depend upon their having been accumulated at a time when the rock was not so deeply weathered and the roots of the plants were still in contact with the parent rock. In totally depleted areas, virgin forest can develop once more if, as a result of soil erosion, the entire soil down to the underlying rock is removed and a new primary succession is initiated. If, however, the parent rock itself is poor in nutrients, as is the case with weathered sandstone or alluvial sand, then the nutrients only suffice to support a rather weak tree or heath population or a sparse savanna. These are a special type of pedobiome, termed *peinobiomes,* and cover very large areas. They have true podsol soils with a raw humus layer 20 cm thick (pH 2.8) and bleached horizons (pH 6.1), or even peat soils. The latter have been found in Thailand and Indomalaya, as well as in Guiana (*Humiria* bush, *Eperua* forest) and in the regions of the Amazon basin drained by the Rio Negro, the waters of which appear to be black owing to the presence of humic acid colloids from the raw humus soil. Peat soils have also been reported from Africa (in the Congo basin) and in the heather moors of the island of Mafia, and have been studied in great detail in northwestern Borneo, where extensive (14,600 km^2) raised forest bogs (helobiomes) with *Shorea alba,* etc., occur. These bogs commence immediately beyond the mangrove zone, and their peat deposits are up to 15 m thick (pH approx. 4.0). Heath forests *(Agathis, Dacrydium,* etc.) with *Vaccinium* and *Rhododendron* are also found on raw humus soils. The total area occupied by tropical podsol soils is estimated at 7 million ha.

At the other extreme, are the tropical limestone soils, i.e., *lithobiomes,* associated with peculiar topographical conformations such as are found in Jamaica. Limestone dissolves easily in damp tropical climates, the softer parts disappearing completely, with the harder parts left in the form of sharp ribs or ridges. The entire area turns into karst, and a honeycomb formation develops out of the original plateau. Circular depressions, or dolines, up to 150 m in depth, develop from the sinkholes caused by subterranean streams. If erosion continues still further, as it has in Cuba, all that finally remains of the network of ridges are solitary towers with almost vertical faces such as the "mogotes," or organ-pipe hills, of Cuba or the "moros" in northern Venezuela. The floor of the dolines is covered with bauxitic "terra rossa" soil upon which a wet evergreen forest develops. The bare limestone rock of the honeycomb ridges presents a heterogeneous habitat, depending on whether an alkaline soil (pH 7.7) can accumulate in the few depressions or not. This explains the extremely interesting flora, ranging from rain-forest species to cactus stands. In the above-mentioned areas, the annual rainfall is less than 1000 mm. A "limestone" vegetation has not been reported from true rain-forest regions.

The subject of halobiomes (mangroves) will be treated on p. 99.

3. Vegetation

a) Structure of the Tree Stratum, Periodicity and Flowering

The most conspicuous feature of a tropical rain forest is the large number of species constituting the tree stratum. As many as 40, or even 100, species can be counted on 1 ha, most of them belonging to different families. On the other hand, there are also forests containing species of only a few families, as in Indomalaya, where Dipterocarpaceae frequently dominate, and Trinidad, where the upper tree story consists of *Mora excelsa* (Leguminoseae). Large floristic differences exist between the forests of South America, Africa, and Asia (palms, for instance, are almost completely absent from African rain forests, although they are abundant in wet habitats in South America), and correspondingly, the forest types are also very dissimilar, but we shall only be able to discuss the features which are more or less common to them all.

The tree stratum reaches a height of 50–55 m, occasionally even 60 m, and three stories – an upper, a middle, and a lower story – are sometimes recognizable, although these are by no means always distinguishable. As a rule, the upper tree story is not compact, but consists of solitary giants which reach far above the other trees. It is the middle or lower stories which form a dense leaf canopy, and in such cases the trunk region is relatively free owing to lack of light and thus of undergrowth. This makes walking in the forest quite easy. However, the detailed structure of such forests varies greatly (Figs. 26 and 27), and generalizations should be treated with caution. The

Fig. 26. A rain-forest profile taken through the Shasha forest in Nigeria. The strip involved is 61 m long and 7.6 m wide. All trees taller than 4.6 m shown. The letters indicate different tree species. (After Richards; from Walter 1964/68)

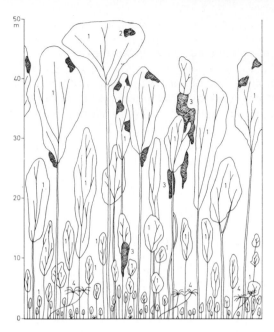

Fig. 27. Schematic profile through the Dipterocarpaceae rain forest in Borneo, length 33 m, width 10 m. (1) Tree species; (2) epiphytes; (3) lianas; (4) *Pandanus*. Herbs are entirely lacking. (After Vareschi; from Walter 1973)

trunks of the trees are usually slender with a thin bark; the crowns begin high up and are relatively small as a result of crowding. It is difficult to judge the age of the trees since annual rings are not present, but estimates based on rate of growth put them from 200 to 250 years. The roots reach further down than was hitherto assumed: 21–47% are in the upper 10 cm, and the rest is mostly below this and down to 30 cm, but 5–6% reach down as far as 1.3–2.5 m (Huttel 1975). A root mass of 23–25 t/ha was calculated (49 t/ha by other methods).

The giant trees achieve stability by means of enormous *plank-buttress roots* which reach, pillarlike, as far as 9 m up the trunk. From the base of the trunk, they radiate outward, much reduced in thickness, perhaps another 9 m.

The wetter and warmer the climate, the larger the leaves on the trees, although in a given species, the leaves that are exposed to light are always much smaller. For example, in an east African rain forest, *Myrianthus arboreus* showed a ratio of 8:1 (largest leaf 48 × 19 cm, smallest leaf 16 × 7 cm), and *Anthocleista orientalis* was 28:1 (largest leaf 162 × 38 cm, smallest 22 x 10 cm). Both belong to the lower tree story.

Any form of bud protection seems to be unnecessary for the trees of the rain forest, although the young leaves are sometimes protected by hairs, mucus, or succulent scales, or even by specially modified accessory leaves. Even though conditions are always favorable, growth is periodic. The growing ends of the twigs often give the appearance of "nodding foliage." This is because at such a rapid rate of growth no supporting tissue is formed at first and the young leafy shoots droop downwards. They are white or bright red initially and turn green only later as they become stronger. The rapid differentiation of the leaf tips results in the formation of "drip tips," which are found, for example, in 90% of the undergrowth species in Ghana. Experiments carried out in the forest have shown that leaves with drip tips are dry within 20 min of a rain shower, whereas those without are still wet after 90 min (Longman and Jenik 1974).

A special problem is presented by the *periodicity of development* in the continuously wet tropics, where there is no annual march of temperature. As has been mentioned, there is a periodicity in growth, and this also holds true for flowering. These phenomena are, however, not confined to a particular season since external conditions are always constant. In Malaya, it is reported that, in wet weather, the old leaves drop after the appearance of the young ones but that, in dry weather, they drop even before new leaves have sprouted so that the tree may be bare of leaves for a short time. Trees which shed their leaves developed this trait in a climate zone with a dry period. It can happen that, of two adjacent trees of the same species, one is bare while the other is in leaf or that the branches of one tree behave differently and put out leaves at various times. The same is true of the flowering period: individuals of the same species may bloom at different times or the branches of one tree may be in bloom at different times. These are all manifestations of an *autonomous periodicity* which is not bound to a 12-month cycle. Periods of 2 to 4 months, 9 months, and even 32 months are observable. This means that a rain forest has no definite flowering season but that there are always some trees in bloom. Although large and beautiful, the flowers are relatively inconspicuous against the predominating green background.

Many European tree species (beech, oak, poplar, apple, pear, almond) have been transplanted to seasonless tropical mountains. The general result has been that at first the trees retained their annual periodicity of leafshedding, growth, and flowering. With time, however, aberrations appeared among the inflorescences, the various branches behaved differently, and finally, on one and the same tree, every stage could be found: leafless, sprouting, blooming, and fruit-bearing branches.

The tropics differ from temperate latitudes in that there is always a short 12-hour day. Temperate-zone species, however, are long-day plants which bloom only when the days are long, as they are in summer. This explains the failure of these plants, as a rule, to bloom in the tropics, although lower night temperatures can replace the effect of the long day. In Indomalaya, *Primula veris* grows only vegetatively at an altitude of 1400 m, but at 2400 m it blooms and bears fruit abundantly. *Fragaria* spp. do not bloom at low altitudes, but produce many runners. In upland regions, runner formation is suppressed, and the plants bloom and bear fruit. *Pyrethrum* plantations are cultivated at 2000–3000 m in Kenya for their flowers, which do not develop at lower altitudes.

This endogenous rhythm of the plants adapts itself immediately to the climatic rhythm wherever there is one–as in the wet tropics, which have a short slightly drier season. On the mango tree, which is cultivated throughout the tropics, the few paler sprouting twigs on the otherwise dark crown are very obvious, but as soon as a dry season occurs, the growth and flowering of all the twigs and trees adapts to it. The teak, or djatti tree *(Tectona grandis)*, is never bare in western Java, which is always wet, but in the dry season of eastern Java it loses all its leaves.

But even in the wet tropics, there are species, such as the orchid *Dendrobium crumenatum*, which bloom on the same day over a large area. In order to open, the buds need the sudden cooling that ensues after a widespread storm. The buds of the coffee tree, too, open only after a short dry spell, and the reproductive organs of the bamboo develop only after a dry year. In such a uniform climate, certain species are extremely sensitive to even small changes in the weather.

Tropical tree species often exhibit the phenomenon of *cauliflory,* where flowers develop on older branches or on the trunk. This occurs in about 1000 tropical species as well as in the Mediterranean *Ceratonia siliqua* and *Cercis siliquastrum.* Most of these species belong to the lower tree story and are either chiropterogamous or chirpterochorous. The fruit-eating or insect-eating bats by which their flowers are pollinated or their seeds distributed can easily reach the cauliflorous flowers and fruits.

The question of the regeneration of virgin forest in the tropics has scarcely been investigated. When a giant tree falls, a large gap is left in which rapidly growing species of the secondary forest (balsa, = *Ochroma lagopus, Cecropia* and *Cedrela* in South America, *Musanga* and *Schizolobium* in Africa, *Macaranga* in Malaya) immediately develop. *Ochroma* puts out annual shoots 5.5 m in length, with light wood; those of *Musanga* are 3.8 m, and of *Cedrela,* 6.7 m. These trees are later ousted by the uppertree-story species. It has been observed that, in the rain forest, a tree will often have no saplings of its own species growing beneath it. This led to the conclusion that such a forest is mosaiclike in composition, each tree species being replaced by a different one. Only after a lapse of several generations can the species return to its original site so that a sort of rotation or *cyclic replenishment* occurs. A similar process can be observed in meadows of central Europe, although it is not certain whether this phenomenon is generally valid for all species-rich plant communities. It would, however, provide an explanation for the observation that none of the species competing is able to attain absolute dominance and a species-rich, mixed population is the rule.

b) Herbaceous Layer

About 70% of all species growing in a rain forest are *phanerophytes* (trees), and besides this, phanerophytes are quantitatively absolutely dominant. It is difficult to distinguish the shrub and herbaceous layers from each other since herbaceous plants such as bananas and various other Scitamineae can reach a height of several meters. Even in the presence of good light conditions on the ground, undergrowth is often lacking, perhaps because of competition with the roots of the trees nitrogen and other nutrients. The lower herbaceous plants have to make do with little light, and as indoor plants in Europe, they also manage with very weak illumination (*Aspidistra, Chlorophytum, Saint-paulia*, the African violet).

The frequent occurrence of velvety or variegated leaves with white or red patches or a metallic shimmer deserves mention.

At such high humidities, guttation plays an important role in the herbaceous layer, and the hydrature of the protoplasm in correspondingly very high (cell-sap concentration only 4–8 bar). The cell-sap concentration in ferns, which have a less efficient conducting system, is 8–12 bar. Heterotrophic flowering plants, saprophytes or parasites, occur but play only a minor role, whereas lianas and epiphytes are particularly important.

A large variety of synusiae is undoubtedly present in connection with the different light and water conditions, but so far they have not been investigated. The groups discussed below form a number of typical synusiae.

c) Lianas

In dense tropical rain forests, autotrophic plants have to struggle primarily for light. The higher a tree, the more light its leaves receive and the greater its production of organic matter. But to reach the light, a trunk has to be formed over many years, a process involving the investment of large quantities of organic substance. Lianas and epiphytes attain favorable light conditions in a much simpler manner. The former exploit the trees as a support for their rapidly upward growing, flexible shoot instead of developing a rigid stem (Fig. 28), whereas epiphytes germinate in the topmost branches of the trees, which thus serve as a base (Fig. 29).

Lianas attach themselves to the supporting tree by various means. The scrambling lianas climb among the branches of the trees using divaricating branches armed with spines or thorns which prevent them from slipping, as for example the climbing palm *Calamus* or the *Rubus* lianas. The root climbers put out adventitious roots to fasten themselves to cracks in the bark or encircle the trunk (many Araceae). The winding or twining plants have long, rapidly-growing, twining tips to their branches and very long internodes upon which the leaves are at first underdeveloped, and the tendril-climbing lianas possess modified leaves or side shoots which are sensitive to touch stimuli and serve as grasping organs.

Light is essential for the growth of lianas, and for this reason they are common in forest clearings and grow simultaneously with the trees of the forest regrowth until they finally reach the level of the forest canopy. Tropical lianas are long-lived compared with those of temperate latitudes. Their axial organs are equipped with secondary thickening, but their stems remain pliable, enabling them to follow the growth of the supporting trees. The woody structure is not compact: the ligneous portions are permeated by strands of parenchymatous tissue or broad medullary rays (anomalous

Fig. 28. Virgin forest in Siam. Liana on the trunk is *Rhaphidophora peepla*. (Photo J. Schmidt)

Fig. 29. Epiphytes on the branch of a tree in the rain forest in Brazil. Rosettes of Bromeliaceae on the extreme right, three *Rhipsalis* spp. hanging down, and the lanceolata leaves of *Philodendron cannaefolium*. (Photo H. Schenck)

thickening). In cross section, the vessels are large and have no dividing walls so that, despite the small diameter of the pliable stem, the crown of the liana receives a sufficient supply of water. Should the supporting tree die and decay, the lianas still remain fastened to the tops of the other trees and their stems hang down like ropes. They often slip down and lie on the ground in loops around their own base, but the shoot tips work their way up again. If this happens several times, the stem can attain a great length: a total of 240 m has been measured in *Calamus*. Complete deforestation is especially favorable for the development of lianas, which for this reason are more numerous in secondary than in virgin forests, where they prefer the margins.

Ninety percent of all liana species are confined to the tropics; in western India, 8% of all species are lianas. In central Europe, few species of woody lianas are found: the root-climbing ivy *(Hedera helix),* the scrambling traveler's joy *(Clematis vitalba),* the twining wild grape *(Vitis sylvestris),* and the winding *Lonicera* species. Although European blackberries (*Rubus* spp.) do not grow high above the ground, in New Zealand they grow thick as a man's arm and reach up to the treetops. The difficulty of water transport is probably what confines lianas mainly to the wet tropics. In a dry climate, the high suction tension (low water potential) built up in the leaves causes disruption of the water column by overcoming the cohesion in the wide vessels. In temperate climates, lianas are at their most abundant in wet floodplain forests.

d) Epiphytes and Hemiepiphytes (Stranglers)

Particularly characteristic of the tropical rain forest are the epiphytic ferns and flowering plants. These plants only occur, however, in forests which are frequently wetted; high air humidity is not enough. A large number of interesting adaptations can be found (Fig. 29), and in Liberia, 153 species have been investigated ecologically (Johansson 1974).

The fact that epiphytes germinate high up on the branches of trees provides favorable light conditions but brings with it the problem of water supply, since the constant water reservoir otherwise provided by the soil is lacking. Epiphytic habitats can be compared with rocky habitats, and epiphytes do, in fact, grow well on rocks if light conditions are satisfactory. Since they can take up water only while it is actually raining, the frequency with which they obtain water is of greater importance to them than absolute quantity of rain. On windward, mountainous slopes, frequency of rainfall is greater than on flat land on account of the ascending air masses. This is why montane forests, especially the cloud forests (Fig. 31), where leaves are constantly dripping, are rich in epiphytes. To be able to withstand the large intervals between rain showers, epiphytes either must be capable of resisting desiccation without undergoing damage, as is the case with many epiphytic poikilohydric ferns, or must store water in their organs, as do the succulents of dry regions. A whole series of cacti has changed over to an epiphytic way of life (*Rhipsalis, Phyllocactus, Cereus* spp.). Epiphytes conserve their water economically just as the succulents; many orchids possess leaf tubers as water reservoirs, and the majority of orchids, bromeliads, peperomiads and other epiphytes have succulent leaves. The velamen of the aerial roots of orchids ensures rapid uptake of water during showers; bromeliads are equipped with water-absorbing scales which take up the water collecting in the funnels formed by the leaf bases or serve to retain the water by capillary forces and then suck it in.

The roots in epiphytic bromeliads serve only as adhesive organs and are completely absent in *Tillandsia usneoides,* which appears superficially similar to the lichen *Usnea. Myrmecodia, Hydnophytum,* and *Dischidia* spp. develop special cavities, sometimes inhabited by ants. Ferns, which cannot tolerate dehydration, can produce their own soil by collecting litter and detritus between their funnellike, erect leaves *(Asplenium nidus)* or with the aid of special overlapping "niche" leaves *(Platycerium).* In this way, a soil is formed which is rich in humus and which retains water so that the roots growing into it are well provided for. In a forest densely populated by epiphytes, epiphytic humus can amount to several tons per hectare. In this way, a new biotope is created far above ground level, which can even be considered an ecosystem. Dripping water and dust bring in nitrogen and other nutrients, and ants colonize the area, build their nests, and also drag in seeds which then germinate and grow into flowering plants. Such "flower gardens" occur in South America and harbor a special fauna and microflora. Mosquito larvae, water insects, and protista inhabit the funnels of the bromeliads, which often achieve considerable dimensions. It should be mentioned that the insectivorous *Nepenthes* (pitcher plants) as well as certain *Utricularia* species can also grow epiphytically.

Epiphytes are distributed by means of spores (ferns), dustlike seeds (orchids), or berries (cacti, bromeliads) that are eaten by birds, the seeds reaching the branches of trees in the droppings. Many epiphytes are able to endure a long period of drought. Among them are orchids, which lose all their leaves, densely scaled *Tillandsias,* and poikilohydric ferns. Epiphytes are also found in the dry type of tropical forest.

Coutinho (1964, 1969) demonstrated the occurence, in some epiphytes in Brazil, of the diurnal oxygen exchange (*Crassulacean acid metabolism,* or CAM) i.e., the *nocturnal absorbtion* of CO_2 through open stomata and its binding as organic acids (carboxylation). The organic acids are decarboxylated during the day, and the carbon dioxide thus set free is immediately assimilated while the stomata are still closed. By means of this process, loss of water due to daytime transpiration can be avoided. Crassulacean acid metabolism has been frequently observed in the succulents of dry regions. Medina (1974) studied this phenomenon in the Bromeliaceae.

In semiarid wooded regions there are no epiphytes, but in their place, hemiparasites such as mistletoes *(Loranthaceae)* are often found on the trees. Mistletoes are even to be seen on the stem-succulents on the edge of the desert. Mosses and hymenophyllous filmy ferns, on the other hand, require constant humidity and are thus the typical epiphytes of the mist forests, in addition to the *epiphyllic,* i.e., microscopic, algae and mosses that grow on leaves.

Hemiepiphytes occupy a position intermediate between lianas and epiphytes. Many Aroideae germinate on the ground and then grow upward as lianas, usually as root climbers. In time, the lower part of the stem dies off, so that the hemiepiphytes turn into epiphytes although still remaining in contact with the ground by means of their aerial roots. Even more striking are the "strangling" trees, of which the numerous "strangling" figs are the best known (*Ficus* spp.) Such "stranglers" occur in many different families, as for example the *Clusia* spp. (Guttiferae) of South America, the New Zealand *Metrosideros* (Myrtaceae), and many others. They germinate as epiphytes in the fork of a branch and at first put out only a small shoot and a long root. The latter rapidly grows down the trunk of the supporting tree and enmeshes it. Only when the root reaches the ground does the shoot begin to grow; at the same time, the roots thicken and prevent secondary thickening in the supporting tree. The tree is thereby "strangled" and dies and the wood rots. The root mesh of the "strangler" unites to form a trunk bearing a broad crown. Such trees can attain enormous dimensions, and it is no longer apparent that they began life as epiphytes. Palms, which do not have secondary growth, are not strangled and may survive, until their leaves are too greatly overshadowed by the crown of the "strangler".

4. Anomalies in the Equatorial Zone

Typical climate diagrams for Zonobiome I show a perhumid diurnal climate with two equinoctial rain maxima, the rains occurring when the sun reaches its zenith, exactly at midday. This type of climate, however, is only found in some parts of the equatorial zone. Regions with wet monsoon winds (Guinea, India, Southeast Asia) exhibit a particularly well-developed rain maximum only in summer and, in addition, have a brief period of dryness or even drought (trend toward ZB II). The vegetation still consists of rain forest, although leaf fall and flowering are clearly connected with a particular season; we can speak of this vegetation type as *seasonal rain forest*.

Ghana, which is not affected by the monsoon, still has two rain maxima with an intervening period of drought, a situation similar to that seen in eastern Africa, where the monsoon winds are dry and rain falls when the wind changes. A distinction is drawn between a longer (and heavier) and a shorter (and lighter) rainy season. In coastal regions and in Somaliland, the rainfall decreases to such an extent that, in places, no wet season can be detected on the climate diagram and the vegetation is desertlike. This is a Zonoecotone I/III.

Trade winds also affect the character of the climate. The southeast trade wind is wet and is responsible for the rain-forest climate in southeastern Brazil, eastern Madagascar and in northeastern Australia from the equator up to 20 °S and beyond. On the other hand, the northeast trade wind in the southern Caribbean brings rainfall only where the wind is confronted with an obstacle such as mountains. Thus mountainous Venezuela exhibits considerable variety in climate and vegetation (Fig. 30).

Venezuela lies between the equator and 12 °N and provides examples of every altitudinal belt from sea level up to the glaciated Pico Bolivar (5007 m).

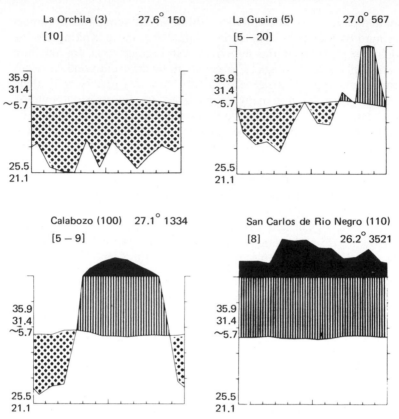

Fig. 30. Climate diagrams of a north-south profile through Venezuela: (1) offshore islands; (2) coastal stations; (3) typical trade-wind climate (rainy season, 7 months); (4) perpetually wet climate of the Amazon basin. (According to Walter and Medina 1969)

The northern part of the country is exposed to strong trade winds from November until March and it rains in the lowlands only during the 7 calm summer months with their ascending air masses and numerous thunder storms.

Only in the south, in the Amazon basin, does no month have rainfall less than 200 mm. In Venezuela, the annual rainfall rises steadily toward the south, from 150 mm (on the island of La Orchila) to more than 3500 mm. In mountainous regions, precipitation increases rapidly on the windward side up to cloud level, but decreases again above this. At the same time, the mean temperature falls by 0.57 °C per 100 m increase in altitude. The inner valleys of the Andes, situated in a rain shadow, are extremely dry.

Figure 31 shows schematically the changes in vegetation with increasing altitude. The sequence is determined by the amount of rainfall; decreasing temperature has no marked effect below 2000 m. The following vegetational sequences, in ascending order, are found: cactus semi-desert, thornbush, deciduous forest, cloud forest, high montane forest, with abundant

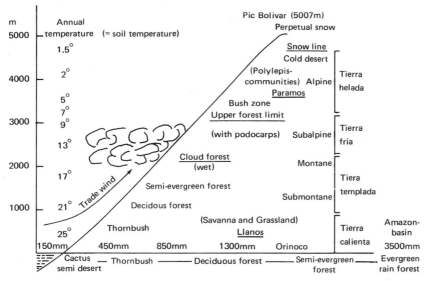

Fig. 31. Schematic representation of the additudinal belts in Venezuela. Man annual temperature is in degrees Centigrade. The abscissa shows change in vegetation from north to south with increasing rainfall (in millimeters)

Podocarpus, upper forest limit, and the alpine belt (páramos, cold desert, permanent snow).

In the driest areas is cactus semidesert; these succulents store so much water that they can easily survive a dry period of 6 months or more (Fig. 32). With a slight increase in rainfall, thornbushes and ground bromeliads occur, and impenetrable thickets similar to the Brazilian caatinga are found. At an annual rainfall of 500 mm, umbrellalike thornbushes predominate *(Prosopis, Acacia)* and are accompanied by *Bursera, Guaiacum, Capparis,* and *Croton* spp., as well as by *Agave* and *Fourcroya.* The tree cactus *Peireskia guamacho,* which has normal leaves and is considered to resemble the ancestral form of all cacti, grows here. During the dry season, these plants are leafless. These cactus semideserts are used only as grazing land for goats.

As the rainfall increases, so too does the number of arboreal species, until true deciduous forests, extremely rich in species, commence. The tree layer is 10–20 m high, and only the Bombacaceae *(Ceiba,* etc.), with their thick, water-storing trunks, and the lovely flowering *Erythrina* spp. extend above this level. During the dry season, such a forest presents the appearance of a European or eastern American deciduous forest in winter, although a few of the trees are already beginning to bloom at this season. [A distinction must be made between dry and wet tropical-deciduous forests; the latter, with a rainfall of up to 2000 mm, attains a height of over 25 m and contains valuable timber species such as *Swietenia* (mahogany) and *Cedrela.*]

The deciduous forests are sometimes cleared for coffee plantations under shade trees and for the cultivation of maize, sugarcane, pineapple, etc.

Fig. 32. Cactus-thornbush semidesert with *Cereus jamaparu,* between Barquisimeto and Copeyal (Venezuela) in February (the dry season). (Photo E. Walter)

Pastureland can also be produced by sowing *Panicum maximum.* Although the forests are poor in lianas, they are rich in epiphytes (drought-resistant ferns, cacti, bromeliads, and orchids).

In the regions with more rain and a still shorter dry season, semievergreen forests occur, in which only the lower bush and tree layers consist of evergreen species. And with even more rain, the evergreen-tropical rain forest commences.

A peculiar situation exists in Venezuela in the Llanos region of the Orinoco basin, which extends far into Colombia. Instead of the deciduous forests expected in this climatic zone, grasslands suddenly occur, dotted with small groves or solitary trees. These are true grasslands or savannas. Although the grassland, nowadays used for grazing, is regularly burned, fire cannot be regarded as the primary reason for the absence of forests. The peculiar soil conditions prevailing in this area will be discussed later (pp. 88–91. The following vegetational formations of Venezuela are determined edaphically (pedobiomes), or by the relief of the terrain, rather than climatically: the mangroves on the seacoast and in the mouths of rivers, the beach and dune vegetation, the freshwater swamps and the aquatic plant communities, the floodplain forests and the vegetation of shallow rocky soils. In addition, the different altitudes in the mountains (orobiomes) must also be considered.

When the path of the trade winds is interrupted by a mountain range, condensation takes place due to cooling of the ascending air masses, and clouds and rain result. Because the force of the trade winds lets up in the late evening, the nights and early hours of the morning are clear, but for the rest of the time, a layer of cloud is present at a certain altitude, so that this altitudinal belt is shrouded in mist during the day. This means that, apart from the orographic rain, the trees are also loaded with drops of water from condensing mist. Because of the saturation of the atmosphere with water vapor, transpiration is entirely lacking. This extremely damp climate, which is cooler, too, on account of the altitude, favors the development of hygrophilic tropical mist forests, which are typical of all wind-exposed tropical mountain regions. In Venezuela, the mean annual temperature decreases by 0.57 °C per 100 m increase in altitude, while the diural climate remains in effect; i.e. the daily temperature fluctuations are larger than the difference between the mean temperatures of the coldest and warmest months of the year. In the mountains, therfore, it is not possible to speak of temperature-dependent seasons. The cooler mountain climate does not correspond to the climate of the temperate zones of the earth. Daily temperature fluctuations are minimal on clouded rainy days, while on clear days they increase with increasing altitude. Above a certain altitude, every day of the year is subject to alternating frost: it is warm during the day and falls to below 0 °C at night. Plants growing at these altitudes must be able to tolerate this change between freezing at night and thawing during the day. The cold desert begins at even higher altitudes, where the daytime temperature does not rise above 0 °C, and where the soil is subject to only little thawing. Where the soil is always frozen, the ground is covered with perpetual snow.

It is clear that, in Venezuela, the altitudinal belts of the orobiome exhibit certain peculiarities. The deciduous forests are an extrazonal phenomenon resulting from the dry trade winds and will be dealt with in more detail in connection with ZB II. Small areas of Zonobiome I/II (semievergreen forest) occur in mosaiclike distribution, and in addition, there is the extensive peinobiome in the catchment area of the Río Negro, the rivers carrying black water containing humus and colloids (heath and moor vegetation, pp. 49–50).

A wide variety of conditions can also be found in equatorial eastern Africa.

5. Orobiome I – Tropical Mountains with a Diurnal Climate

a) Forest Belt

The succession of altitudinal belts seen in tropical mountains rising from an extrazonal dry plain has been shown in Fig. 31. Elsewhere, too, ascending air masses bring increased precipitation to mountain slopes (unless they are sheltered from the wind), and even if there is a dry period in the lowlands, it becomes shorter with increasing altitude, or even disappears altogether. It is therefore not surprising that the montane forests are particularly luxuriant and rich in epiphytes which obtain plentiful water. Tropical mountain slopes are generally very steep, so that the soil is well drained and swamps are absent. In ascending mountains of Zonobiome I, the change in vegetation with decreasing temperature is initially scarcely noticeable. Finally, at cloud level, where a state of maximum humidity prevails, the *cloud forests* commence. They are not connected with a definite altitude but rather with the cloud level itself, which, in turn, is dependent upon the humidity at the foot of the mountain. The greater the humidity at the base of the mountain, the lower the cloud level. In a climate with both a wet and dry season, the clouds are higher during the dry season. Cloud forests can be found between 1000 and 2500 m above sea level, or even higher still, and the variety of temperature conditions which can prevail accounts for the floristic differences exhibited by such forests. Even the height of the tree stratum decreases with increasing altitude, and the trees of very high forests are gnarled and stunted by the wind (elfin forest). But the common feature of all cloud forests is their profusion of epiphytes. Whereas the number of warmth-loving, epiphytic flowering plants decreases with increasing altitude, the ferns, lycopods, and, above all, the filmy ferns (Hymenophyllaceae) and mosses become more abundant. The branches of the trees are typically draped with curtainlike mosses, themselves covered with water droplets, and filmy ferns, which roll up as soon as the humidity drops slightly below 100%, form a green covering on the trunks and branches. The forest floor is often carpeted with bright-green *Selaginella* spp., and tree ferns, which prefer a damp, cool climate, are numerous. In tropical mountain regions, the wettest altitudinal belt is often characterized by palms (South America) or dense bamboo groves (eastern Africa).

Increasing altitude also brings with it changes in the soil. The reddish-brown loams of the lower belts are gradually replaced by more yellow types; at the same time, a humus horizon appears, and the clay content decreases. Further up, a slight podsolization is detectable, and eventually true podsols, with a leached horizon and raw humus, occur. In the perhumid cloud belt, gley soils are found.

b) The Upper Forest Limit

Precipitation decreases rapidly above the cloud belt, and if the forest extends further upward, the leaves of its trees become smaller and more xeromorphic. Conifers of *Podocarpus* species, which possess narrow, leafshaped structures in place of needles, are present. Mosses are replaced by beardlike lichens, and finally, at the upper forest limit, a shrub zone commences. This occurs at a much lower level in the tropics than in the subtropics: an altitude of from 3100 to 3250 m above sea level is given for the Venezuelan Andes. The shrub zone is narrow, and the shrubs themselves get smaller the higher up they grow. In Venezuela, the last are found, sheltered by rocks, at 3600 m.

The question of what factors are responsible for determining the timberline in the tropics is still unresolved. The fact that precipitation decreases with increasing altitude would suggest that aridity sets the limit. On the other hand, it might well be that frost is the limiting factor since at this elevation temperatures can drop below the freezing point. The author's own investigations in Venezuela, however, suggest that in all probability it is the soil temperature that is of ultimate importance, although, as always with such phenomena, a variety of other factors is involved. In the equatorial zone, where the climatic fluctuations are of a diurnal rather than an annual nature, the temperature variations do not penetrate very deeply into the ground. On shady ground, the temperature at a depth of 30 cm is constant throughout the year and is the same as the mean annual air temperature measured by the meteorologists. With only a spade and a thermometer, it is possible to determine the annual mean temperature at any spot in the tropics in a few minutes. In dense forests, the temperature is constant immediately below the ground surface, and it is this temperature that is decisive for the germination of tree seeds and for the root system of the plants. Although the minimum temperature requirements for root growth of tropical trees are unknown, it is known that the enzymes responsible for protein synthesis in the roots have a temperature minimum well above 0 °C. This means that at temperatures considerably above the freezing point, tropical species can be "chilled" and die. *Ceiba* seedlings, for instance, only grow at temperatures above 15 °C. Assuming the temperature minimum of germination or root growth of trees situated at the timberline to be around 7–8 °C, this would coincide exactly with the temperature of the soil at the timberline in Venezuela, where the vegetation is made up of typical tropical species and is completely lacking in holarctic species. If this assumption holds true, it would explain the higher timberline in the subtropics. In these regions, there is already an annual march of temperature, and in the summer the temperature of the soil rises considerably above the mean annual temperature. Arboreal species are able to exploit this favorable season, which is lacking in climatic zones where the fluctuations in temperature are of a diurnal rather than an annual nature. At an altitude of 3864 m in the Pamir, for example, the August temperature at a depth of 1 m remains above 10 °C, and the maximum temperature at a depth of 40 cm exceeds 20 °C. The temperature minimum of the roots of species typical of a tropical shrub zone must be lower than that of forest trees, an explanation apparently supported by the presence of many holarctic species of the family Ericaceae and the genera Hypericum, Ribes, etc.

c) Alpine Belt

In the wet tropics, the alpine belt is termed the *páramo*. It is described as being perpetually wet, misty, desolate, and cold, although annual temperature and precipitation curves are not given. For Venezuela, such curves have been made available to the author, and it can be seen from Fig. 33 that there is little rain during the trade-wind period, between November and March. In January, the author experienced a whole week in the páramo under a cloudless sky when the cloud layer was situated at a lower altitude. The hourly

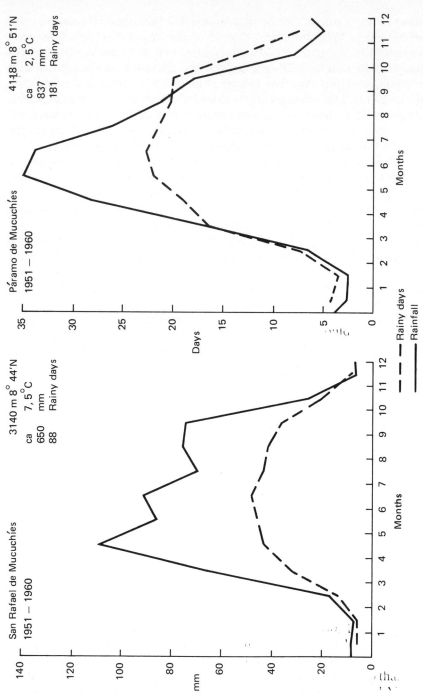

Fig. 33. Rainfall distribution in the páramo belt. The drier season is from November to March

temperature values (Fig. 34) reflect the lack of incoming radiation or the intense incoming radiation (February 10) and high net outgoing nocturnal radiation (February 12). The coldest day of 1967 followed the warmest almost immediately. The air at 3600 m usually reaches a daytime temperature of 10 °C during the dry season, although the temperature goes down to freezing at night. Furthermore, the plants themselves are exposed to much greater extremes than are the thermometers in the shelters. Apparently, this continuous freezing and thawing does the plants no harm, because it coincides with the main flowering season. The upper soil layers harboring the roots of the páramo plants warm up during the day at this season to temperatures above the annual mean. A rocky habitat is apparently more favorable than a wet, cold soil. On the basis of soil measurements, the mean annual temperatures can be given as follows: at 3600 m, 5.0 °C (agrees with

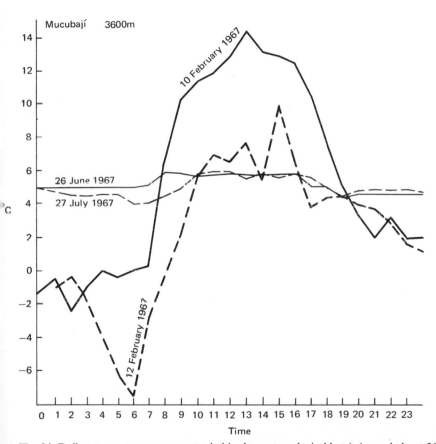

Fig. 34. Daily temperature curves, recorded in the meteorological hut (páramo belt, at 3600 m above sea level) on June 26 and July 27 during the rainy season (diurnal fluctuations only 1.6° and 2.0 °C, respectively) and on February 10 (hottest day) and February 12 (coldest day) during the dry season (fluctuations were 17.0° and 17.5 °C, respectively). Temperature maximum, 14.5 °C; temperature minimum, − 7.5 °C

meteorological data); at 3950 m, 3.9 °C; and in the firn, or permanent snow, at 4765 m, −1.5 to −3.5 °C. A decrease in temperature forces the plants to spread their roots nearer the surface, and this inevitably leads to sparser vegetation, until vegetation finally ceases altogether some 100 m below the permanent snow line. This *cold desert belt,* with frost-rubble soils due to freezing and thawing, is typical of tropical mountain regions. In higher latitudes (in the Alps), the plants take advantage of a favorable season to grow in snow-free places even above the snow line. The soil in the páramos itself remains moist during the dry season, so that the vegetation does not suffer from drought and gives a hygromorphic impression. In Columbia, besides Páramos with dry periods, constantly wet Páramos with low ground cover vegetation, dwarf bamboo, grasses and mosses were investigated also.

The floristic composition of the páramos in South America, Africa, and Indonesia varies greatly, each area having its own peculiarities. It is striking that, apart from the low plants which give the impression of hugging the ground, there are also tall species, mostly Compositae, with a proper stem and large, bushy, upright leaves covered with thick, white hairs (Fig. 35). These plants are, in the Andes, *Espeletias* (27 species), in equatorial African regions, *Senecio* tree species, and, in Indonesia, species of *Anaphalis.* The woolly, candlelike form of *Lupinus* and *Lobelia* is also conspicuous. On Kilimanjaro, extremely hairy species of *Helichrysum* grow as far up as 4400 m, and it is assumed that their hairiness serves as a heat insulation providing protection against sudden variations in leaf temperature. At such altitudes, the passage of a cloud on a sunny day invariably leads to a sharp drop in temperature. The upper vegetation limit lies at about 4000–4500 m and probably coincides with a mean annual temperature of about 1 °C.

Fig. 35. Páramo de Mucuchíes, 4280 m above sea level (Venezuelan Andes). In the *foreground,* a cushion of *Aciachne pulvinata* in the process of disintegrating. *Behind,* numerous *Espletia alba.* (Photo E. Walter)

It is especially remarkable that in the Venezuelan Andes, in the middle of the alpine belt, at an altitude of approximately 4200 m and with a mean annual temperature of 2 °C, small stands of *Polylepis sericea* trees (Rosaceae) occur. They are invariably found on steep east- or west-facing talus slopes consisting of large boulders, and are exposed to the sun in the morning or afternoon. The roots of *Polylepis* sometimes go down as far as 1.5 m, and the only possible explanation for this isolated occurrence of trees 1000 m above the timberline is that the talus slopes provide particularly favorable temperature conditions. Incoming radiation warms up the air layer nearest the ground very strongly; the heavier cold air between the boulders flows out at the lower end of the talus, so that warm air is drawn in between the boulders in the upper part. This explanation finds support in the observation that the lower part of the talus slope is devoid of trees and is often even completely bare. Exact temperature measurements on talus slopes in Mexico have proven this to be the case.

The altitudinal belts on the African volcanoes are slightly less humid since they rise from a savanna zone (Mt. Elgon, Mt. Kenya, Kilimanjaro). The soil temperature at the upper limit of the forest *(Hagenia, Podocarpus)* is similar to that in Venezuela. *Erica arborea* plays a considerable role in the lower alpine belt.

6. The Biogeocenes of Zonobiome I as Ecosystems

The tropical evergreen forest is one of the most complicated of all plant communities. The individual biogeocenes are still totally uninvestigated. In one of the most thoroughly studied forests, in northeastern Australia, 18 lists of forest stands reveal 818 species taller than 45 cm, 269 of which were trees. For the fieldwork itself, a whole year was required, and the determination of the species took again as long. The computer, however, was able to show within the space of 13 min that 6 floristic groups were represented in the 18 lists, with 3 stands in each group. These floristic groups are the result of climatic differences, although nutrients, water supply, and altitude also play a role. From a further 70 forest-stand lists, spread across 20 degrees of latitude, 24 life-forms and structural characteristics were considered. The computer revealed an arrangement corresponding to geographical regions (which were already known) and also showed that the combinations in which the plants occur are continually changing.

A classification into ecosystems is obviously an extremely complicated matter.

The luxuriance of the vegetation would suggest a very high primary production. The preliminary estimates of about 100 t dry substance/ha proved to be too high. It must be borne in mind that the phytomass of tropical rain forests has a very high water content (75 to 90% for the herbaceous parts) and that, although the green leaves are able to assimilate CO_2 throughout the year, nocturnal respiration losses are especially large owing to the high temperatures. Wood production in tropical forest plantations can attain values of 13 t/ha, which is only about twice that of a good European beech forest and occurs only because the vegetational period is twice as long. The leaf area index (LAI), or the ratio of total leaf area to the total ground area covered in a particular ecosystem, is of great significance for the productivity of the biogeocene. This index has been determined for a tropical rain forest on the Ivory Coast, with the very low result of 3.16. However, this experimental plot cannot be considered to be representative. The gross production proved to be very large (52.5 t/ha), but 75 % of the organic substance produced is lost by respiration: respiratory losses of the leaves, 16.9 t/ha; of the axial organs, 18.5 t/ha; and of the roots (estimated), 3.7 t/ha (total, 39. 1 t/ha). The losses due to respiration in a central European beech forest, on the other hand, amount to only 10.0 t/ha, i.e., 43% of the gross production of 23.5 t/ha. The primary production of this tropical rain forest is therefore no higher than that of a well-managed beech forest in central Europe. The values for primary production (gross production minus respiration) are as follows:

$52.5 - 39.1 = 13.4$ t/ha   Tropical rain forest
$23.5 - 10.0 = 13.5$ t/ha   Beech forest

In tropical forests, small mosaics consisting of three phases can be distinguished. The first is a young phase with rejuvenating stock and an increasing phytomass. The next is the optimum phase with a maximum, constant phytomass. Finally, there is the ageing phase with decreasing phytomass. The stand on the Ivory Coast was clearly a thin juvenile phase.

Data on the optimum phase of a luxuriant tropical rain forest reveal that the total phytomass amounts to 350–450 t/ha, with an LAI of 12–15 and an annual gross production of 120–150 t/ha. This corresponds to a primary production of 30–35 t/ha, of which 10–12 t/ha are accounted for by litter. Soil respiration corresponds roughly to the quantity of litter, but the larger part of the primary production is probably mineralized aboveground (dead, upright trees and epiphytes). About 106 kg of nitrogen/ha are returned to the soil in the Amazon region annually, as compared with only 2.2 kg of phosphorus/ha. Impoverishment of the secondary forest is probably mainly a question of phosphorus deficiency. Nitrogen is also added from the atmosphere in the course of the frequent thunderstorms.

A forest investigated in Thailand, with 2700 mm annual rainfall and an annual temperature of 27.2°, had an aboveground phytomass of 325 t/ha, which probably indicates a total of 360 t/ha. This value increased still further by 5.3 t/ha per year, over the 3 years of the study. The LAI was 12.3; gross production was 124 t/ha, and respiratory loss was 95 t/ha (= 76%), which gives an annual primary production of 30 t/ha.

Characteristic of the fauna is that more than half the mammals inhabit the crowns of the trees and posses prehensile tails. Very large numbers of birds, as well as invertebrates (termites), are found both below- and aboveground. Nevertheless, the zoomass is never large.

The typical tree-crown herbivores of the Neotropical realm are the sloths *(Cholopus, Bradypus)*. Their habits have been studied in great detail by Montgomery and Sunquist (1975). The total zoomass of these animals was 23 kg/ha, and the quantity of leaves consumed annually amounted to 53 kg/ ha, i.e., 0.63% of the total leaf production. Their excrement decays slowly and represents a nutrient reserve in the soil.

The leaf-cutting ants *(Atta)* exert an influence on the ecosystem by reason of their selective habits (Haines 1975). They drag material from the trees of the secondary forest (up to 180 m aboveground) into their underground nests, which have a diameter of 10 m. Here they cultivate the fungus gardens that provide them with their food.

Zonoecotone I/II – Semievergreen Forest

The semievergreen tropical rain forest is the zonoecotone between Zonobiome I, evergreen rain forest, and Zonobiome II, the tropical summer-rain region with deciduous forests. It is thus a transitional zone with a diffuse mixture of both types of vegetation.

In discussing the vegetation of Venezuela, the following zonation was mentioned in connection with decreasing annual precipitation and lengthening dry season: evergreen rain forest, semievergreen forest, and deciduous forest. This zonation is only rarely observable within the equatorial climate zone, since a stepwise increase in rainfall, such as is found in Venezuela, is exceptional. As one leaves the tropics, however, such zones can be distinguished. In moving away from the equator, the tropical climate zone with zenithal summer rains is left, absolute rainfall decreases continuously, and the rainy season becomes shorter. In contrast to Venezuela, the annual march of temperature becomes more marked, and the dry season occurs at the cool time of year. But since the latter means a rest period for the vegetation, the temperature variations are of little significance from this point of view.

It has already been mentioned that, when a short dry period occurs in the very wet tropical region, the endogenous rhythm of tree species adapts itself to the climatic rhythm. The general character of the forest remains unchanged, but many trees lose their leaves at the same time, or sprout or flower simultaneously, so that the vegetation does in fact exhibit definite seasonal changes in appearance (seasonal rain forest).

If the dry season becomes even longer, the type of forest changes. The upper tree story is made up of deciduous species (in South America, these are the large, thick-trunked Bombacaceae and the beautiful flowering *Erythrina* species), whereas the lower stories are still evergreen, so that it can be termed *tropical semievergreen forest*. These forests can be seen by boat in passing through the Panama Canal.

With a further decrease in rainfall and a lengthening of the dry season, all of the arboreal species are deciduous, so that the forest is bare for long or short periods of time. These are the moist or dry deciduous tropical forests. Figure 36 shows climate diagrams for forests of this type in India, where, in the monsoon-rain region, the transition can be particularly well observed.

The question now arises as to whether the amount of rain or the duration of the dry season determines the structure of the forest. The diagram in Fig. 37

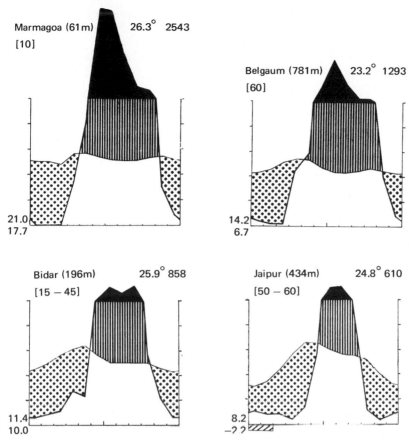

Fig. 36. Climate diagrams of Indian stations in the evergreen, semievergreen, moist- and dry-monsoon-forest. (From Walter and Lieth 1967)

shows that both factors are ecologically important and that neither should be considered alone. The course taken by the limiting lines indicates that for the wet type of forests, the duration of the dry season is more important, where as the absolute annual rainfall is of greater significance for the dry types.

In Africa, the above-mentioned succession of forests is difficult to recognize. Dense population and shifting cultivation have led to the deforestation of semievergreen and wet deciduous forest. This vegetation lends itself better to clearing than the rain forests since it can be burned during the dry season. Besides this, the rainfall is sufficient to ensure an annual crop yield, which is not the case with the dry forest.

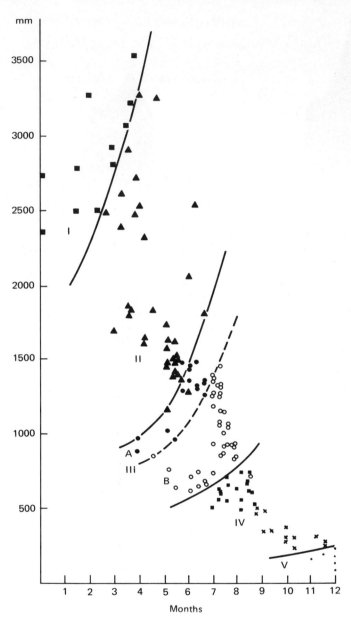

Fig. 37. The relationship of annual rainfall (ordinate) and length of drought in months (abscissa) to forest vegetation in India: I, Evergreen tropical rainforest; II, semievergreen tropical rainforest; III, monsoon forest (A, moist; B, dry); IV, savanna (thornbush forest); V, desert. Details in text. (From a study carried out for UNESCO by H. Walter)

II Zonobiome of the Humido-Arid Tropical Summer-Rain Region with Deciduous Forests

1. General

The tropical and frost-free Zonobiome II shows a distinct annual temperature cycle with heavy zenithal rains in the warm perhumid season and an extremely dry cool season.

In the Americas, this zonobiome covers a large area south of the Amazon basin and smaller areas extending beyond 20 °N in Central America. It is also found extrazonally in Venezuela.

In Africa, zonobiome II is found on both sides of the equator, extending particularly far to the south. In the uplands south of the Zambesi, severe frost damage has been observed in the cold season, and this limits further extension of the zonobiome to the south. The cool tablelands near Johannesburg are largely grassland.

In Asia, the main areas occupied by Zonobiome II are India and Southeast Asia. In Australia, this zonobiome is limited to the northernmost part of the continent (cf. Figs. 17–22).

The humido-arid climate of Zonobiome II determines the euclimatotopes on the zonal soils. These store such a large amount of water during the rainy season that they do not completely dry out during the dry season. This is a prerequisite for the growth of the zonal deciduous forests which, although they significantly reduce transpiration losses in the dry season by shedding their leaves, still must take up a certain amount of water from the soil. Even the leafless twigs and branches lose so much water that the water stored in the trunk does not suffice for the entire dry season.

A distinguishing feature of Zonobiome II, however, is the absence of the zonal forest vegetation in large areas where it is replaced by the vegetation type of the savannas. There are several different reasons for this (p. 80), one of the most important being the presence of water-impermeable barriers (laterite crusts, etc.) at various depths in the ground. Although their presence has been known for some time, their extensive distribution was first demonstrated by exact soil profile measurements by Tinley (1982) on a 200-km-long profile in East Africa. He was able to determine the position of the impermeable layers with 7-m-deep ditches. These impermeable crusts alter the water balance of the soil to such a degree that the development of the zonal forest vegetation becomes inhibited (Fig. 43). The savannas and grasslands are determined by the soil (edaphically) rather than by the climate and are, therefore, to be regarded as pedobiomes. A second edaphic determinant is the low level of nutrients often found in the soils of Zonobiome II. The land surfaces in Africa, as well as in Australia, Anterior India, and especially in the Brasilian Plate of South America are all parts of the Gondwana shield, the ancient continent which separated into the respective continents millions of years ago. The land has never been covered by oceans and the soils are ancient, never having been rejuvenated by ocean sediments. The exposed rock surfaces have been constantly subject to erosion, resulting in a graduated landscape with various ancient plateaus. Wherever recent volcanic rock is absent, the weathered products of which the soil is composed have been leached out substantially so that the soils are too deficient in the essential minerals required by plants (phosphorous, trace elements) for a forest to be able to develop (p. 92, Campos cerrados).

The scarcely noticeable deeper areas on the large plateaus are flooded during the rainy season and their soils are waterlogged. Wooded groves only grow on the somewhat higher surfaces which are not flooded, while tropical grasslands occupy the wet areas. The result is a mosaiclike parkland with wooded groves and grasslands which are not savannas in an ecological sense. Savannas are ecologically homogenous plant communities of scattered woody plants within a relatively dry grassland (p. 80). Most geographers employ the term "savanna" much more freely. Three types of vegetation, therefore, are characteristic for Zonobiome II: (1) zonal deciduous forests (2) relatively dry savannas and (3) the parklands with a wet rainy season. Many laterite crusts are of a fossil nature, having been formed in the Pleistocene, the geological period before the present, which was subject to several glaciations from the North Pole southward to below 50 °N. In the desert zone of the Sahara, these ice ages resulted in pluvial periods with heavy rainfall. Recent pollen analyses have revealed that, at the same time, the tropical zones experienced dry periods extending into Zonobiome I. This resulted in the formation of laterite crusts and relict savannas which can still be found surrounded by evergreen forests.

2. Zonal Vegetation

Zonobiome II is divided into dry and wet subzonobiomes, depending on the length of the dry season and the amount of annual precipitation. Respective climate diagrams are presented for India in Fig. 33. It is not possible to specify certain limiting climatic values for each of the continents, since the respective conditions are quite variable.

Corresponding to the climate, zonal tropical deciduous forests are distinguished as either wet or dry types. The zonal soils are not yet well enough understood to specify general characteristics for the differentiation between the wet and dry forests. Like the soils of Zonobiome I, they belong to the red ferrallitic group. The leaching of SiO_2 from the soil, which is wet only during the warm rainy season, however, is not as acute. In Zonobiome I, the ratio of SiO_2 to Al_2O_3 is below 1.3, while in Zonobiome II it is between

1.7 and 2.0. The adsorbability of the zonal soils is also somewhat greater, meaning that they are more capable of retaining (by adsorption) ions necessary for the nutrition of the plants and are, therefore, not quite so poor in nutrients.

The most conspicuous difference between the zonal vegetation of Zonobiomes II and I is the shedding of foliage. In every climatic zone, a type of foliage has developed which guarantees the greatest production under the respective climatic conditions. Leaf organs are always short-lived structures, since they age rapidly, and soon lose the ability to assimilate CO_2, which is their main purpose. The rapid aging process is probably caused by the accumulation of ballast, which reaches the leaves dissolved in the transpiration current, and by metabolic side products, such as tannins, alkaloids, terpenes, etc. The evergreen trees of Zonobiome I also soon shed their old leaves after the younger leaves have become functional. Some species of Zonobiome I have been found to be evergreen in years of plentiful rain, while during unusually dry periods they shed their leaves before the new leaf buds have sprouted, and are therefore leafless for a short time. In Zonobiome II, a long dry period is normal and the rainy season, in contrast, is extremely wet. Correspondingly, the tree species develop very drought-sensitive large thin leaves at the beginning of the rainy season, for which less material is required per unit leaf surface than for the thick leathery leaves of the species of Zonobiome I, which can only tolerate very short periods of drought. Although the thin leaves assimilate CO_2 only during the rainy season, the saving of material is advantageous considering the annual production balance. Besides the leaf surface area, the intensity of assimilation (which is higher in thin leaves) is also decisive for the assimilation of CO_2 (i.e., the production of organic material).

The water budget of the trees of Zonobiome II is well balanced during the rainy season, since the diurnal transpiration curve and the evaporation curve run parallel courses and have almost no midday depressions, which are always an indication of a beginning water deficiency. The osmotic potentials of the leaves of the different species are low, ranging from -7.5 to -18.9 bar. At the onset of the dry season, the sugar concentration in the cells of the leaves increases sixfold (an increase of 2 bar). Soon afterwards, the leaves turn yellow and become dry.

Frost damage has been observed in especially cold years in the southern limits of Zonobiome II in southern Africa (Ernst and Walker 1973). The growth of the annual sprouts and the development of the leaves begin after the arrival of the first rains (Fig. 38). A striking factor, however, is that the buds of the blossoms of many tree species open before the first rain. Since the petals possess only a very low cuticular transpiration, they are subject to an only insignificant amount of water loss. On the other hand, insects are able to pollinate the blossoms of the leafless forest more easily.

The factor responsible for the induction of blossoming is most probably the maximum of the temperature curve, which occurs at the end of the dry season and before the onset of the rainy season.

Fig. 38. *Colophospermum mopane* forest coming into leaf at the beginning of the rainy season near Victoria Falls (Rhodesia). (Photo E. Walter)

Fig. 39. A very large baobab *(Adansonia digitata)* in Krüger National Park. South Africa. (Photo E. Walter)

The most expansive forests of Zonobiome II are situated south of the equator in the less populated regions of Africa. These are the "Miombo" forests, located on the watershed between the Indian and Atlantic Oceans, and on the Lunda threshold south of the Congo basin, where enough water necessary for settlements is lacking during the dry period. A conspicuous sight in the dry limit of Zonobiome II is the baobab *(Adansonia digitata)* with its bizarre trunk which may attain a circumference of more than 20 m (Fig. 39) and is able to store up to 120,000 l of water[1]. It may, therefore, be assumed

---

[1] An age of 1000 years was determined by the ^{14}C method for a large, not hollow tree

that such plants are able to survive the dry season in a leafless condition without the uptake of water from the soil. In South America, the same family (Bombaceae) is represented by the "bottle trees" (Flaschenbäume).

References on the production of the sparse Miombo forests with a leaf area index of only 3.5 were presented by Cannel (1982):

1. Miombo forest in Zaire (11°37'S, 27°29'E, 1244 m above sea level).
   Tree species: *Brachystegia, Pterocarpus, Marquesia,* etc.
   Soil: latosols
   Phytomass (aboveground):   144.8 t/ha (leaves: 2.6 t/ha)
   Phytomass (underground):   25.5 t/ha (estimated)
   Net production: litter, 4−6 t/ha/yr
   The wood production was not determined.

2. Dry monsoon forest in India (24°54'N, 83°E, 140−180 m above sea level).
   Tree species: *Anogeissus, Diospyros, Budenania,*
                                   Pterocarpus, etc
   Soil: reddish-brown, lessivated sandy clay
   Phytomass (aboveground):   66.3 t/ha (leaves: 4.7 t/ha)
   Phytomass (underground):   20.7 t/ha (estimated)
   Net production: trunks and branches: 4.40 t/ha/yr
                                   leaves   4.75 t/ha/yr
                                   underground   0.35 t/ha/yr
                                   roots (estimated)   3.40 t/ha/yr

Medina (1968) determined the soil respiration in a deciduous forest at 100 m above sea level (mean annual temperature: 27.1 °C, annual precipitation: 1334 mm) in Venezuela. The respiration was three times as intensive during the rainy season as in the dry season. This corresponds to an average amount of catabolized organic substance of 11.2 t/ha/yr. The litter amounted to 8.2 t/ha/yr. The difference may be attributed to root respiration.

Data from Thailand is given by Ogawa et al. (1971) for the following forest types:

1. A dry forest of Dipterocarpaceae at an altitude of 300 m with a thin growth of trees about 20 m tall and a covering of grass 20−30 cm in height and
2. a mixed wet-deciduous forest with trees 20−25 m tall and a sparse covering of grass.

The following values were obtained for the phytomass and the annual primary production in t/ha (LAI, leaf area index):

Type of forest	Phytomass	LAI	Production
1	65.9	4.3	7.8
2	77.0	4.2	8.0

The deciduous forests are exploited by the population for shifting cultivation, 3–5 years at a time. A secondary forest grows up on the fallow patches within about 10–20 years. The trees apparently do not live more than 100 years.

3. Savannas

As previously mentioned, the only ecosystems to be properly designated as savannas are tropical grasslands in which scattered woody plants exist in competition with the grasses (Fig. 40).

Grass and woody species are ecologically antagonistic plant types, one usually excluding the other. Only in the tropics, where both summer rain and a deep, loamy sand coincide, are they found existing in a state of ecological equilibrium. The cause of the antagonism is to be sought in differences in (a) their root systems and (b) their water economy.

Grasses possess a very finely branched, *intensive root system* with which a small volume of soil is densely permeated. Such a root system is specially suited to fine, sandy soils with an adequate water capacity in regions of summer rain where the ground contains plenty of water during the growing season. Woody species, on the other hand, have an *extensive root system* with coarse roots which extend far into the soil in every direction, thus penetrating a much larger soil volume but less densely. This type of root system is well suited to stony soil where the water is not uniformly distributed, not only in summer-rain regions, but also in the winter-rain regions where the water which has seeped down to great depths has to be drawn up again by the roots in summer. For this reason, grasses are unsuited to the latter type of climate.

As far as the water economy is concerned, it is characteristic of grasses that, given sufficient water, they transpire very strongly and photosynthesis proceeds very intensively. Within a short period of time their production is very large. At the end of the rainy season when water becomes scarce, transpiration is not slowed down but continues until the leaves, and usually the entire aerial shoot system, dies. Only the root system and the terminal growing point of the shoot survive. The meristem tissue is protected by many layers of dried-out leaf sheaths and is capable of surviving long periods of drought, even though the soil itself is practically desiccated. Growth recommences after the first rain.

The water economy of woody plants, on the other hand, with their large system of branches and many leaves is well regulated. At the first indication

Fig. 40. *Acacia mellifera* ssp. *detinens* savanna in Southwest Africa. The grass layer is dried out after the rainy season. It appears as if there were a forest growing in distance, although it is actually only the same savanna. (Photo E. Walter)

of water scarcity, the stomata close and transpiration is radically reduced. If the lack of water becomes acute, then the leaves are shed; during the dry season, only the branches and the buds remain. Although well protected against loss of water, these remaining parts have been shown to exhibit a very small but demonstrable loss of water over the course of several hours. The water reserves in the wood are insufficient to compensate for water losses over a lengthy dry period, which means that woody plants are obliged, under such conditions, to take up a certain, albeit small, quantity of water. If the soil contains no available water, however, then they dry out and die.

With these differences in mind, we are in a position to understand the ecological equilibrium of the savannas. As an example, we have chosen an area in Southwest Africa with gradually increasing summer rainfall, uniform relief, and a fine, sandy soil which takes up all of the rainwater and stores the larger part of it (see Fig. 41). This transitory region between Zonobiome II and the desert with summer rain is designated as Zonoecotone II/III, in which climatic savannas occur with an annual precipitation of 500–300 mm and an 8-month-long dry season. Where the annual rainfall amounts to only 100 mm (a) and the water cannot penetrate very far down into the ground, the roots of the small tufted grasses which permeate the upper, wet soil layers use up all of the stored water and then dry up at the end of the rainy season, with the exception of the root system and the apical meristem. *Woody plants cannot survive because the soil offers no available water during the dry season.* At a rainfall of 200 mm (b) the situation is similar: the soil is wet to a greater depth and the grasses are larger but still use up all of the water. Only when the rainfall reaches 300 mm (c) is some water left in the soil by the grasses at the end of the wet season, and although this is insufficient to keep them green, it

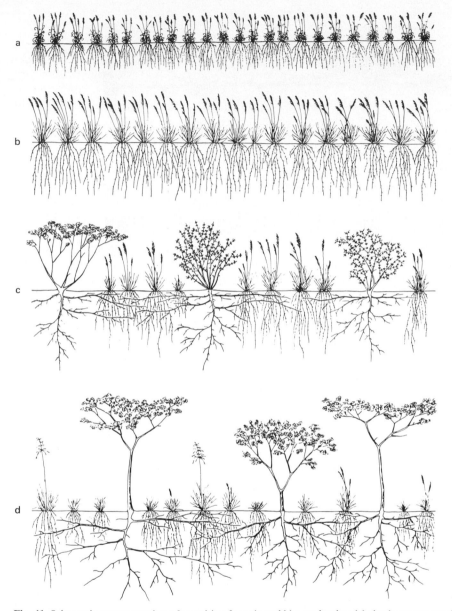

Fig. 41. Schematic representation of transition from (a and b) grassland to (c) shrub savanna and (d) tree savanna. For explanation, see text. (From Walter 1939)

is enough to enable small woody plants to survive the dry season and to form a shrub savanna. If the annual rainfall is 400 mm (d), then the larger amounts of water remaining in the soil at the end of the rainy season support solitary trees, and a tree savanna is formed. But it is still the grasses which are the dominating partner; they determine how much water is left over for the woody plants.

Only when the rainfall reaches a level where the crowns of the trees link up to form a canopy whose shade prevents the proper development of grasses is the competitive relationship reversed. It is now the woody plants which are the dominant competitors in the savanna-woodland or dry tropical-deciduous woodland, and the grasses are obliged to adapt themselves to the light conditions prevailing on the ground.

Such a labile equilibrium in the savanna is readily disrupted by man when he begins to utilize the land for grazing purposes. Water losses due to transpiration cease when the grass is eaten off, so that more water remains in the soil, to the advantage of the woody plants (mostly *Acacia* species), which can consequently develop luxuriantly and produce many fruits and seeds. Their seeds are distributed in the dung of the grazing animals, and the tree seedlings are not exposed to competition from grass roots. The predominantly thorny shrubs grow so densely that thorny shrubland is formed, which is then useless for grazing purposes.

In all grazing areas that are not rationally utilized, there is a great danger of such a *bush encroachment,* and it is for this reason that thornbush as a substitute plant community is nowadays more widespread than the climatic savanna in, for instance, the arid parts of India and in northern Venezuela and the offshore islands (Curaçao, etc.). If the area is more densely populated, and if the woody plants are used for fuel or for making protective thorny hedges around the kraals, then a man-made desert is often produced, which has only a covering of annual grass during the rainy period. The cattle starve during the dry season, with nothing but the straw remnants as fodder. Such is the situation, for example, in the Sudan. The only remaining natural savannas seen by the author were in central Argentina at a rainfall of 400–200 mm with *Prosopis* as the woody species (see Fig. 128).

The transition between Zonobiomes II and III is of a different nature on stony ground. The grasses are unable to compete with the woody plants and are absent. With decreasing precipitation, the woody plants become smaller and are further apart. Each plant requires a larger root area, and the roots run near the surface since rain moistens only the upper layers of the soil. In approaching Zonobiome III, only a few small dwarf shrubs showing xerophyllous adaptations remain.

A special situation is encountered in two-layered soils such as are found in Southwest Africa, where a shrub savanna is found, although annual precipitation is only 185 mm (Fig. 42). If the soill were deep and sandy, pure grassland would be expected at such low rainfall. In fact, the soil profile reveals that beneath the 10–20 cm of sand there is sandstone of the "Fischfluss" formation, which either is arranged in thin layers with small cracks or forms thick banks with larger crevices. The upper sandy layer cannot retain all the rainwater, and part of it seeps into the cracks in the sandstone. The grasses can utilize the water from the sandy layer, and the roots of the shrubs penetrate into the sandstone layer and take up the water

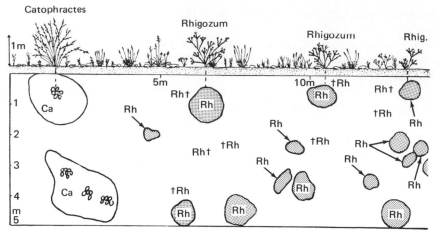

Fig. 42. Transect (1 m wide) through a typical patch of vegetation near Voigtsgrund (Southwest Africa). The grasses dry out during the dry season. Below: groundplan of plant cover (without grasses): Ca, *Catophractes;* Rh, *Rhigozum* († dead). (From Walter 1939)

from its cracks. The water reserves in the cracks of the thin-layered sandstone are sufficient only for the small *Rhigozum* bush, whereas the larger *Catophractes* shrub flourishes in the crevices of the thicker-banked sandstone. The distribution of the bushes reflects the structure of the sandstone even in places where the covering layer of sand is missing. The bushes compete with one another, and although both types can germinate in the larger crevices, the smaller bushes are in time ousted by the larger ones, and only their dead remains are left. There is no competition between grasses and woody plants in this case.

In Zonoecotone II/III, zonal savannas occur in place of deciduous forests where the annual precipitation is too low for the latter. In Zonoecotone II, however, they are found wherever (in spite of sufficient precipitation) the soil does not contain enough water for the survival of a forest during the dry season. On the other hand, too much water during the rainy season results in a waterlogged soil and excludes the presence of woody plants. In such a case, a pure grassland develops which can dry out during the dry season and which is typical of the parklands. This is illustrated in Fig. 43. The rainy season usually commences with storms. The abundant dry grass present at this time of year easily becomes ignited by lightning. The large number of pyrophytes (fire-resistant woody plants) attest to the frequency of such fires. These trees or shrubs often possess thick bark which becomes scorched while the cambium remains protected. Many shrubs have dormant buds located below the soil, which sprout when the parts of the plant aboveground are burned off. A number of plants have underground storage organs capable of lignifying (lignotubers) which allow for rapid regeneration.

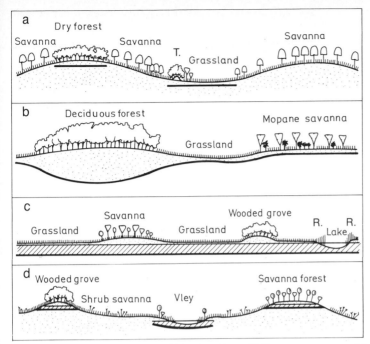

Fig. 43. The types of landscape within Zonobiome II (forest, savanna, parkland) as determined by positioning of an impermeable barrier (laterite crusts, etc.) beneath the soil (Tinley 1982)

a Slightly hilly area of the northern Kalahari with deep sand. Relatively little water is retained at field capacity, so that a large portion of the rain seeps through the soil and the retained water is only sufficient for savanna vegetation. Occasionally, laterite crusts (black lines) are present as barriers, thereby preventing water seepage. If the crust is relatively deep (left), a dense wooded grove or dry forest develops on the moist sand. If the crust is positioned less deeply in a lowland, the ground above it is waterlogged, only allowing the growth of grass [with the exception of the termite mound (T), which is more adequately drained and suitable for the growth of trees].
b Sandy soil with an uninterrupted laterite crust with a basin on the left which collects water by lateral seepage. The soil is well-aerated and is wet deep underground, providing favorable conditions for the zonal deciduous forest. In the center is a lowland with a grassland which is flooded during the rainy season. On the somewhat higher level on the right, on an alkaline clay soil, the grassland is interspersed with species of woody plants adapted to heavy soils (*Mopane, Balanites,* flute acacia).
c The uninterrupted crust is at a uniform distance below the surface. The soil above it is saturated with water during the rainy season (diagonally shaded) and covered with a grassland. The soil somewhat better drained only in smaller elevations, which are covered with a tree savanna, or with wooded groves if there is sufficient root space. On the outer right is a basin with an open groundwater surface and a zone of reeds (R).
d A relatively dry sandy region with savanna vegetation. Woody plants dry out down to the soil during the drought season and grow again from the base of the stem or sprout from the roots during the rainy season. Depending on the availability of water, laterite crusts at various depths below higher ground support the development of wooded groves or savanna forests. Valleys are characterized by vleys with swamp vegetation, which may dry out during the drought season, and sparse tree growth on the edge of the crust along laterally flowing water veins. This illustrates how the presence of forest and savanna is dependent on the position of the crusts, and that the development of parklands depends on water-saturated soils. Later, the presence of savannas on nutrient-poor soils will be discussed (p. 92). Other factors favouring the development of savanna vegetation are fire, herds of large herbivors and various human activities. In Zonobiome II, fire was already an effective natural factor before the appearance of man

In prehistoric times, primitive man almost certainly started grass fires in order to protect his settlements from the effects of fire caused by lightning. The tall grass of the wetter zones permits fire to spread much more rapidly and with a greater ferocity than elsewhere. Nowadays it is common practice to burn the grass in the dry season to facilitate big-game hunting, to destroy vermin such as snakes, etc., or simply for the sake of an exciting evening's entertainment.

After a grass fire, the grasses sprout earlier, which is advantageous for grazing. Grass fires only penetrate dry forests with an undergrowth of grass, but they can also push back the limits of the wet forests. Above all, they prevent the forest from winning back ground that has been cleared and turned into grassland. Thus the forests are continually becoming smaller, except where special protective measures are taken. Grazing of big game on the savannas represents an important factor (Anderson and Herlocker 1973).

The growth of young trees is inhibited by their being bitten off and trodden upon. Elephants are especially destructive to forests. They pull out trees by the roots or strip off the bark. Elephant paths cause clearings, allowing grass fires to penetrate into the forest. On the average, one elephant is capable of destroying four trees per day. The loss of trees in Miombo forests can be as high as 12.5% annually. The number of elephants in the national parks is increasing rapidly. Murchinson Park, on Lake Albert is becoming increasingly deforested by elephants. In Serengeti Park, however, the damage caused by wildlife and the regeneration of the vegetation appear to be in equilibrium.

It is interesting to note that many woody plants of the savannas of Africa, with their high wildlife population, are thorny, while this does not hold true for South America, where wildlife is less abundant. This indicates the selection of species protected from grazing by wildlife.

Wildlife paths influence the vegetation indirectly by exposing surfaces to channel erosion. This is especially true for hippopotamuses, which climb up the banks of the rivers at night to graze on the grass.

The erosion channels may lead to the drainage of a wet grass area, making the advance of woody plants possible. A summary of the complex effects of big game was presented by Cumming (1982). The influence of man is even greater, whether he functions as animal breeder or farmer. All types of intervention, such as fire, grazing, forest clearing for shifting cultivation and the gathering of fire wood, are activities to the detriment of the forest. The use of the savanna north of the equator for grazing began at least 7000 years ago. Forests remain only as small relicts in this region, and it may be assumed that a large portion of the savannas are of a secondary nature (Hopkins 1974).

In conclusion, the following types of savanna are to be distinguished:

1. *Fossil savannas*, which arose earlier under other circumstances, such as in the range of Zonobiome I.
2. *Climatic savannas* in the range of Zonoecotone II/III, with an annual precipitation of less than 500 mm.
3. *Edaphic savannas* which are determined by the soil conditions in Zonobiome II
   a) on soils in which the water balance is altered by the presence of impermeable barriers (laterite crusts, clay layers, dense silt or sand layers), being less advantageous than they would be based on the amount of rain alone;
   b) on soils which are primarily so deficient in nutrients that they cannot support a forest (p. 92);
   c) within parklands where soils are saturated with water during the rainy season, resulting in palm savannas (p. 88).
4. *Secondary savannas* caused by fire, the effects of large wildlife species and various interventions by humans.

The determination of the respective type of savanna cannot be achieved by superficial observation. Instead, a thorough investigation is necessary.

4. Parkland

Parkland usually develops in very flat country of Zonobiome II. Its formation is due to the minute differences in relief, which are inconspicuous during the dry season. During the heavy rains of summer, all the slightly deeper depressions are flooded, the water taking months to drain away. The grey clay soil of this type of biotope (termed "mbuga") is covered by grassland, whereas the slightly more elevated areas which are not flooded, and to which the woody plants are confined, have a thick layer of red, sandy loam.

Here the watershed begins, in barely perceptible grass-covered narrow valleys which gradually unite and become deeper and deeper, with an increasing gradient, to form streams and river beds (easily recognizable from the air).

A peculiar variant is seen in the so-called "termite" savanna (Fig. 44), which consists of large grassy hollows from which the deserted termite heaps rise like islands. Since these mounds are never submersed, they are covered with trees, so that the scenery is a mosaic of different plant societies, and is, therefore, not a true savanna.

The deeper hollows with black clay soils, designated as "mbuga", are a special type of amphibiome with an alternately wet and dry soil and a hard layer of iron concretion at a depth of 50 cm. Since the potential evaporation by far exceeds the 1000 mm of yearly rainfall, the clay soil dries out in August-December to a depth of 50 cm and is split polygonally by deep cracks. Biotopes of this kind are unsuitable for trees, which only thrive where the ground water table does not rise above a depth of 3 m. At this depth the laterite crust, to which the roots of the trees extend is formed.

Fig. 44. "Termite savanna", periodically flooded grassland with trees on old termite mounds, in northwestern Kenya. (Photo E. Walter)

In contrast to the "termite savanna", the "palm savanna" is a homogenous plant community. Palms as woody monocotyledonous plants, have a bundle-like root system with uniform, almost branchless roots which extend radially, so that palms are situated individually in the grassland. They tolerate occasional flooding. Although there is no available information on the subject, the soils of palm savannas probably dry out as much as those of pure grasslands (see "Palmares", p. 93).

5. Some of the Larger Savanna Regions

In South America, expansive savanna-like vegetation communities are found on the Orinoco of central Brasil and in the Chaco region.

a) The Llanos on the Orinoco

The Llanos, on the left bank of the Orinoco in Venezuela, occupy a basin 400 km in width that was left by a Tertiary sea and extend for a distance of 1000 km, into Columbia. The inflowing rivers have filled the basin with weathering products from the Andes. In the central Llanos around Calabozo (see diagram in Fig. 30), the climate is very typical of Zonobiome II, i.e., annual precipitation amounts to more than 1300 mm, the rainy period lasts 7 months, and the dry period lasts 5 months. With a climate of this type, wet-deciduous forests would be expected, and they are in fast present as "matas", which are small, scattered woods. Apart from this, the Llanos bordering the river, which are flooded in the rainy season, are pure grassland, as is usual in Zonobiome II (with trees only on the banks, as gallery forests).

Fig. 45. Savanna with *Curatella americana* in the Llanos near the Estacion Biologica Calabozo (Venezuela) during the dry season. (Photo E. Walter)

Apart from this, the area is covered with grass about 50 cm in height, scattered with small trees *(Curatella, Byrsonima, Bowdichia)* and is in fact a true savanna (Fig. 45). Since the savanna cannot be climatic in origin (the rainfall is too high), edaphic factors such as soil conditions must be responsible.

The assumption has often been made that this is an anthropogenic savanna resulting from the use of fire, but this explanation is too simple and uncritical since the savanna existed long before the arrival of the white man. It was neither cultivated nor used as grazing land by the Indians. Fires caused by lightning are common occurrences in grasslands, and although the Indians probably often burned down the dry grass, this was only possible because natural grassland already existed. Fire has certainly played its part in forming the savanna, insofar as only fire-resistant woody species could survive in the grassland and on the fringe of the matas, but it is by no means the primary reason for the existence of these enormous grassy expanses. In the central Llanos, it has been shown that, at a time when the groundwater table in the basin was very high in this region, a laterite crust was formed which was cemented by ferric hydroxide. This is called "arecife" (Fig. 46).

Arecife runs beneath the surface at varying depths (mainly between 30 and 80 cm), rarely below 150 cm, and can sometimes also be seen on the surface. The statement that arecife is impermeable to water cannot be correct in this case, since the 750 mm of rain falling during only 3 months of the summer rainy season cannot possibly be absorbed by the soil overlying the arecife. In such a completely flat region, flooding would inevitably result, but this does not occur. The reddish color of the soil is another factor which speaks against prolonged water-logging. The groundwater table beneath the

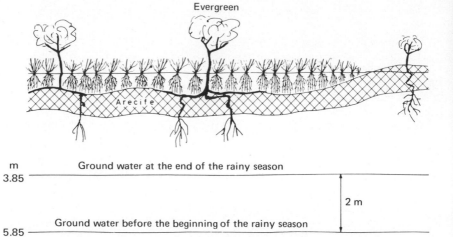

Fig. 46. Scheme to illustrate the situation in the Llanos north of the Orinoco. Further details in text

arecife has in fact been found to rise from –575 to –385 cm, a rise of almost 2 m, by the end of the rainy season. Assuming a pore volume of about 50% for the alluvial deposits, this would mean that about 300 mm is retained by the soil above the arecife and 1000 mm seeps through. Arecife laid bare by erosion on the river banks shows quite clearly that there are irregular channels penetrating the hard crust in places. This suggests that, the grasses can take root in the fine soil above the arecife and use up the 300 mm of stored water in the course of their development. Woody plants, however, grow wherever their roots can find a way through the arecife layer to the damp, underlying rock layers where adequate water supplies are available. Groups of trees can grow wherever the channels through the arecife are large enough or where a number occur close together. But small woodlands exist only in isolated patches where the arecife is either entirely absent or extremely far down. In such places, deciduous forests (matas) are found, which are, in fact, the type of vegetation corresponding to the regional climate. These savannas can thus be considered as a stable, natural plant community in which the distribution of the trees is a reflection of the arecife structure. The following facts add weight to our point of view:

1. Wherever the arecife is on the surface, a grass cover is completely wanting, but solitary trees grow on it at rather large intervals. The roots must in such cases reach down below the arecife through existing channels.
2. *Curatella* remains green throughout the dry season, in contrast to other woody plants in the typical savannas. This indicates that the water supply is sufficient throughout the year. Transpiration measurements have demonstrated that a single tree transpires approximately 10 l per day during the dry period. Since the topsoil is dry at this time, the water must be provided

from the layer below the arecife. The same holds true for other species of woody plants.
3. The small woods (matas) grow in places where arecife is locally absent and their roots can penetrate unhindered into the ground.

Final proof would be afforded only by excavating the roots over larger areas, which would be difficult to do. The use of dynamite on the arecife would certainly favor woodland expansion.

Scattered over the savannas of the Llanos are slight depressions into which the water drains after a heavy downpour (1961, 38 mm in 20 min) and in which sediments of gray clay collect. The water in the depressions reaches a depth of about 30 cm during the rainy season, but toward the end of the dry season, the gray bottom dries out completely.

This alternating wetness and dryness is well tolerated by certain grasses *(Leersia, Oryza, Paspalum, etc.),* but not by tree species other than palm trees. This results in the formation of "palmares," i.e., palm savannas, (also widespread in Africa), which are grasslands containing the palm *Copernicia tectorum.* Such areas, too, often burn, but the palms are resistant to fire (as are the tree ferns) because they possess no damageable cambium. The dead leaves sheathing the trunk are burned, and the outermost vascular bundles are charred, but the resulting layer of carbon acts as an insulation against later fires. The apical meristem, surrounded by young leaves, survives. If old leaves are completely missing from the trunk, this is a sign that the palm savanna has recently been burned. If the leaves sheath the trunk down to the ground, then the palm has not been exposed to fire at all. And if only the lower portion of the trunk is bare, this means that the palms has had a chance to grow for several years since the last fire.

Some of the water must drain away from the areas on which palms grow because otherwise, with a rainfall of 1300–1500 mm and a potential evaporation of 2428 mm (i.e., a negative hydrological water balance), the soil would become brackish.

On permanently wet areas, the palm *Mauritia minor* is found. Black, acid peaty soils are formed, on which a few grasses, *Rhynchospora, Jussiaea (= Ludwigia), Eriocaulon,* and the insectivorous *Drosera* spp. (sundews) are found. Areas of this kind, just as the alternately wet and dry grassland mentioned above, have to be considered as amphibiomes, a special form of helobiome.

Further east, the Llanos are succeeded by a plain with sandy deposits from the Orinoco, which at one time changed its course here to the north, flowing through the Unare lowlands and emptying into the Caribbean Sea. The often white quartz sands are weathered products of the quarzite sandstones of the Guayana Table Mountains, and correspond to those of the Brasilian shield, which are also low in nutrients. These savannas, and to some extent pure grasslands, were probably induced by the same factors as the Campos cerrados, which will be discussed next.

b) Campos Cerrados

The Campos cerrados is a region with a savanna-like vegetation covering an area of two million km^2 in central Brasil (Eiten 1982). The 4–9 m high trees cover between 3 and 30% of the total surface. The climate is characterized by a five month long dry season and an annual precipitation of 1100–2000 mm. Rawitscher (1948) presented the first information on the water budget of these savannas and demonstrated that the deep soil is constantly wet at a depth of only 2 m. This provides the deep-rooting species of woody plants with plentiful water, so that they remain evergreen and transpire heavily during the dry season. Only the grasses and shallow rooting species dry out or shed their leaves during the dry season. The soils are weathered products of the granite and sandstone of the Brasilian shield and are very poor in nutrients, especially phosphorous (but also potassium, zinc and boron). Various mineral fertilizer compositions are used to yield such crops as cotton, corn and soybean. That fact that the nutrient deficiency, rather than the water factor, is responsible for inhibiting the growth of the zonal deciduous forest is demonstrated by the observation that a semi-evergreen forest may be found on a basaltic soil in the vicinity of São Pãulo. The Campos cerrados were burned regularly and the presence of a large number of pyrophytes is an indication that fire has been a natural factor in this region for ages. Although they are not the actual reason for the lack of forest vegetation, fires do somewhat reduce the density of the wooded stands (Coutinho 1982).

c) The Chaco Region

The Chaco region is situated in the westernmost part of Zonobiome II in South America – an expansive plain between the Brasilian shield in the east and the Andean foothills in the west, which spreads across southern Bolivia, most of Paraguay and far into western Argentina, thereby covering an area of 1500 km from north to south with an average width of 750 km (Hueck 1966). The central region of the plain is only 100 m above sea level.

During the heavy summer rains, large portions of the plain – especially in the east – are flooded (annual precipitation: 900–1200 mm). Here, a parkland with forests, expansive periodically flooded grasslands, palm savannas and swamps may be found. The western section, in Argentina, is heavily overgrown with brush, and saltpans with the halophytes *Allenrolfea* and *Heterostachys* are also present. The southern Chaco is succeeded by the pámpas. The relief is quite level and there are water-impermeable layers in the soil. The vegetation is primarily a *Prosopis* savanna with a grass cover of *Elionurus muticus* and *Spartina argentinensis*. The predominant tree species of the chaco forests are the tannin-rich Quebracho species, *Aspidosperma quebracho-blanco*, *Schinopsis quebracho-colorado* and *S. balansae* etc. *Trithinax campestris* is a common palm species, while *Copernicia alba* is typical in wet depressions. The mammalian fauna is relatively sparse. *Myrmecophaga tridactyla* and *Tamandua tetradactyla* are present as termite eaters. The carnivores are represented by the jaguar *(Leo onca)*, the puma *(Felis concolor)* and a number of smaller species. Rodents are numerous. The trees are inhabited by the tree sloth *(Bradypus boliviensis)*, three species of monkey (Cebidea), the tree porcupine *(Coenda spinosus)*, the mustelid *Eira barbara*, a number of insectivores and fruit or nectar-eating bats, as well as the vampire bat *(Desmodus rotundus)*.

The only bird species to be mentioned here is the large flightless *Rhea americana*. The reptiles are represented by two rare cayman species, three turtle species, several poisonous snakes (25 snake species in all) and various lizards. Up to 30 species of anurans have also been identified.

It is not possible to mention the innumerable invertebrates. Research on the ecosystem is still lacking. The main influences by humans are deforestation and grazing, which may lead to brush overgrowth. A brief summary with references is presented by Bucher (1982).

d) The Savannas and the Parklands of East Africa

This huge area at the base of the large volcanoes with the giant crater Ngoro-Ngoro, the East African Rift Valley and the expansive Serengeti region is well known, especially because of its abundant wildlife, which is supported by the nutrient-rich volcanic soil and its superior vegetation.

This equatorial region, with its diurnal climate and monsoon climate, however, has two rainy seasons – a long one and a short one. These are usually only separated by a short dry season and have the same effect as a summer rainy season, so that with an annual precipitation of 800 mm, savannas and parklands similar to those in Zonobiome II may be found here. Forest clearing, annual fires and overgrazing have affected the plant cover substantially, resulting in various stages of degradation. Although this is a typical tree savanna, it is often referred to as an "orchard steppe". Where the climate is drier, and in dry stoney regions, the large candelabrum euphorbia and aloe species are found.

e) The Vegetation of Zonobiome II in Australia

This vegetation type exists in only a few small relicts of deciduous forest in northeastern Australia with Indo-Malaysian floral elements and in several practically insignificant deciduous species of *Eucalyptus* in northern Australia. Parklands with palms and evergreen *Eucalyptus* species are distributed within the range of Zonobiome II. Somewhat more to the south, where the annual precipitation is lower, savannas occur with a grass cover of *Heteropogon contortus* and evergreen *Eucalyptus* species.

The term "savanna" is not used in the extensive vegetation monography of Beadle (1981). In contrast, the Australian researchers Walker and Gillison (1982) designate as savannas all sparse forests in which grasses and the herbaceous layer comprise more than 2% of the ground cover. According to this criterium, most sparse *Eucalyptus* forests would have to be considered savannas.

6. Ecosystem Research

The following two savanna ecosystems were investigated ecologically: the Lamto savanna in West Africa, a relict savanna in a rainforest region, and the Nylsvley savanna in South Africa, bordering on the Kalahari in the west.

a) The Lamto Savanna

This relict savanna is situated in the Guinea forest zone (in the Ivory Coast) at 5° W and 6° N, and is, therefore, still within Zonobiome I. It is burned annually to prevent the advance of the neighboring rainforest, irregardless of whether the soil conditions are capable of accommodating it.

The mean annual precipitation is 1300 mm and the climate diagram indicates a dry period of only 1 month in August. The weather varies greatly from year to year, however, and the annual rainfall fluctuates between 900 and 1700 mm. The higher levels of the relief are occupied by tree or brush savannas on red savanna soil with laterite concretions; on the lower levels, however, palm savannas are found on soils saturated with water not far below the surface. The various plant communities were studied by Menaut and Cesar (1982). Here, only the extreme values for a low brush savanna and a dense tree savanna are presented:

Number of woody plants per ha	120−800
Area covered by woody plants	7%−45%
Leaf area index	0.1−1
Phytomass (aboveground) (t/ha)	7.4−54.2
Phytomass (underground) (t/ha)	3.6−26.6
Net wood production per year (t/ha)	
(aboveground)	0.12−0.76
(underground)	0.05−0.37
Net production of leaves and green sprouts	0.43−5.53
Net production of the grass layer per year (t/ha)	
(aboveground)	14.9−14.5
(underground)	19.0−12.2

Lamotte (1975) recorded observations on the consumers and decomposers in these savannas. Large animals appear only sporadically; the zoomass of birds is 0.2–0.5 kg/ha, of the 12 rodent species 1.2 kg/ha, of the earthworms 0.4–0.6 kg/ha. It was not possible to determine the mass of the grass, humus and wood-eating termites or of other invertebrates. The soil respiration, which serves as a criterium for the amount of microorganism activity, was calculated at 8 t CO_2/ha per year. An attempt to determine the energy flow of decomposition (Lamotte 1982) yielded the following results:

1. The annual fires result in the mineralization of approximately 1/3 of the primary production.
2. Probably less than 1% of the primary production is eaten by the consumers. The decomposition by detritus eaters, including the earthworm as the most important group, is significant.
3. 80% of the primary production is decomposed by microorganisms, making the representation of the energy flow in the form of a pyramid quite doubtful. The author's opinion is herewith confirmed that the long cycle by way of the consumers is quantitatively insignificant (p. 9). A large amount of faunistic information on the individual animal groups of the savannas is presented by Boulière (1983).

b) The Nylsvley Savanna

This region lies at 24° S, north of Johannesburg in the Nylsvley Nature Reserve, and comprises 745 ha, 130 ha of which are rocky soils (Huntley and Morris 1978, 1982).

The climatic conditions are illustrated in the climate diagram in Fig. 47. Here, we have a relatively dry tropical climate with summer rains (Zonobiome II), an annual precipitation of 610 mm and rare frosts of down to
– 6 °C in the months of May to September. The nutrient-poor soils are sandy latosols (pH = 4), and are derived from the weathering of bedrock of the Waterberg Series (B-horizon at a depth of 30–130 cm, and a clay content of 6–15%).

The vegetation constitutes a *Burkea africana* + *Eragrostis pallens* tree savanna with trees of up to 14 m in height. Besides *Burkea*, other dominant deciduous trees are *Terminalia sericea* and *Combretum molle*. Among the brush species are *Ochna pulchra* and *Grewia flavescens*. The crowns of these plants cover 27.5% (20−60%) of the surface. In areas which were inhabited by natives until 50 years ago, the soil is more compacted, more abundant in the elements N, P and K and bears a secondary growth consisting of an *Acacia tortilis* + *A. nilotica* + *Dichrostachys cineraea* thorn savanna (covering 10%) with *Eragrostis lehmanniana* as the dominant grass species.

The aboveground phytomass of the woody producers in the *Burkea* tree savanna amounts to 16.3 t/ha, 14.9 t/ha of which is contained in trunks and branches, 0.3 t/ha in twigs and 1.1 t/ha in leaves. Another 1.9 t/ha are present in the dead wood. The leaf area index is 0.8.

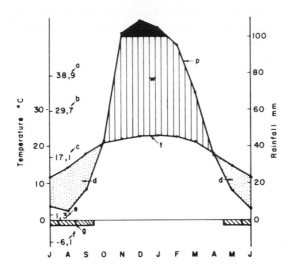

MOSDENE 1097m (ppt 40yrs) 609,5mm
NYLSTROOM 1143 m (temps, 27yrs) 18,3°

Fig. 47. Combined climate diagram of Mosdene and Nylstrom near the research area. (a) Absol. maximum; (b) mean daily minimum of the warmest month; (c) mean daily temperature fluctuation; (e) mean daily minimum of the coldest month (+ 1.3°C); (f) absolute minimum, otherwise as usual. (According to Huntley and Morris 1978)

Herbaceous plants are of no special importance in the grass layer. The grasses are less dense beneath the trees than in between them. The phytomass of the grass layer fluctuates significantly within a small area, and varies from year to year according to the amount of precipitation. Three years of measurements have revealed maximum values of 235 g/m² between the trees, and of 62 g/m² beneath them. Minimum values of 141 and 16 g/m² were determined respectively.

The underground phytomass is calculated at 15.5 t/ha. Between the trees, half of this mass is attributed to the roots of the grasses. 75% of the root mass is located in the upper 20 cm of the soil. In the summer, 13% of the roots are dead, compared to 30% in the winter. Between April 7 and November 14, 1977, the amount of litter in the *Burkea* savanna totaled 160 g/m² (84.8% leaf litter, 9.4% twigs, 5,5% fruits and seeds, 0.3% bark and bud scales). 35% of the litter originated from each *Burkea* and *Ochna*. The average annual litter production is estimated to be 170 g/m². The total litter on the ground on October 18, 1976 was 1853 g/m² and decreased to 1342 g/ha by July 12, 1977. Both the vertebrate and invertebrate fauna of the *Burkea* tree savanna are notably different from those of the *Acacia* thorn bush savanna. In all, 18 species of amphibians occur in the National Park (in the Nyl River valley), 11 of which are found in the research area. The toad, *Bufo garmani*, and the frogs, *Breviceps mosambicus* and *Kassina senegalensis*, may be found far from the nearest source of water. The reptiles are represented by 3 species of turtles, 19 lizards and 26 snakes.

325 different bird species are found within the National Park; 197 of them permanently. 120 species inhabit the research area [14 raptors, 71

insectivorous species, 4 baccivorous (berry eating) species, 10 granivorous species and 26 omnivorous species]. 46 of the 62 mammalian species in the park are found in the research area. The most numerous group is that of the rodents. One species each of porcupine, warthog and jackal and two species of monkey inhabit the area.

The most important artiodactyls are the koodoo *(Tragelaphus strepsiceros)*, the impala *(Aepyceros melampus)*, the deuker *(Sylvicapra grimmia)* and the steenbok *(Raphicerus campestris)*. Since it is so difficult to determine the number of individuals or the living zoomass, this was achieved rarely and then only as an approximation. The number of snakes was estimated at 3/ha, and the most frequent reptile, the gecko *(Lygodactylus carpensis)*, at 195–262/ha. The common lizard, *Ichnotropis capensis*, was found with 7–11 animals/ha.

The living zoomass of birds on 100 ha of the *Burkea* savanna was 40 kg, which is reduced by 25–30% when the migratory species depart for the winter. Mammals were captured so seldom and so sporadically that the results are of little meaning. Captures of *Dendromys melantois*, for example, averaged approximately 5 (0–15) animals/ha, and of other rodents only 2/ha.

The following mean values were given per 100 ha for the artiodactyls: impala 13, koodoo 2, warthog 1, dcukcr 2, stccnbok 1–2 (rictbok, rarc).

According to the previous owner of Nysvley, grazing was only allowed in the area between the months of January to April during the last 40 years. Otherwise, losses would occur due to the poisonous geophytic relative of the Euphorbiaceans, *Dichapetalum cymosum*. The biomass of the cattle during those 4 months was approximately 150 kg/ha. In 1975, effects of overgrazing became evident, so that the herd was reduced to almost half in the following ycar.

The number of invertebrates is so large that it was necessary to limit the study to certain groups which play important roles in the ecosystem: wood-eating Colepterans, Lepidopterans, social insects, root-eaters and spiders. The zoomass of the invertebrates in dry weight averaged 135 g/ha (minimum in August = 60 g/ha, maximum in March = 300 g/ha) on woody plants. The dry weight of the insect zoomass is higher in the grass layer. The following figures give the dry mass in kg/ha for the *Burkea* savanna; the values in parentheses are for the *Acacia* thorn savana:

Acridoidea	0.76 (2.32)
Other Orthopterans	0.06 (0.02)
Lepidopterans	0.05 (0.03)
Hemipterans	0.08 (0.08)
Others	0.05 (0.05)
Total insect mass	1.00 (2.50) kg/ha

Occasionally, large numbers of catepillars *(Spodoptera exempla)* or beetle larvae *(Astylus atromaculatus)* were seen on the grass species *Cenchrus ciliaris*. During the warmer season, 77% of the dung is removed within a day by dung beetles (Coprinae, Aphodiinae), which dig below the deposit, while tumblebugs (*Pachilomera* spp.) distribute the dung over a large area. These coprophags lead us to the following group.

The decomposers are the small saprophagous animals of the soil and within the litter layer, which feed on dead plant and animal remains, thereby breaking them down further in size. Finally, protozoans, fungii and bacteria provide for their complete mineralization. The most important saprophages are the termites. Oligochaetes, Myriapods and Isopods are less significant. Acarids and Collembola eat bacteria and fungii.

Of the 15 termite species, the most common are *Aganotermes oryctes, Microtermes albopartitus, Cubitermes pretorianus* and *Microcerotermes parvum*. Four of these 15 species are humus-eaters, while others eat dead wood or grass litter. More detailed studies on these termites are not available. An average of 2540 termites were found under 1 m² ground surface (maximum: 8204 in November; miniumum: 596 in July.

Microorganisms were determined by the usual counting methods on agar media, whereby the number of *Actinomycetes* was conspicuously high. The activity of the soil organisms was quantified by ATP determinations and soil respiration measurements. A mean of 2.33×10^6 fg ATP/g soil, and a daily (24 h) respiration of 1866 mg CO_2/m^2 (minimum: 226 mg in August, maximum: 4367 mg in January). The methods used for determining these values are not identified. Parallel laboratory experiments were performed on soil samples to determine the formation of CO_2.

Eight contributions by Huntley and Walker (1982, pp. 431–609) give further information on the South African savanna. These concern methodical problems in the calculation of energy flow, photosynthesis, the nitrogen cycle, formation of zoomass, decomposition of litter and ecosystem stability. Final results are not yet available. The wood mass of various *Burkea* savannas was ascertained by Rutherford (1982).

7. Tropical Hydrobiomes in Zonobiomes I and II

High rainfall combined with relatively low potential evaporation accounts for the large surplus of water in the wet tropics. San Carlos de Rio Negro in southern Venezuela, for example, with a rainfall of 3521 mm, has a potential evaporation of only 520 mm. In flat, poorly drained areas, therefore, extensive swamps have developed. In Uganda, these areas cover 12,800 km², about 6% of the entire country. The drainage basins of the river systems are not separated by watersheds but are connected with one another by a network of swamps. The flight from Livingstone to Nairobi provides excellent views of the large Lukango swamps as well as those surrounding Lakes Kampolombo and Bangweulu. The largest swamp area of all is formed by the White Nile in the southern Sudan, which, together with its left tributary the Bar-el-Ghasal, fills an enormous basin lying 400 m above sea level. This region is known as the "Sudd" and extends 600 km from north to south and from east to west at its widest and longest. The total area covered is estimated at 150,000 km² and varies according to the water level. Half of the water of the Nile is lost by evaporation in the Sudd region. Seen from the air, the water appears to be dotted with small islands barely extending above the surface, but these are in fact swimming islands or floating lawns formed by the shoots of *Vossia* grass and *Cyperus papyrus,* as well as grassy rafts of the South American *Eichhornia* and *Pistia.* From the air, it is possible to distinguish free waterways and small stretches of water. When the water level sinks, part of the land emerges, and grassland consisting of tall *Hyparrhenia rufa* and *Setaria incrassata* develops. The wettest parts are covered with *Echinochloa* species, *Vertiveria,* and *Phragmites* (reed).

It was previously assumed that the Great Pantanal in the Mato Grosso in Brazil, bordering Bolivia and Paraguay, was a similar type of swamp, from which the southern tributaries of the Amazon and the right tributaries of the upper Paraná arose. But this region is flooded only during the rainy season and is used as grazing land in the dry season, although ringlike lakes bordered with woodland remain. Swamp areas and watery basins are also widespread in the rest of the wet tropics. The aquatic vegetation consists of some cosmopolitan and pantropical species, but each region also has its own floristic peculiarities.

8. Mangroves as Halohelobiomes in Zonobiomes I and II

Approaching a tropical coast with its protective coral reefs from the sea, one's attention is caught by partially submerged mangroves. At high tide, the crowns of the trees barely extend above the surface of the saltwater, and only at low tide are the lower portions of the trunks and the pneumatophores visible. Such forests are found in the tidal zone in saltwater (Fig. 48), where the salt concentration is about 35 parts per thousand, or 3.5%, which corresponds to a potential osmotic pressure of 25 atm. They are composed of about 20 woody species in all. Distinction must be made between the species-rich Eastern mangroves on the coasts of the Indian Ocean and the western coast of the Pacific, and the Western mangroves on the American coasts and the east coasts of the Atlantic Ocean, which are poorer as regards number of species. The best developed mangroves are found near the equator in Indonesia, New Guinea, and the Philippines, but with increasing latitude, the number of species decreases until only one species of *Avicennia* remains. The last outposts are to be found at 30 °N and 33 °S in East Africa, 37–38 °S in Australia and New Zealand, 29 °S in Brazil, and 32 °N in the Bermudas. Thus, although the mangrove is at its best in equatorial regions, it also extends throughout the tropical and subtropical zones almost as far as the winter-rain regions or the warm-temperate zone (Chapman 1976).

Fig. 48. Outer mangrove zone with *Rhizophora mangle* near Marina/Bahia de Buche (Venezuela). (Photo E. Walter)

Mangroves are an azonal vegetation confined to the saltwater of tidal regions. The chief genera of mangrove are *Rhizophora,* with stilt roots (Fig. 48) and viviparous seedlings, and the nonviviparous *Avicennia,* the white mangrove, which has thin pneumatophores growing up out of the ground. *Laguncularia* is a Western mangrove. *Conocarpus* grows only where the salt concentration is low. In the Eastern mangrove vegetation, species of the genera *Bruguiera* and *Ceriops* (both viviparous, with kneelike pneumatophores) occur, as well as *Sonneratia* (nonviviparous with thick pneumatophores) and species of *Xylocarpus, Aegiceras,* and *Lumnitzera* etc. The different mangrove species usually grow in distinct zones, and only seldom are they mixed. This zonation is dependent upon the tides, since the nearer a species grows to the outer edge of the mangrove, the longer and deeper does it stand in saltwater (Fig. 49).

The tidal range (the difference in depth between high and low tides) not only differs from place to place on the coast but also varies periodically according to the phase of the moon and position of the sun. It is at its maximum at new and full moon (spring tide) and at a minimum midway between the two (neap tides). The spring tides are at their highest twice a year, when day and night are equally long (equinoctial spring tides).

Besides the coastal mangroves growing on flat shores, often in a belt many kilometers wide and with no influx of fresh water from the land, there are also the estuary mangroves, extensive in the delta regions of rivers, as well as the less common reef mangroves growing on dead coral reefs protruding above the surface of the sea. Coastal mangroves have been investigated in most detail in eastern Africa, with special regard to salt relationships (Fig. 50).

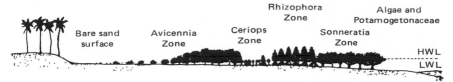

Fig. 49. Zonation of the east African coastal mangroves. HWL, high-water limit; LWL, low-water limit. (From Walter and Steiner 1936)

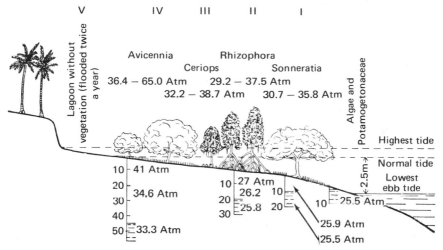

Fig. 50. Concentration of the cell sap in atmospheres (smallest and highest) of the leaves of mangrove species and of the soil solutions at various depths (in centimeters). Coastal mangroves of eastern Africa, arid type. (From Walter 1960)

The east African coast in the neighborhood of Tanga has a relatively dry monsoon climate. Potential evaporation is equal to or greater than annual rainfall. Apart from a short dry season, there is also a pronounced period of drought which is responsible for the fact that, within the tidal region, the further inland and, consequently, the shorter the period of time during which the ground is flooded, the higher the salt concentration of the soil. The most extreme conditions in the mangrove zone are to be encountered at the inland margin, which is reached only by the equinoctial spring tides. Here the saltwater in the soil is strongly concentrated by evaporation during drought, whereas during the rainy season, the soil may be completely leached. Since no plant can tolerate such drastic variations in salt concentration, these areas, found wherever the climate prevailing at the inner boundary of the mangroves is such that a period of drought occurs, are devoid of vegetation. In northern Venezuela, however, small clumps of salt-sensitive plants such as columnar cacti and *Opuntia* or bromeliads do, in fact, grow under such conditions. It is known, though, that the bromeliads take up water through

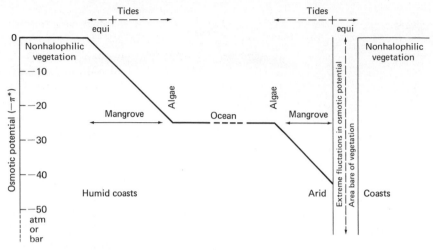

Fig. 51. Scheme showing the salt concentration in the soil and the mangrove formations on the humid and arid coasts (equi-equinoctial)

their leaves and do not root in the soil. The cacti invariably grow on small sand heaps from which the salt has been washed away in the rainy season. They obtain their water by means of shallow roots spreading in the sand and are therefore unaffected by the underlying salty soil. The tissues of both cacti and bromeliads are free of salt, and they are not halophytes – a further example of the observation that obvious soil properties may not always be an indication of the ecological conditions under which plants grow.

In very humid regions, on the other hand, the exposed areas are continually being washed by rainwater, so that the salt concentration of the water in the soil must decrease landward. This is also true, moving upriver, of the estuary mangroves which, via a brackish zone (where the fern *Acrosticum,* the *Nipa* palm, *Acanthus ilicifolius,* and many other species are found), give way to a freshwater community without the interpolation of a barren zone (Fig. 51). In spite of the fact that mangroves are an azonal vegetation, their zonation is determined by the climate and they are, therefore, different in the humid Zonobiome I than in climates with pronounced periods of drought.

Plants rooting in saline soils take up a certain amount of salt and store it in the cell sap. In the case of mangroves, the salt concentration of the cell sap in their succulent leaves is roughly equal to that of the soil. Apart from this, nonelectrolytes are present in a concentration usual for tropical species. Figure 50 illustrates the typical zonation and the potential osmotic pressure in the soil and the mangrove leaves, and Fig. 51 shows the differences between mangroves in arid and humid regions.

Zonation in the mangroves results from competition between the various species. Investigations carried out in eastern Africa suggest that the salt factor is decisive in determining this zonation. *Avicennia* is the weakest competitor but has the highest salt resistance, so that stunted individuals of this species constitute the landward limit of the mangrove. *Sonneratia* is the strongest competitor but seems not to tolerate a salt concentration above that of seawater and is therefore confined to the outer fringes. In continuously humid areas, the situation is more complicated: *Avicennia* appears to be confined to sandy ground, whereas *Sonneratia* prefers silty soil. In such regions, type of soil, duration of inundation, water movement, decrease in saltiness, and variations in salt concentration seem to be of greater significance.

The salt economy of the mangroves presents an interesting problem. Mangroves are unable to take up seawater in its original form because the salts which would be left behind after the loss of water by transpiration would soon lead to the formation of a saturated salt solution in the leaves. It has recently been shown by direct measurement that suction forces of from 30 to 35 atm, or bar, can be produced in mangrove leaves. This is higher than the osmotic pressure of the soil solutions. These suction forces are transmitted to the roots by the cohesion tension in the conducting vessels. The roots act at the same time as an ultrafilter, permitting only the passage of almost pure water, which is then transported to the leaves. Only the very small amount of salt necessary for producing suction tension enters the plants, and the salt is stored in solution in the vacuoles of the leaf cells.

The mechanisms whereby the salt concentration is regulated have not been completely elucidated. Some of the salt from the old, fading leaves can be transported to the young, developing leaves, or an excess of salt can be eliminated when the old leaves drop. *Avicennia* has salt glands, situated on the underside of its leaves, for regulating the salt concentration. This species can excrete a 4.1% salt solution, which is more concentrated than seawater; the proportion of NaCl to KCl is the same as in seawater (90 and 4%). No excretion takes place at night, but it reaches a maximum at midday. In 24 hours, 0.2–0.35 mg of salt per 10 cm^2 leaf area is excreted. In the dry season, the salt crystallizes on the underside of the leaves during the daytime, and at night, when the humidity is greater, the salt dissolves and drips away.

It is interesting to note that the viviparous seedlings are almost free of salt and have a potential osmotic pressure of only 13–18 atm. This means that water must somehow reach them with the help of glandular tissue in the cotyledons. But as soon as the seedlings drop off and take root in the salty ground, their salt concentration increases, and the potential osmotic pressure attains the normal level. The radicle appears, initially at least, to be permeable to salt.

The function of the pneumatophores has also been elucidated. They are equipped with lenticels with minute openings which permit the entry of air but not water. When the pneumatophores are completely submerged in water, their intercellular oxygen is used up in respiration and a negative pressure develops because carbon dioxide, being very soluble, escapes into the water. As soon as the roots emerge from the water, a pressure compensation takes place, whereby air (and oxygen) is drawn in. The oxygen content of the intercellular spaces in such roots therefore varies periodically between 10 and 20%.

Together with their fauna–the numerous small fiddler crabs and the mangrove fish *Periophthalmus,* which can be seen crawling out of water and up the trees–the mangroves present a highly interesting ecosystem belonging to neither the sea nor the land.

9. Shore Formations – Psammobiomes

The sandy shore formations of tropical coasts offer fewer peculiarities. Beyond the barren zone resulting from exposure to the force of the breakers, sand plants with long runners can be found, among them the widespread *Ipomoea pes-caprae* and the halophytic *Sesuvium portulacastrum* and *Sporobolus virginicus.*

Still further inland, beyond the influence of saltwater, the sand in the tropics very soon becomes covered by shrubs and trees, the floating fruits of which can be seen in the drift on all tropical shores. Typical representatives are *Terminalia catappa* and coconut palms, although nowadays the palms are nearly all planted by man. *Barringtonia, Calophyllum, Hibiscus tiliaceus,* and *Pandanus* are characteristic of the Eastern oceans, and *Coccoloba uvifera* (Polygonaceae), *Chrysobalanus icaco,* and the poisonous *Hippomane manicella* (Euphorbiaceae) of the Western oceans.

No large dune areas occur in the tropics, with the exception of the northern coast of Venezuela, near Coro, where a semidesert type of climate prevails. As a result of the continual trade winds blowing in from the northeast or east-northeast, large quantities of sand drift landward from the beach and are trapped by *Prosopis juliflora.* The dunes thus formed continue to grow in the direction of the wind and are soon covered by *Prosopis* bushes. In this manner, a series of dune ridges is established, running parallel to one another and to the wind direction and reaching a considerable height. Migrating dunes, or barchanes, are found in one part of the dune region, probably resulting from wood-clearing. They join up to form ridges at right angles to the wind direction.

10. Orobiome II – Tropical Mountains with an Annual Temperature Periodicity

Whereas a brief dry period in the alpine belt has no effect on the water supply of the plants in Orobiome I, the dry period in ZB II, even at considerable heights, has very obvious consequences, depending upon its duration.

Precipitation increases in the montane belt, and cloud cover reduces the number of hours of sunshine to such an extent that an evergreen montane forest develops. Cloud forests may even be found above this, in the tradewind or monsoon region (Figs. 31 and 52).

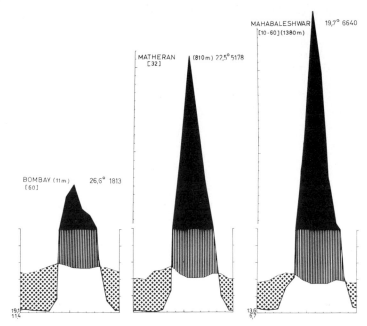

Fig. 52. Increase in quantity of precipitation in the monsoon region of India. Climate diagram of Bombay and two mountains stations above it. At the highest station, 1380 m above sea level, almost 3000 mm of rain falls in July. The rainy season is only one month longer, although the annual precipitation exceeds 6000 mm

The entire sequence of the altitudinal belts of Orobiome II is best seen in the southern slopes of the eastern Himalayas, by following the extremely humid Sikkim profile from Darjeeling toward the north. The forest belts are not easy to distinguish from one another, however, and a further complication is provided by the fact that the tropical floral elements are increasingly supplanted by holarctic elements.

The foot of the mountains is taken up by wet-deciduous forest with *Shorea robusta* or with bamboos and palms on wet ground. An evergreen tropical-montane forest *(Schima, Castanopsis)* with tree ferns begins at an altitude of about 900 m above sea level. In its higher regions, holarctic genera *(Quercus, Acer, Juglans)* and *Vaccinium,* among others, are well represented. Above this is a cloud forest with Hymenophyllaceae and mosses. The proportion of holarctic genera *(Betula, Alnus, Prunus, Sorbus)* increases with increasing altitude.

The frost limit lies between 1800 and 2000 m above sea level. In the belt immediately above this, there are large numbers of tall *Rhododendron* and *Arundinaria* species, which are replaced further up by conifers *(Tsuga, Taxus,* etc.). Between 3000 and 3900 m, is an *Abies densa* fir forest containing broadleaf species. The timberline is composed of *Abies* and *Juniperus.*

The subalpine belt is again characterized by high *Rhododendron* species, which gradually diminish in size in the alpine belt, with its flower-filled meadows, until, at 5400 m, *Rhododendron nivale* is merely a tiny shrub. This orobiome system is especially complicated in the Himalayan Mountain range (Troll 1967, Meusel and Schubert et al. 1971).

From 5100 m upward, the plants are mainly hemispherical and cushionlike in form (*Arenaria, Saussurea, Astragalus, Saxifraga,* etc.). The snow line is reached at 5700 m.

In the Andes, the order of the altitudinal belts on the eastern slopes differs from that on the western slopes, and is different again in the mountainous valleys of the interior. A brief schematic survey has recently been published by Ellenberg (1975). The high plateau of the Altiplano has undergone anthropogenic changes due to human settlements and grazing of llama herds. In accordance with the climate, the altitudinal belts on the western slopes become increasingly xerophytic toward the south. In addition, the deciduous-forest belt, in leaf only during the rainy season, extends farther and farther upward, and the evergreens become increasingly sclerophyllous with smaller leaves. The occurrence of a warm season pushes the forest limit up to 4000 m, with scattered stands of *Polylepis* even at 4500 (4900) m. The páramo is replaced first by puna of the wet type with cushionlike plants, then further south by dry puna with tussocks of xerophytic grasses (*Festuca orthophylla, Stipa ichu,* etc.), and, finally, in Orobiome III, by desertlike puna with *salares* (saltpans). The soils of the alpine belt change correspondingly, from peat soils to chestnut-brown soils and sierozems, and even solonetz and solonchak.

The Puna of northwesten Argentina (between 22° and 24$^1/_2$° S) has been investigated in detail, including microclimatic studies (Ruthsatz 1977).

Zonoecotone II/III – Climatic Savannas

This zonoecotone includes the climatic savannas, such as those of Southwest Africa, which have already been discussed. Similar conditions prevail in the Sahel zone, south of the Sahara, where the transition to the summer rain region of the Sudan (Zonobiome II) occurs.

The Sahel zone has become completely degraded by overpopulation and overgrazing as a result of the typical and recurring years of drought in this zone. Due to the limited number of natural water sources in this area, it was only capable of supporting a very small population and a correspondingly small amount of cattle until foreign aid programs attempted to develop the land by drilling a large number of wells. This made water available for larger herds of cattle, and resulted also in a rise in the human population as long as the annual precipitation remained above the long-term average. The occurrence of several years of drought led to a catastrophe. Although enough water was available for both humans and the animals, the parched pastures yielded no more grass. The cattle died of starvation and the humans were forced to leave the area or were supported by foreign aid projects. The pastures, however, were irreparably damaged and were converted into a man-made desert. In Southwest Africa, with its similar climate, several successive years of drought also prove disastrous, although the small number of farmers is able to survive by reducing the size of their herds in time. The economy recovers quickly after a few good years of rain and the natives in the reservation receive the necessary aid. A further Zonoecotone II/III is situated in the border region of India and Pakistan in the Thar or Sind desert. This homogenous arid region between the Aravalli Mountains in the east and the Baluchistanian Heights in the west is also referred to as the Great Indian Desert (Fig. 53). The aridity increases from east to west.

The reference in the literature to a Saharo-Sindic desert zone is incorrect. The Sahara, a rainless region (or one with scant winter rain) is floristically mainly part of the holarctic, and extends eastward into the Egyptian-Arabian desert as far as Mesopotamia. The Sind desert, however, is the final arid extension of the Indian Monsoon region and floristically, it must be considered part of the paleotropics. Climatically, the Indian Thar Desert is a Zonoecotone II/III, which may be compared to the "Sahel", the transitional region between the Sudan and the southern Sahara. Both receive light summer rains, although the Indian region lies north of the Tropic of Cancer and the mean annual temperature is $2°-3$ °C lower than in the Sahel, meaning that frosts may occur between the months of December and February (Fig.

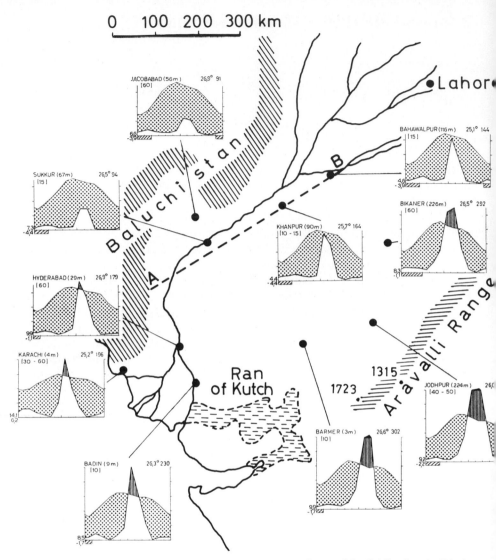

Fig. 53. Climate diagram map of the Sind-Thar Desert. Northwest of the dividing line *A* − *B* is the extremely arid region

53). Only the Indus valley has an actual rainfall of less than 100 mm and would, therefore, qualify as a desert. Because of the Indus and its tributaries, however, it is a water-rich irrigation region.

On the other hand, the Great Indian Desert is a man-made desert. The area was settled 4000 years ago, but since the campaigns of Alexander the Great, the population has increased to an extent that overgrazing, wood clearing and, to some degree, farming have completely degraded the

Fig. 54. Barchane that has started to move recently in the region between Jaisalmer and Jodhpur with single specimens of *Prosopis, Acacia,* and *Calotropis.* (Photo courtesy of M. P. Petrov)

countryside (Mann 1977). In its natural state, the region was a *Prosopis* savanna with an annual precipitation of of 400–150 mm on a deep reddish brown sandy savanna soil, as can be seen today in the area around Jodhpur, which has been under protection for the past decades (Rodin et al. 1977).

Here, the following thorny shrubs are found: *Prosopis cineraria (= specigera), Ziziphus nummularia, Capparis decidua (= C. aphylla)* etc. With an annual precipitation of 500–600 mm, *Prosopis* grows to a height of 8 m, forming stands of 150–200 individuals/ha. With a precipitation of 300–400 mm, it grows to 5–6 m in stands of 50–100 individuals/ha and with 200 mm it only grows to 3–4 m in stands of 25–30 plants/ha. Decreasing amounts of rain also result in the replacement of high grass species *(Lasiurus, Desmostachys)* by lower growing species *(Aristida)* (Gaussen et al. 1972). The conditions are, therefore, similar to those in Southwest Africa (pp. 81–82).

In regions with over 250 mm annual precipitation, the savannas are used for grazing and are degraded as a result of overstocking with cattle, whereby the annual grass species *Aristida adscensionis* becomes dominant in the pastures.

The soils are extremely sandy in the Bikaner District. As a result of overgrazing in the vicinity of the villages, moving barchanes (dunes void of vegetation) give the impression of an extreme desert (Fig. 54). Actually, however, the water content in the sand of such naked dunes is much higher than of overgrown dunes as is illustrated by the following figures from an area with an annual precipitation of 260 mm:

Water content (in mm) in the sand of naked (I) and overgrown (II) dunes, according to Mann (1976)

Depth (in cm)	March I	March II	June I	June II	September I	September II	January I	January II
0–105	41	10	33	17	45	10	34	7
0–210	106	39	94	48	120	33	105	28

This is explained by the fact that a growth of *Prosopis* requires approximately 220 mm of water for transpiration. The often-planted grass species, *Pennisetum typhoides*, requires 160–180 mm. The inhabitants take advantage of the water content of the sand in the naked dunes by planting watermelons every 2 m. They use branches to prevent the sand from blowing away. No information is available on the natural vegetation of the driest parts of the Sind Desert in the Indus Valley. This irrigation region is densely populated and there are no areas with natural vegetation to be found. Due to irrational irrigation practices, the water table is rising, resulting in a secondarily increased salt concentration in the moist soil. This has caused the loss of 40,000 ha of agricultural land annually, and means that the food production will not be able to keep pace with the growing population. A rehabilitation of the brackish soil would be an extremely costly endeavor in this level landscape.

Naturally salty soils are widely distributed in the southern part of the Thar Desert on the Gulf of Kutch. Mangroves grow in the tidal zone, followed inland by salt marshes with *Salicorina*, *Suaeda*, *Atriplex* and the salt grass species *Urochondra*. In the region of Ran of Kutch, with its high water table, nearly sterile salty clay soils are found with halophytes *(Haloxylon salicornicum, Aeluropus, Sporobulus),* and *Chenchrus* spp., *Cyperus rotundus,* or scattered woody plants on good locations (Blasco 1977).

A difficult region to cataegorize ecologically, is the arid "Polygono da Sêca" in the Caatinga of northeastern Brasil. It is characterized by a precipitation which may vary extremely from year to year. In the driest location in Cabaceira, for example, the years 1940–1946 with plentiful rainfall (664–150 mm) were followed by the drought years of 1948–1958 with precipitation below 80 mm (1952 only 24 mm, 1958 only 22 mm), excepting the years 1954 (170 mm) and 1955 (187 mm). In such an unreliable climate large succulent columnar cacti and large thorny bromeliads, which grow close to the ground survive best, as well as bottle trees (Ceiba) or deciduous shrubs, which are leafless for long periods of time. This region is difficult to use agriculturally and is only sparsely settled since the drought periods are unpredictable and force the inhabitants to leave the area. Similar conditions are found in the trade wind deserts on the Venezuelan-Columbian border on the north coast of South America and on the Galapagos Islands. These dry regions also experience years with very high precipitation.

Fig. 55. The tuberlike trunk of *Aderia globosa* (Passifloraceae) from the top of which the bow-shaped branches armed with green thorns and which normally envelop the trunk, were cut. In the foreground lie the cut-off thorny-tipped succulent leaves of *Sanseviera*. Thorny brush on the southern foothills of the Pare Mountains in East Africa. (Photo H. Winkler)

Finally, and still belonging to the Paleotropic realm, there are the extensive arid regions in the tropical parts of eastern Africa, as well as a small area in the rain shadow between the Pare and West Usambara mountains where very odd succulents are found [*Adenia globosa* (Fig. 55), the boulderlike *Pyrenacantha, Euphorbia tirucalli, Caralluma, Cissus quadrangularis, Sansevieria*, etc.]. This is probably the driest region near the equator, with an annual mean temperature of 28 °C and a rainfall of only 100−200 mm. In northern Kenya, western Ethiopia, Somaliland, and on Socotra, there are

Fig. 56. *Adenium socotranum* (Apocynaceae) with a diameter around the trunk of 2 m on western Socotra. (Photo F. Kossmat)

even more extensive arid regions where *Adenium socotranum* (Apocynaceae), a plant with bizarre succulent stems achieving diameters of up to 2 m, is found (Fig. 56), and *Dracaena cinnabari* with a trunk diameter of 1.6 m. The driest southeastern corner of Madagascar is distinguished by Baobab trees and plants of the columnar cactuslike family Didieraceae, which only occurs her.

III Zonobiome of the Subtropical Arid Climate with Deserts

1. Climatic Subzonobiomes

Deserts are arid regions where potential evaporation is very much higher than annual precipitation. These regions can be further subdivided into semiarid, arid, and extremely arid, which together, cover 35% of the earth's land surface. The cold winter period which is typical of the arid regions of the temperate zone is lacking in the subtropical desert zone (see Sect. VII).

The term desert is a relative one. Contrasted with the humid eastern part of North America, the Southwest looks like desert, although Tucson (Arizona) has an annual rainfall of 300 mm. On the other hand, the Mediterranean coast of Egypt, with barely 150 mm of rainfall, is not considered to be desert by an Egyptian from Cairo.

In general, a subtropical region is termed desert when the annual rainfall is less than 200 mm and the potential evaporation more than 2000 mm (up to 5000 mm in the central Sahara).

The sparse precipitation of arid regions falls at different times of year, thus providing a basis for a subdivision of Zonobiome III into subzonobiomes, as follows:

1. sZB with two rainy seasons (Sonoran desert, Karroo);
2. sZB with a winter rainy season (northern Sahara, Mohave desert, Middle-Eastern deserts);
3. sZB with a summer rainy season (southern Sahara, inner Namib);
4. sZB with sparse rainfall occurring at any time of year (central Australia);
5. sZB of the coastal deserts with almost no rainfall but much fog (north Chilean-Peruvian desert, outer Namib);
6. sZB of the rainless deserts devoid of vegetation (central Sahara).

Figure 57 shows the climate diagrams of the various subzonobiomes, with the exception of sZB 5 (since fog cannot be measured as precipitation and thus does not show up on the diagrams (see Fig. 79). A very distinctive feature of all arid regions is the large variability in amount of rain falling in different years. This means that average figures are of little value. Although years in which the rainfall is below average are most frequent, the water reserves of the soil can be replenished for decades in the few years in which precipitation values are high.

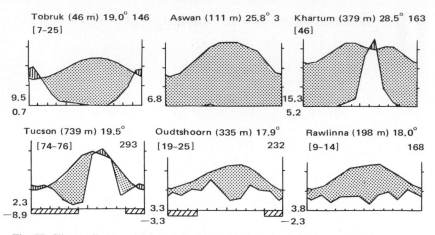

Fig. 57. Climate diagrams of desert stations. *Above:* northern Africa, with winter rain, no rain, and summer rain. *Below:* with two rainy seasons (Sonoran Desert and the Karroo) and with rain that may fall at any season (Rawlinna, Australia). Compare also Fig. 79. (From Walter and Lieth 1967)

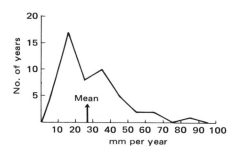

Fig. 58. Curve showing variability in annual precipitation near Cairo, between 1906 and 1953. (From Walter 1973)

The variability curve for Cairo (winter-rain region) is shown in Fig. 58. That for Mulka, the most arid station in central Australia, is similar, except that the mean value is 100 mm and the highest and lowest values are 18 and 344 mm. In Swakopmund (outer Namib) the mean precipitation is 15 mm the highest and lowest values being zero and 140 mm.

Ecological conditions differ so much from year to year that an accurate picture of desert ecosystems can only be formed on the basis of long-term observations, and each desert has to be considered individually. The few features that they have in common will be discussed first.

In all deserts (except in the fog variety), the air is very dry. Both incoming and outgoing radiation are extremely intense, which means that the daily temperature fluctuations are large. In the rainy season, however, the extremes are greatly reduced.

2. The Soils and Their Water Content

Desert soils are not soils in the true sense of the word, but rather lithosols (syrozems), consisting of the erosion products of the underlying rock, altered by the action of wind and water. Therefore, it is the properties of the frequently loose bedrock that are decisive. In other words, we cannot speak of climatic soils, but only of soil texture. Further, instead of euclimatopes supporting a vegetation typical of the climatic zone, there are pedobiomes (lithobiomes, psammobiomes, halobiomes).

The water supply of plants in arid regions depends upon the soil texture (particle size). Quantity of rain is only of indirect importance; the amount of water remaining in the soil, and thus available to plants, is far more important. Part of the rainwater runs off, and a further portion evaporates (Fig. 59). How much of the water remains in the soil, and is thus available to the plants, is determined by the texture of the soil. In humid regions, the sandy soils are dry because they retain only small amounts of rainwater, whereas the clay soils are wet. The reverse is true for arid regions.

On flat ground in arid regions, water does not sink down to great depths and thus does not reach the groundwater. Only the upper soil layers are damp, and the depth to which the water penetrates depends upon the field capacity of the soil. Let us assume that 50 mm of rain falls upon a dry desert soil and that it completely soaks into the ground. If the soil is sandy, then the upper 50 cm is wetted to field capacity. If the soil is a finely granulated clay with a field capacity 5 times as large, the water can only penetrate to a depth of 10 cm. On rocky ground with small cracks, the water goes down much further, possibly 100 cm (Fig. 60).

Evaporation follows the rain. If the upper 5 cm of a clay soil dries out, then 50% of the water originally entering the soil is lost. Sandy soils do not dry out so much, and even if the upper 5 cm were to dry out, this would

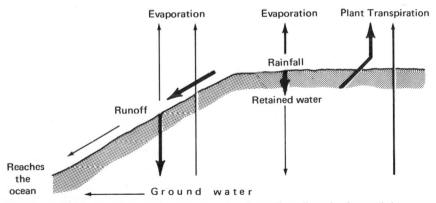

Fig. 59. Diagram showing the fate of rain in arid regions. The soil-retained water is important to plants. The runoff water seeps down to the groundwater in the dry valleys and is only seldom reached by roots. (From Walter and Volk 1954)

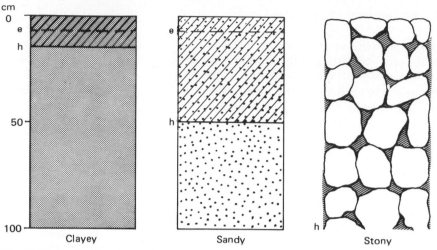

Fig. 60. Diagramatic representation of water retention in various kinds of soils after a rainfall of 50 mm in arid regions. h, Lower level of moistened soil; e, Lower level to which the soil dries out again. The clay soil retains 50%; the sandy soil, 90%; and the stony soil, 100%

involve a loss on only 10% of the water. On rocky ground, there is almost no evaporation, and nearly all of the water is retained. This means that, *contrary to the situation in humid regions, the clay soils form the driest habitats, whereas the sandy soils offer better water supplies. Fissured, rocky ground provides the wettest habitat* if there is no runoff from the rock surface and if there is enough fine soil in the cracks to retain the water.

Such considerations have been confirmed by measurements in the Negev desert. Comparison of areas with identical absolute rainfall revealed that loess soil offered the plants the equivalent of 35 mm of rainfall in available water; rocky habitats with a relatively high run-off, 50 mm; sandy soils, 90 mm; and dry valleys with a large inflow, 250–500 mm.

That the sandy soils are favorable habitats for plants in arid regions can be seen from the fact that the same type of vegetation occurs on sand at a lower rainfall than on clay. In the Sudan, *Acacia tortilis* semidesert is found on sandy soil in a zone which has a rainfall of 50 to 250 mm. On clay, however, this species is found only in a zone with a rainfall of 400 mm. The *Acacia mellifera* savanna begins on sandy soill at a rainfall of 250–400 mm, but on clay soil at 400–600 mm. In the shortgrass prairie region of the Great Plains in western Nebraska, a tallgrass prairie occurs on sandy soil although, otherwise, it only occurs farther east at a higher rainfall. The favorable water supply on rocky ground is often marked in arid regions by the occurrence of trees in the midst of the lower-statured vegetation supported by fine-grained soils.

If sandy soil or the soil in rocky clefts is wet down to the groundwater, the roots of the plants also grow down this far, thus securing their water supply. The following example is worthy of mention.

North of Basra in Mesopotamia, groundwater is present at a depth of 15 m and is constantly replenished from the Tigris and Euphrates via gravel strata. Since, however, the rainfall amounts to only 120 mm annually, the upper soil layer alone is damp, and the plants are unable to reach down to the groundwater. As a result, there is only a sparse ephemeral vegetation after the winter rains. The native population has dug wells and uses the water for cultivating vegetables, which they plant in furrows and irrigate several times daily since the temperature can reach 50 °C. The soil rapidly turns brackish due to the high evaporation, so that vegetables cannot be planted in the same spot for more than one year. Between the vegetable plants, however, *Tamarix articulata* cuttings are planted and rapidly take root. In the second year, the furrows are not irrigated but the soil is still damp down to the groundwater from the previous year. The tamarisk roots can therefore grow deeper and deeper in the ensuing years until they finally reach groundwater, after which their water supply is secure and they can develop into large trees. They are cut down every 25 years for fuel and shoot up again from the stumps; in this way, the farmland is transformed into tamarisk forest. *Deserts with deep-lying groundwater can be converted into forest* if the soil is irrigated to such an extent in the first year after the trees have been planted that it is wetted down to the groundwater.

This example provides the explanation of the fact that phreatophytes, which are dependent upon groundwater, can reach down to it with their long roots, even through several meters of overlying dry soil. This can only be accomplished, however, if the soil is wet from the surface down to the groundwater, as it is after several years in succession with good rains. Once the groundwater has been reached, however, woody plants can attain their usual lifespan. It need not necessarily be groundwater that the plants seek; often it is merely the ground moisture, i.e., interstitial water stored in the soil. Once this water has reached a depth of 1 m or more, it remains there for long periods if there is no vegetation or if only a few plants have succeeded in tapping it with their roots.

The desert vegetation has larger water reserves at its disposal than is at first sight apparent to the observer.

Of very frequent occurrence in the desert, and particularly in depressions, are saline soils, which will be dealt with separately on p. 125.

Desert biomes can be classified on the basis of soil texture into the following biogeocene complexes. Since these soils were first studied in the Sahara, they bear the local descriptive names.

1. *The rocky desert, or hamada,* is mainly found on plateaus of the table mountains (mesas), from which all the finer products of weathering have been blown away and where the exposed rocks have undergone severe wind erosion due to sandblast. The surface is covered with a pavement formed of hand-sized stones, darkly stained by desert-varnish (Mn oxides), which lend a forbidding aspect to the landscape. Beneath the stony pavement, there may be a water-repellent, dusty soil, rich in gypsum and salt if it has originated in marine sediments, which prevents the development of a plant cover. Hamada

Fig. 61. Great "Fischfluss" canyon in the desert of southern Southwest Africa. (Photo E. Walter)

areas are cleft by deep erosion valleys with steep, rubble-covered slopes (Fig. 61). In the cracks and crevices of the rocks, a few xerohalophytic species can take hold.

2. *The gravel desert, or serir (reg),* arises from heterogeneous (e.g., conglomerate) parent rock. The cementing substance is readily weathered and removed by the wind, and the harder pebbles collect on the surface. Such autochthonous gravel deserts are in contrast to the allochthonous ones, which consist of alluvial deposits of earlier rainy periods and from which, again, the finer material has been blown away. Under the darkly stained gravel layer, there may also be a crust, cemented hard with gypsum. It is a particularly monotonous type of desert, slightly undulating, with broad, shallow, sandfilled valleys offering better growth conditions for plants typical of sandy soil and a few xerohalophytes.

3. *The sandy desert, erg (or areg),* has formed in the large basin areas by the deposition of sand blown off the raised ground. Sand dunes are formed in this region. If there is a prevailing wind direction, then sickle-shaped dunes are formed (barchanes), gently sloping on the convex, windward side, and steeply sloping on the concave, lee side. The dunes move in the direction of the wind, but if the wind direction varies periodically, the crest of the dunes alters while the base remains fixed. A thin covering of iron oxide on the sand grains accounts for the bright-red color of the dunes in hot, dry regions. Near the coast, where the air is more humid, the color changes to yellowish brown.

These mobile, and therefore barren, dunes can store water because the rain sinks in readily and does not evaporate. Even at an annual rainfall of only 100 mm, a fresh groundwater horizon is present, so that water can be obtained by sinking wells.

If the sandy covering is not very deep, colonization by plants is possible (nonhalophytes such as dune grasses, *Ziziphus,* etc.). Perennial species, including shrubs, serve as sand catchers and grow up through the sand that has accumulated around them, thus trapping still more sand. In this manner, each plant can form its own dune-hillock (several meters high), called a nebka. These miniature dunes lend a characteristic note to the entire landscape.

4. *The dry valleys, or wadis (oueds),* known in Southwest Africa as riviers and in America as washes or arroyos, are an important feature in all deserts. They mostly originated in the past (during pluvial periods of the pleistocene), when the rainfall was higher. The dry valleys commence as scarcely noticeable erosion gulleys which then unite to form deep ditches or small valleys and often end in deep canyons. Gravel and sand are deposited by the water as it drains off after a shower. Some of the salt is washed out, and the soil is soaked to a considerable depth, which provides favorable conditions for the growth of halophytic plants *(Tamarix, Nitraria).* The beds of the larger dry valleys bear no vegetation owing to the redistribution of the soil by occasional floods. Vegetation is confined to the valley sides, which are safe from floods, and the degree of luxuriance depends upon the amount of water held in the alluvial deposits. There is often a permanent underground flow of water, and in such cases, dense, nonhalophytic woods are present as extrazonal vegetation.

5. *The pans, dayas, sebkhas, or shotts* are hollows or larger depressions in which alluvial silt or clay particles are deposited. If there is subterranean drainage (in karst areas), they do not turn brackish. This is also true of the takyr, or deltalike formations at the valley exits, from which a part of the water drains off after a particularly heavy rainfall. The heavy clay soils, however, provide unfavorable habitats since the water can scarcely penetrate the soil and the ground rapidly dries out again after a flood. For this reason, mainly algae, lichens, and ephemeral species grow on takyr soil. If there is no outflow and all the water evaporates, then salt concentration takes place, and in such saltpans (sebkhas or shotts), or halobiomes, compact layers of salt form in the deepest places. On the edges, where the salt concentration is lower, hygrohalophytes take hold. The salt content of the groundwater is often low, and a salt crust forms only on the surface. If a thin layer of sand is deposited on the surface of such a saltpan, then there can be no capillary rise of water and hence no salt concentration. Plants soon establish themselves on the sandy deposits and serve to trap even more sand, so that a hillocky, or nebka, type of landscape is formed around the pans.

6. *Oases* are those sites in the desert with a dense vegetation, where water of a low salt concentration reaches the surface, either by means of normal springs or artesian wells. As has already been mentioned, hygrophilic species can grow here. Such oases are nowadays densely populated, and the natural vegetation has been replaced by cultivated plants or weeds.

Oases with abundant water are often fringed by saltpans (shotts), where the excess water collects and evaporates (southern Tunisia, Algeria).

3. The Water Supply of Desert Plants

The extreme dryness of arid regions has led to the false assumption among investigators with no personal knowledge of the desert that desert plants possess special physiological properties (a physiological resistance to drought) enabling them to grow under such conditions. In particular, an (allegedly) high cell-sap concentration is often mentioned in connection with the ability of the plants to take up water even from almost dried-out soils. However, detailed ecophysiological investigations over the past decades have shown that this view is incorrect. The water supply of desert plants is not so poor as would be suggested by the low rainfall. Rainfall measured in millimeters is equivalent to liters of water per square meter of ground surface. In order to judge how much water is available to the plants, the transpiring surface per square meter of ground surface must be calculated.

Although there are many different kinds of deserts, they are all alike in the sparseness of their plant cover, so that the character of the landscape is determined by the naked rock and not by the plants. In order to study the exact relationship between rainfall and density of vegetation, identical lifeforms must be compared (e.g., grasses or trees with similar foliage), and a region must be chosen in which rainfall varies over a relatively short distance and temperature conditions remain more or less constant. Furthermore, euclimatopes should be chosen, where the vegetation has in no way been disturbed by human action.

Suitable regions are to be found in Southwest Africa, with grass cover and an annual rainfall of 100–500 mm, and southwestern Australia, with *Eucalyptus* forests and a rainfall of 500–1500 mm. Such investigations have revealed a linear relationship between amount of rainfall and production of plant mass or transpiring area (Fig. 62). This also holds true for the creosote bush desert *(Larrea divaricata)* in southeastern California as well as in areas with rainfall of 0–100 mm, as Seely has recently demonstrated in the Namib. Grass seedlings require 16–17 mm for the process of germination, and are less economical with water than older grasses, so that the curve rises less steeply.

From this it can be deduced that *the water supply per unit of transpiring surface is more or less the same in arid and in humid regions* (annual rainfall of 100–1500 mm). The drier the region, the further apart the plants grow, thus leaving a greater area from which the individuals can take up water. This has been confirmed in northern Africa in olive plantations. The number of trees per hectare decreases with decreasing rainfall until, finally, only 25 trees per hectare are left. But since the individuals bear approximately the same amount of fruit, it is evident that the water supply is unaltered. In cereal-growing areas, too, it is known that crop-plant density should decrease with decreasing rainfall.

To take up water from a larger soil volume, *a correspondingly larger root system has to be developed, and this, combined with a steady reduction of the transpiring surface, ensures an adequate water supply in the face of increasing*

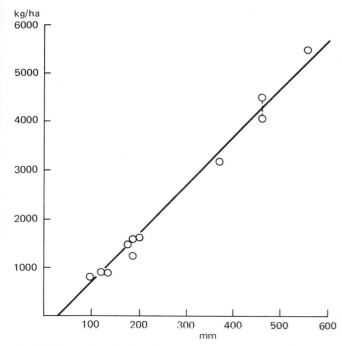

Fig. 62. Material production (aboveground dry weight, in kilograms per hectare) of grassland in Southwest Africa, in relation to the annual precipitation (in millimeters). (From Walter 1939)

aridity. A negative water balance in the plants and a rise in cell-sap concentration have been shown to be accompanied by an immediate and marked inhibition in shoot growth, whereas roots at first grow even longer. In wet regions, the larger part of the phytomass is aboveground, and in arid regions, it is underground. This does not mean that the roots penetrate deeper in dry regions as is usually suggested. Rather, the opposite is true: the root system flattens out. The scantier the rainfall, the less deeply the soil is wetted, and beneath the upper, watercontaining layer, there is no water available at all for the plants. Long taproots have been observed only in plants that are dependent upon groundwater. These are, however, special cases from which generalizations should not be made.

In extremely arid regions with a rainfall below 100 mm, a change in type of plant cover is noticeable. Instead of being *diffuse,* or evenly distributed, over apparently flat areas, the vegetation changes to a so-called contracted type (*Végétation contractée* of Monod 1954), the plants growing only in the often barely noticeable erosion gulleys or depressions, and the raised areas having no vegetation at all. This is connected with the distribution of the water in the soil.

In extreme desert regions, the soils, apart from the shifting sands, usually have a surface crust which can be moistened only with difficulty. The rare, but usually torrential, rains therefore scarcely penetrate the soil but to a large extent run off the surface. The sandy erosion gulleys and depressions therefore receive much more water than the rainfall would suggest, and this runoff water penetrates deeply into the soil. In such places, the plants develop roots that reach down as far as there is any water, and this is sometimes several meters. In some places, groundwater even collects in the valleys. With a rainfall of only 25 mm, vegetation is present in all of the valleys in the desert near Cairo-Heluan. Assuming that 40% of the rainwater runs off into the deeper parts of the relief and that these depressions account for only 2% of the total area, then at a rainfall of 25 mm, the amount of water available to the plants in these biotopes is the same (thanks to the inflowing water) as if they were growing on level ground with a rainfall of 500 mm. Water losses due to transpiration of the plant cover in such a habitat near Heluan were, in fact, found to be 400 mm. Even in a rainless summer, the cell-sap concentration of these plants does not rise, thus confirming that they are well supplied with water. The sandy depressions in the gravel desert along the Cairo-Suez road permanently contain 2.5% water at a depth of only 75 cm (wilting point, 0.8%), so that they never dry out and are capable of supporting a sparse perennial vegetation. In some erosion gulleys, roots descend as far as 5 m, depending upon the depth to which the soil remains damp. Despite the extreme aridity, 200 vascular plant species can be found in the neighborhood of Cairo.

The water supply of plants in extreme deserts is therefore not so poor as is usually assumed. However dry the soil surface may appear to be, there is always some water available, at least at certain times, wherever plants are found growing in the desert. These plants must, of course, be able to resist long periods of drought, and this they achieve chiefly by means of special morphological adaptations. There is no significant protoplasmic resistance to drought, and the cell-sap concentration is generally low (with the exception of halophytes).

The Berber population of southern Tunisia has exploited the principle of "contracted vegetation" for countless centuries in order to obtain crops at an annual rainfall of 200 mm or less. Each small gulley is provided with a dam to prevent the water from draining off, and date palms or barley can be cultivated on the damp soil caught behind the dam.

It has been established that a similar type of *runoff farming* was practiced in pre-Arabian times in the Negev desert. The old dams have been renovated in recent times, and experiments with various cultivated plants have been successful (Evenari et al. 1971).

4. Ecological Types of Desert Plants

All plants growing in arid regions have been called "xerophytes." This is incorrect, because in every such region there are habitats, such as the oases, where plants are well supplied with water. Species typical of the humid tropics even grow in such habitats. In the rainless Aswan desert, on an island in the Nile, coconut palms, mangoes, papaya, maté, sweet potatoes, manioc, camphor trees, mahogany trees, coffee, pomegranates, and many other species typical of the Indian monsoon forests are cultivated with the aid of artificial irrigation. The microclimate in the dense plantations is less extreme than in the open desert. In dry valleys with groundwater, plants can grow under natural conditions without suffering any water shortage and without any sign of adaptation to the dryness. Besides, most deserts have at least a brief wet season, with the exception of the rainless central Sahara, the Namib, and the Peruvian-Chilean desert. The species which develop during these damp periods *(ephemerals),* including those which survive the rest of the time as seeds *(therophytes)* or in the ground *(geophytes = ephemeroids),* do not exhibit any particular adaptation to water shortage either.

It would be illogical for an ecologist to draw a distinction between plant species that avoid drought and those resistant to it. All plant species tolerate drought, some as seeds (ephemerals), tubers, or bulbs (ephemeroids); some in a latent condition, like poikilohydric lower plants (algae, lichens), a number of ferns, *Selaginella* spp., and even flowering plants, the best known of which is *Myrothamnus flabellifolia* (Rosales) (Fig. 63); and some in a state of reduced activity (xerophytes and succulents).

The term *xerophyte* is used here to describe ecological groups that require a certain minimum amount of water uptake even in periods of drought, since they possess no adequate water-storage organ. These fall into three subgroups linked by transitional forms: (1) malakophyllous xerophytes, (2) sclerophyllous xerophytes, and (3) stenohydric xerophytes.

Malakophyllous xerophytes are characteristic of semiarid regions. They have soft leaves which wilt under dry conditions while the cell-sap

Fig. 63. *Myrothamnus flabellifolia* in latent state (twigs drawn together, leaves folded) on mica-schist of the steep drop to the Namib desert in Southwest Africa. (Photo E. Walter)

concentration rises steeply. They lose their leaves in lengthy dry periods, and only the youngest of the leaves within the hair-covered buds survive. Typical examples are the many Labiatae and Compositae of arid regions and *Cistus* spp., among others.

Sclerophyllous xerophytes have small, hard leaves and owe their rigidity to mechanical tissue. They are found especially in regions with a long summer drought and are able to reduce their transpiration to a minimum when water is scarce, whereby the cell-sap concentration rises only in extreme circumstances. Typical examples are provided by evergreen oaks, olive trees, etc.

Stenohydric xerophytes are able to prevent a rise in cell-sap concentration by shutting their stomata at any sign of water shortage. Gaseous exchange and photosynthesis are also of necessity brought to a standstill, so that the plants are in a state of starvation. The leaves of such species do not dry out during the long droughts but turn yellow and finally fall off. Some nonsucculent *Euphorbia* species may be cited as examples. Most plants of the extreme deserts belong to this group. Since there is no competition among the aerial parts of the plants in deserts, the only important thing is for them to survive the drought, not to produce large quantities of phytomass. This they achieve with incredible endurance, often as pitiful-looking cripples, and may live for 100 years or more. Although many branches die, enough survive to assure further growth after a future rainfall.

A special group is formed by the *succulents,* water-storing species which use their stored water extremely sparingly in times of drought. Their small absorbing roots die, so that no water at all is taken up from the ground during the dry period. According to the nature of the organ responsible for storing water during the rainy season, the succulents can be divided into the following three groups: (1) plants with succulent leaves, such as *Agave* and *Aloe,* or *Cotyledon, Crassula,* and *Sansevieria;* (2) plants with succulent stems, such as cacti and many species of *Euphorbia* and *Stapelia;* and (3) plants with succulent roots, that is to say, with underground storage organs, such as *Asparagus* species, *Pachypodium succulentum,* as well as some Leguminosae with enormous tubers found growing in the sandy regions of the Kalahari.

The concentration of the cell sap in the succulents is very low and does not rise even during long periods of drought when large amounts of water have been lost. The water content, calculated on the basis of dry weight, remains constant since respiration involves the breakdown of organic compounds (sugars, organic acids, etc.). Many succulents can survive for a year without taking up any water, and in many of them the diurnal acid metabolism (CAM = **C**rassulacean **A**cid **M**etabolism) has been demonstrated. Plants with this type of metabolism open their stomata at night, when water losses due to transpiration are small, and take up CO_2, which leads to the formation of organic acids and thus to a rise in acidity of the cell sap. The stomata are closed in the daytime, and the CO_2 which has been bound during the night can be assimilated in daylight, with an accompanying decrease in acidity. In this manner, the necessary gaseous exchange is effected with a minimum loss of water (Dinger and Patten 1974).

The salt plants, or *halophytes,* constitute a very important group in many deserts. Their occurrence depends upon the presence of a saline soil rather than upon climate, and for this reason the salt factor will be discussed below. Halophytes usually exhibit succulence, *although they should not be classed with the true succulents.* Their succulence results from an intense storage of salt, that is to say of chloride, with the result that the *cell-sap concentration is often very high* and may even exceed 50 atm. The *Mesembryanthemums* represent a link between the true succulents, with their low cell-sap concentration, and the halophytes. They can be extremely succulent and can also occur on nonsaline soils, although their cell sap always contains a certain amount of chloride.

5. Salt Soils – Halobiomes

As a result of the long periods of drought in arid regions, the rivers carry water only periodically or even sporadically. Since potential evaporation is many times larger than annual rainfall, all of the water running down into depressions in the arid regions evaporates. The salts which are left behind concentrate over a period of time, so that a saturated brine may even be formed, from which salt then crystallizes out. The largest part of the soluble salt is NaCl, since the sulphates soon precipitate as gypsum ($CaSO_4$) and the hydrocarbonates as $CaCO_3$ after loss of CO_2.

In the process of weathering of silicates and clay formation, sodium ions are set free. Although almost 20 g of chloride ions is present per liter of seawater (sulphate ions, only 2.7 g), chlorine-containing minerals are extremely rare, so that chloride ions are not set free by weathering. In river water, however, the presence of NaCl can always be demonstrated. The NaCl of the salty soils of arid regions can be of various origins:

1. Salt from rocks that were formed as marine sediments. This salt can be washed out of the rocks by rainwater and carried into the undrained depressions. In deserts with marine sedimentary rocks of Jurassic, Cretaceous, or Tertiary age (the northern Sahara and the Egyptian desert for example), saline soils are common, whereas in arid regions with underlying magmatic rock or terrestrial sandstone, hardly any saline soils are found at all.
2. Brackishness in areas that in the most recent geological past were lake or marine beds. Examples are the areas surrounding the Great Salt Lake in Utah, around the Caspian and Aral seas in Central Asia (see p. 245ff.), and the Tuz Gölü in central Anatolia.
3. Seawater turned into a fine spray by the force of breakers along arid coasts. The small droplets dry and form a salty dust which can be blown inland. This salt is then either washed into the soil by rain or fog or simply

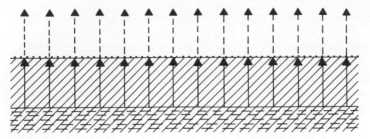

Fig. 64. Formation of a salt crust by means of capillary ascent *(continuous arrows)* of groundwater (horizontal shading) and evaporation of the water *(broken arrows)*. (From Walter and Volk 1954)

deposited. A similar process also takes place in humid regions, but here the salt is continuously being washed out and returned to the sea via the rivers (cyclic salt). In arid regions with no outflow, however, the salt concentrates and in this way leads to such brackishness as is encountered in the outer Namib desert and the arid parts of western Australia. Once salt accumulates on the surface of depressions, it can be blown still further by the wind.

4. Salty water coming to the surface in springs, as it does in the northern Caspian Lowlands. In this case, the salt originated from marine beds which became desiccated in earlier geological times and formed large salt deposits at a considerable depth.

In fact, all of the chloride of the salt found in the deserts came from the sea, where it was deposited in the course of the earth's history, the chlorine being mainly from volcanic exhalations containing HCl.

In the desert, salt is washed from the higher elevations to lower-lying parts with each rainfall so that, for the most part, only the depression soils are saline. If the sedimentary rocks contain much salt and rainfall is very scanty, as it is around Cairo-Heluan, then the soil of the plateau habitats also contains salt. There is no salt transport in the completely rainless central Sahara, so that lower areas do not receive any additional salt.

As far as plants are concerned, it is the salt concentration of the solution surrounding the roots that is important and not the salt content of the soil calculated on a dry-weight basis. In slightly salty but dry soils, the concentration is often higher than in very saline but wet soils.

Evaporation from the soil surface in places where the groundwater is less than 1 m below the soil surface can also lead to salt accumulation. Water rises to the surface by means of capillary forces (Fig. 64), bringing salt to the surface, even if the groundwater contains only minute quantities of salt (Fig. 65). A salt crust forms wherever the capillary water column ends, which is at the highest point of the microrelief (Fig. 66). The occurrence of a salt crust in the dry periods does not interfere with plant growth as long as the roots have access to nonbrackish groundwater.

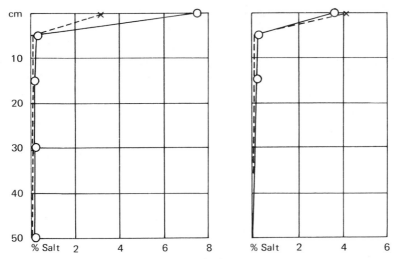

Fig. 65. Salt content at various depths in (left) a watered plot with ascending groundwater and (right) an unwatered plot. Swakop Valley of Southwest Africa. NaCl, solid lines; Na_2SO_4, broken lines. The salt collects only at the surface. (From Walter and Volk 1954)

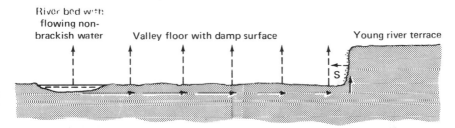

Fig. 66. Salt accumulation in Swakop Walley (Namib, Southwest Africa). Solid arrows indicate direction and magnitude of water flow in the ground; broken arrows represent evaporation. The salt concentration increases toward the sides of the valley; salt efflorescence occurs at S at the foot of the terrace, where the water flow ends

If irrigation in arid regions is not accompanied by at least a certain degree of drainage, the cultivated areas necessarily turn brackish in time, even if the water used for irrigation purposes contains only a small quantity of salt. Extensive cultivated areas in Mesopotamia, the Indus region, have been transformed into salt deserts in this manner. So far, this has not happened in the undrained cotton fields of the Gezira in the Sudan, but only because the water of the Blue Nile, which is used for irrigation, contains hardly any salt. Small quantities of salt are removed in the crop itself with each harvest. An important rule to be followed in irrigation farming, therefore is: no irrigation without sufficient drainage (in order to remove accumulating salt).

6. The Salt Economy of Halophytes

Plants growing on saline soils (halobiomes) have been termed "halophytes." It would be preferable, however, to take the plant itself as a starting point in making any definition. *True halophytes are plants that store large quantities of salt in their organs without thereby undergoing any damage and even benefitting from the salt if its concentration is not too high.* The salts involved are usually NaCl, but can occasionally also be Na_2SO_4 or organic Na salts.

What was said for mangroves (p. 102) holds true for all halophytes. The osmotic effect of the salt concentration in the soil has to be balanced by an equally high salt concentration in the cell sap. Since there are also other osmotic substances present in the cell sap, the transpiring organs can produce a suction tension sufficiently high to extract water from saline soils. The salts in the cells have an effect upon the protoplasm and are toxic to salt-sensitive species, which for this reason cannot survive on saline soils. Facultative halophytes that can tolerate salt nevertheless develop better on a nonsaline soil. But the growth of true halophytes, or *euhalophytes,* is, in fact, stimulated to a certain degree by salt (on normal soils containing only traces of NaCl, such plants take up the available salt so that their salt content remains high). The stimulating effect is due to the chloride ions, which cause a swelling of the proteins and, therefore, lead to a special ionic hydration of the protoplasm. This results in a cell hypertrophy due to water uptake, or in other words a succulence of the organs. The higher the chloride content of the cell sap, the better developed is the succulence. Only the chloride ion has this effect; the sulfate ion has the opposite effect. Certain halophytes that store larger quantities of sulfates as well as chlorides in the cell sap are only barely or not at all succulent. A distinction must be made, therefore, between *chloride-* and *sulfate-halophytes,* although they can exist side by side on the same type of soil. The uptake of salt is thus seen to be species-specific (Breckle, 1976). Investigations on halophytes are quite clearly incomplete if only the salt content of the soil is determined, since, for the plants, only those salts are important that come into contact with the protoplasm. *The concentration and composition of the salts in the cell sap must therefore always be measured.*

Even the euhalophytes have an upper limit, varying from species to species, for the concentration of salt tolerated in the cell sap. If this rises too far, the plants wilt (which, in the case of the Chenopodiaceae, is usually accompanied by their turning red due to the formation of N-containing anthocyanins) and finally die. In yet a further group of halophytes, the sodium in the cell is at a higher equivalent concentration than that of the Cl and SO₄ put together, so that the Na ions must be equilibrated by anions of organic acids. When such plants die, carbonic acid is produced as a result of the breakdown of organic acids, and the sodium reaches the soil in the form of Na_2CO_3 (soda), which leads to an increased alkalinity. These plants are termed *alkalihalophytes.*

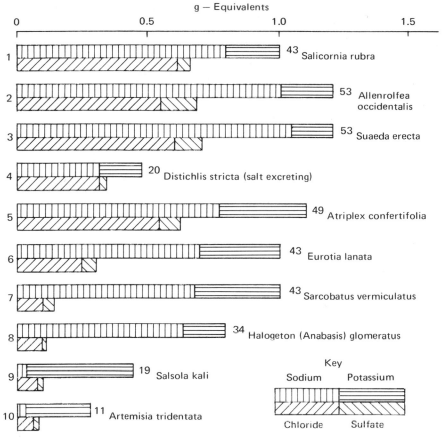

Fig. 67. Concentration of inorganic salts in the cell sap of the transpiring organs of halophytes and nonhalophytes. Figures on the right are concentrations (in atmospheres) of the cell sap. (1 5) Chloride halophytes (all except the salt-excreting grass *Distichlis* are succulent); (6–8) alkali halophytes (cations bound to organic acids, often oxalic acid); (9) represents a transition to nonhalophytic (10) (cations almost exclusively potassium)

There are also *salt-excreting halophytes,* usually nonsucculent species with salt glands, as, for example, *Limonium, Frankenia, Reaumuria,* and halophilic grasses. The well-known tamarisk tree *(Tamarix),* of which there are many species in arid regions, also has salt glands. Salt dust rains down if the branches of these trees are shaken. Since *Tamarix* primarily excretes NaCl, sulfates predominate in the cell sap, and the leaf organs are not succulent.

For most of the halophytes in arid regions, growing on the wet saline soils of the saltpans *(hygrohalophytes),* salt economy is more important than water supply itself. But there are other plants that grow on saline soils and often suffer from water shortage, despite a considerable salt concentration in the cell sap *(xerohalophytes). Atriplex, Haloxylon,* and *Zygophyllum* are of this type, and they often reduce their transpiring surface during the dry season. *Zygophyllum,* for example, sheds its small leaves, and others lose the young terminal shoots or even the green bark of the leafless shoots from the preceding year.

Figure 67 shows the differences in composition of the cell sap of halophytes and nonhalophytes. Characteristic for all halophytes, as was discussed for mangroves (p. 103), is the functioning of the roots as a type of ultrafilter, which only take up almost pure water from the salty soil and convey it to the leaves via the plants' connective tissue. High cohesive tension has been found to exist in the vascular tissue of halophytes.

7. Desert Vegetation of the Various Floristic Realms

At the time when the conquest of the desert by plants took place, during the evolution of a terrestrial vegetation, the floristic realms were already differentiated. Since the plant families, or, speaking more generally, the taxa of the various floristic realms, differ in their genetic constitution, adaptation to life under arid conditions has also taken different directions in the various floristic realms. In addition, the life-forms are not necessarily similar, although convergences do occur (p. 21). These facts have to be borne in mind by the ecologist. Only the northern part of the largest "subtropical" desert, the northern Sahara-Arabian desert, belongs to the Holarctic realm. In the east, this desert borders the Irano-Turanian and central Asiatic deserts, which have cold winters. The northern limit for productive date cultivation forms the border between the two. Chenopodiaceae are especially well represented in the Sahara-Arabian desert, partly on account of the extensive occurrence of saline soils. Succulent species of *Euphorbia* are found only in western Morocco; most of the species are xerophytic dwarf shrubs, some of them broomlike bushes. The only grasses present are xeromorphic with hard leaves: *Stipa tenacissima* and *Lygeum spartum* (transitional zone), *Panicum turgidum, Aristida pungens,* etc. Many ephemeral species appear after a good winter rain. Shrubs confined to wet habitats are *Tamarix, Nitraria,* and *Ziziphus.* Paleotropic elements are numerous, including the species of *Acacia* found in the dry valleys carrying groundwater.

In the United States, only the southern Californian and southern Arizonan deserts can be considered to be subtropical deserts. The arid regions in northern Arizona, Utah, and Nevada already have a very cold winter.

The southern Sahara, with the Sahel providing a transition to the summer-rain region of the Sudan, belongs to the Paleotropic realm. Grasses with less-hard leaves *(Aristida, Eragrostis, Paniceae)* are much more common here. There are also far more shrubs *(Acacia, Commiphora, Maerua, Grewia)* as well as herbaceous plants *(Callotropis, Crotalaria, Aerva,* etc.), which are also typical of the Thar of Sind deserts (Fig. 53). The South African deserts,

Fig. 68. Great Karroo near Laingsburg (South Africa) with succulent *Euphorbia, Rhigozum obovatum, Rhus burchelli,* and dwarf shrubs. (Photo E. Walter)

the Namib and Karroo, are also paleotropic. The Namib extends along the coast of Southwest Africa, and we shall mention it again on p. 142.

The Karroo extends into the Orange Free State. The two rainy seasons favor the development of innumerable succulents: the larger species of *Euphorbia, Portulacaria,* and *Cotyledon* in rocky habitats and many smaller Crassulaceae, succulent Liliaceae, and *Mesembryanthemum* spp. on quartz veins (Fig. 68). The vast flat areas are covered with dwarf shrubs, mainly Compositae. Woody plants such as *Acacia, Rhus, Euclea, Olea, Diospyros,* and even *Salix capensis* grow in the dry valleys. In the transitional region of the Upper Karroo, the grassland of the summer-rain region is found growing on deep, fine-grained soils, whereas on shallow rocky ground, the Karroo succulents still abound (Fig. 69).

In the Neotropic realm, too, there are several semidesert to desert regions. The Sonoran desert (northern Mexico and southern Arizona), although actually in North America, belongs to the Neotropic realm, floristically speaking. Extensive investigations of this desert (or rather semidesert) have been carried out from the Desert Laboratory in Tucson, Arizona. The vegetation, with its tall candelabra cacti, is termed a "cactus forest" (Fig. 70). By means of a sort of bellows mechanism, these succulents can store so much water that they are capable of surviving for more than a year without any further water uptake. They have a shallow root system, but within 24 hours after wetting of the upper soil layer, fine absorbing roots are put out and the water-storing tissue fills up. Apart from succulent cacti, other ecological types are represented here: winter and summer ephemerals, poikilohydric ferns, malakophyllous half-shrubs *(Encelia),* sclerophyllous

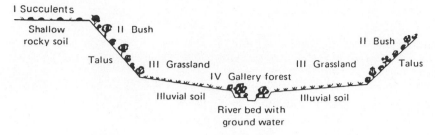

Fig. 69. Vegetation profile of a valley of the Upper Karroo near Fauresmith (South Africa). Distribution of the plant cover is determined by differences in soil. Bush with *Olea, Rhus,* and *Euclea.* (From Walter 1939)

Fig. 70. Sonoran desert near Tucson, Arizona. Slope with giant cacti *(Carnegiea gigantea),* erosion gulley with *Fouquieria splendens* and *Acacia* bushes; in immediate foreground, *Cercidium microphyllum.* (Photo H. Walter)

species, stenohydric plants, and the deciduous *Fouquieria,* which develops new leaves after each heavy rain shower, although they afterwards rapidly turn yellow due to water shortage. Wide, flat, dry areas are covered with the particularly drought-resistant creosote bush *(Larrea divaricata),* which smells strongly of creosote when its leaves are wetted. This species is also characteristic of the Mohave desert, which only receives winter rain and is poor in succulents.

In the lee of the High Andes, along their eastern foot, a *Larrea* desert stretches more than 2000 km, from northern Argentina into Patagonia. The predominating species, *Larrea divaricata,* is probably identical with that found in Arizona (Böcher et al. 1972).

The Peruvian-Chilean coastal desert is, at its most extreme, just as dry as the Namib, although the fog plays a greater role here because of the steepness of the coast in places. *Tillandsia* (Bromeliaceae), the only true fog plant known to exist among the flowering plants, occurs here. Although it cannot take up water from the humid air like the lichens, it sucks in the water drops from condensed fog with the aid of special leaf scales. The rosettes of these plants sit loosely on the sandy ground.

The "garua", as the fog blanket is termed in Peru, hangs at an altitude of 600 m for months on end during the cooler season. The soil on the slopes is so wet that a carpet of herbage grows, called "loma vegetation", which is used for grazing. Although absent nowadays, woody plants formerly grew in these areas. The quantity of water from fog condensations collected under the trees in a *Eucalyptus* plantation amounted to the equivalent of a rainfall of 600 mm. In northern Chile, the slopes exposed to fog are densely covered with cacti which are draped with lichens. Further south, near Frey Jorge, there is a true mist forest immediately adjoining a cactus semidesert on the fog-free slopes. In the neighborhood of the large saltpeter deposits in northern Chile the desert is completely barren, and only along the rivers fed by the snowfields of the High Andes is there vegetation or irrigated farmland.

A very different situation is encountered in the arid regions of the Australian realm. The whole of central Australia is arid; however, it has no climatic deserts. Sand-dune regions (the Gibson and Simpson deserts), although not climatically the driest parts of Australia, are desertlike in character, as are the gibber plains (bare, stony areas produced by overgrazing). The vegetation of the driest parts, with scanty rainfall at any time of year, is composed of saltbush *(Atriplex vesicaria)* and bluebush *(Kochia = Maireana sedifolia)*, both Chenopodiaceae. They occur either in pure populations, or mixed. The soils upon which *Atriplex* grows contain little chloride (about 0.1% dry weight), but since these loamy soils dry out to a considerable extent, the concentration of chloride can in fact be very high. The cell-sap concentration of *Atriplex* is also correspondingly high (usually 40–50 atm, chlorides accounting for 60–70% of the total), and it is in fact an euhalophyte, the growth of which is enhanced by salt. A certain degree of salt excretion is achieved by means of the short-lived vesiculated hairs, which are continuously being replaced. *Atriplex* is a half-shrub, lives for about 12 years, and, like most halophytes, possesses weakly succulent leaves and a root system spreading widely at a depth of about 10–20 cm (above a chalk caliche layer). The bushes, therefore, grow rather far apart.

In contrast, *Kochia (Maireana) sedifolia* is said to be long-lived. Its root system not only penetrates to a depth of 3–4 m into the cracks in the caliche but also spreads equally far laterally. This species grows wherever rainwater percolates to greater depths, such as on a light or stony soil. The cellsap concentration of *Kochia* is only half that of *Atriplex,* and the part played by chloride is also smaller (about 20–40%). This species is thus probably a facultative halophyte. It can attain dominance if the climate becomes more humid.

In the saltbush region, there are scattered sand dunes or sandy areas where moisture conditions are more favorable and where the soil is not saline.

Fig. 71. Mulga vegetation in the interior of Australia near Wiluna after a rain. Large bushes are *Acacia aneura,* smaller bushes *Eremophila* spp. The ground is densely covered with temporarily active everlasting plants *(Waitzia aurea* and white *Helipterum* species). (Photo H. Walter)

Shrubs such as *Acacia, Casuarina,* and *Eremophila* can be found. Treelike species of *Heterodendron* and *Myoporum* as well as species of *Eremophila* and *Cassia* are confined to silty soils.

The most widely occurring species in central Australia is *Acacia aneura* ("Mulga") (Fig. 71). It dominates large areas, which look like a gray sea when seen from the air. The shrub reaches a height of 4–6 m and has thin, cylindrical or somewhat flattened, resin-covered phyllodes. Its root system is well developed and penetrates the hard soil layers to a depth of about 2 m. Owing to the irregular rainfall, flowering is not connected with any particular season, but rather with the occurrence of rain. Fruits and seeds develop after a heavy rain shower, and at the same time the ground is carpeted with white, yellow, and pink everlasting plants, belonging to the Compositae family (Fig. 71).

Acacia aneura is sensitive to salt but can survive long periods of drought. In dry habitats the bushes grow well apart, but in wet depressions they form thickets. *Rhagodia baccata* and *Acacia craspedocarpa* have recently been the subject of detailed ecophysiological studies. The porcupine grasses *(Triodia, Plectrachne)* are another important group, collectively forming what is known as "spinifex" grassland. They are sclerophyllous species with very hard, rolled-up, pointed, perennial leaves with a resin covering, and they form large round cushions, or cupolas in the case of *Triodia pungens,* with a height of up to 2 m.

Triodia basedowii dominates the sandy areas of the most arid part of western Australia. Its dense root system goes straight down for 3 m. Older cushions disintegrate and form garlands. Other characteristic genera, represented by many species, are *Eremophila, Dodonaea, Hakea, Grevillea,* etc. The structure of the vegetation is determined by the kind of soil and by the sheet floods which follow heavy rains, both factors leading to a complicated mosaic of vegetation.

8. Adaptations to Water Stress from the Cybernetic Point of View

The ecological types so far discussed are genetically fixed. For experimental purposes, the species will be regarded as stable entities, although longterm observations reveal that they are in fact highly subject to change. Every plant is continually adapting in many ways – including morphologically – to the prevailing environmental situation, and this is essential if the plant is to survive. Phenomena of this kind, however, are ignored by physiologists because they involve growth processes and only become noticeable weeks or months after the environmental change that has elicited them. Such adaptations are of great ecological significance and can be followed particularly well in arid regions by observing a plant from the end of a rainy period through the succeeding drought, up to the beginning of the next wet season (Walter and Kreeb 1970).

In studying adaptations to water stress, various values relating to the osmotic state of the plant cell should be taken into consideration:

Suction tension (S) = – Water potential (ψ)
Potential osmotic pressure (π^*) = –Osmotic potential (ψ_s)
Turgor pressure (P) = Pressure potential (ψ_p).

These terms can be related by the following equations:

$$S = \pi^* - P$$
$$\psi = \psi_s + \psi_p.$$

The above parameter were previously measured in atmospheres, but this unit has recently been replaced by the *bar*. However, since 1 atm = 1.013 bar, the difference involved is less than the experimental error of ecological experiments, and thus the unit employed is immaterial. The values S and ψ, and π^* and ψ_s, are always numerically equal, differing only in their sign (ψ and ψ_s are always negative).

What is important is that the significance of the various parameter for the water budget should be clearly realized. If only the more physical process of water flow through the plant, from the soil to the atmosphere, is to be studied, then S and ψ have to be measured. If, however, as in this case, the study is concerned with the biological processes connected with growth, then π^* and ψ_s are the appropriate parameter since they are directly related to the hydrature, or state of hydration, of the protoplasm (see p. 37), and this is what governs growth processes in the living plant.

Even though they only become obvious after a considerable time lag, the adaptations to water stress can be regarded from the cybernetic point of view

Fig. 72. Various forms of *Encelia* leaves. Above: sparsely hairy, hygromorphic leaves; below left: mesomorphic leaves; below right: xeromorphic leaves. In the middle: twigs with terminal buds (all leaves have dropped)

as feedback regulation of the water supply under changing conditions of drought. When the drought period begins, the plants lose more water owing to transpiration, and the water balance becomes negative (water loss exceeds water uptake). As a consequence, the cell-sap concentration rises (π^* increases, i.e., ψ_s becomes more negative), which results in a decrease in hydrature of the protoplasm (dehydration), including that of the meristem cells of the shoot, with the leaf primordia. The changed condition of the protoplasm is responsible for the fact that the internodes are shorter and the newly formed leaves are smaller and more xeromorphic and thus transpire less. The water losses of the plant decrease so that the water balance is restored (water losses equal water uptake).

An example from the Sonoran desert provides an illustration. The Compositae *Encelia farinosa* is a semi-shrub growing to a height of about 50 cm. During the rainy season, the plant has large, soft, greenish, hygromorphic leaves with a light covering of hairs, their π^* amounting to 22–23 atm. In the dry season, when water supply becomes a problem, π^* rises to 28 atm, which results in a slight decrease in hydrature of the protoplasm of the meristematic cells. The leaves subsequently formed from this meristem to replace the hygromorphic leaves are smaller and more mesomorphic and have a denser covering of hairs. With continuing drought, π^* rises to 32 atm, and the next leaves are even smaller, thicker, and more densely covered with white hairs (Fig. 72), all of which tends to reduce transpiration even further. If the drought is extremely long, all the leaves are shed as soon as π^* reaches 40 atm. Only the terminal buds containing rudimentary leaves are left. Water

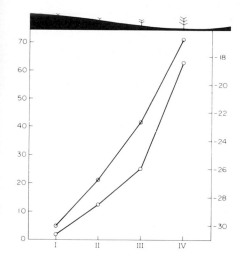

Fig. 73. Relationship between osmotic potential $(-\pi^*)$ in bars, and vertical growth in *Solanum elaeagnifolium* (in centimeters). I–IV, samples from the above-mentioned plants

losses are now reduced to such an extent that the plants are in a state of water balance, even with the minimum amount of water available from the soil.

As soon as the next rainy season begins, the potential osmotic pressure (π^*) drops to the original value of about 20 atm, the hydrature of the meristem cells rises, and the newly formed leaves are large and hygromorphic. As a result of the ensuing intense photosynthesis, rapid growth takes place, transpiration values are high, and water balance is maintained on a higher level. In this way, the ever-recurring cycle of events is completed.

Figure 73 shows the close relationship between size and ψ_s $(-\pi^*)$ for *Solanum elaeagnifolium*, an annual species that develops in the Sonoran Desert on clay soils after a heavy rainfall. In small hollows where water collects and the soil is wet down to a considerable depth, the plants achieve a height of 60 cm. Toward the perimeter of the hollows, as the soil gradually becomes increasingly dry, the plants become smaller and smaller, down to a dwarf form only 1 cm high. The differences in water supply are reflected in the ψ_s values. The two curves run almost parallel (Fig. 73). Dwarf ephemerals of this type are seen everywhere after a poor rain season.

It is interesting that a decrease in hydrature of the meristematic cells elicits different reactions in root and shoot: growth of the shoot is always inhibited (Fig. 74), whereas the roots become thinner and longer without the development of branches and are only inhibited if the decrease of hydrature becomes more pronounced (Fig. 75). *Brassica* seedlings reared in sands of different water content show this very clearly (Fig. 74). This kind of reaction is also a means by which ephemerals keep their water balance steady in order to survive. A smaller shoot means less water loss, and elongated roots enable the plants to reach down to the lower soil levels that may still be wet even if the rains have been light.

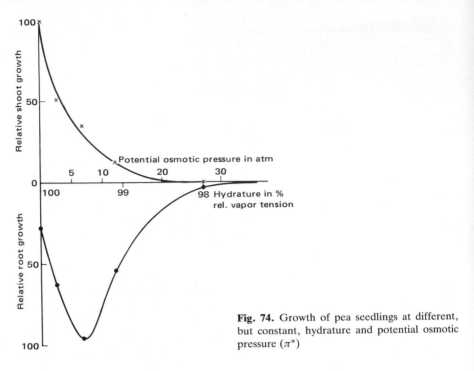

Fig. 74. Growth of pea seedlings at different, but constant, hydrature and potential osmotic pressure (π^*)

A most remarkable phenomenon is seen in the barrel cacti, the various sides of which react differently according to exposure. The southwest side receives more warmth than the rest of the plant and transpires the most, whereas the exact opposite holds for the northeast side. An example is provided by *Ferocactus wislizenii*, which is widespread in the Sonoran desert. A transverse section shows that the iso-osmoses (lines of equal π^*) clearly reflect the asymmetrical water requirements. The π^* value is highest on the southwest side and lowest on the northeast side. Correspondingly, the southwest side is the more xeromorphic (narrower, closer ribs, woody parts thicker) (see Fig. 76). Vertical growth is also less on the southwest side, so that the plant bends in this direction (Fig. 77) and older specimens often tip over.

Fig. 75. *Brassica* seedlings grown in sand of various degrees of wetness (5 days after germination). Water content of sand (a) 15.5%; (b) 6.7%; (c) 4.3%; (d) 2.5%; (e) 1.3%. (Three seedlings from each group are shown)

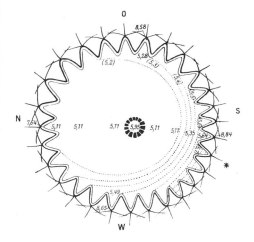

Fig. 76. Distribution of potential osmotic pressure (π^*) in the cross section of *Ferocactus wislizenii*. Isosmoses, lines of equal pressure (cell-sap concentration, in atmospheres). The highest pressure is on the southwest side (0 = East)

Another interesting point is that often the first flower buds form on the southwest side and none at all on the northeast side. This is shown for *Pachycereus pringlei* in Fig. 78. It can be concluded from this that higher π^* (or lower ψ_s) favors generative growth at the expense of vegetative growth. This is seen in the ephemerals, since dwarf plants with their higher π^* always come into flower first. This confirms the observations of gardeners that a poor water supply results in a larger number of blossoms whereas growth is mainly vegetative if water is plentiful.

Fig. 77. *Ferocactus wislizenii* in the Sonoran desert, leaning towards the southwest. (Photo E. Walter)

9. The Productivity of Desert Vegetation

If a plant lowers its transpiration and photosynthesis during times of drought by means of a reduction of active surface, its production is also reduced or may even come to a standstill if dryness continues for longer periods.

Fig. 78. *Pachycereus pringlei* with flower buds on the southwest side (one flower already open). (Photo E. Walter)

In years with plentiful rainfall, on the other hand, the plants develop more luxuriantly, but they cannot use up all of the available water. The surplus is exploited by the ephemerals, which under such circumstances develop particularly well and can be considered as fulfilling a buffer role in smoothing out the larger fluctuations in annual precipitation. In years with meager rainfall, ephemerals scarcely develop at all or are, at best, represented by dwarf forms.

If surface reduction in perennial forms is insufficient to maintain water balance, the larger part of the plant dies off because the maximum π^* has been exceeded. For survival, it suffices if the shoot meristem of only one branch remains alive to sprout after a rain. All woody plants in the desert bear large numbers of dead branches, indicating past years of drought. Reproduction by means of seeds only takes place after a good rain-year or several successive good years, which seldom happens more than once in a century. Young plants are therefore hardly ever found. Under such circumstances, it is impossible to obtain mean values for production. The leaf area index for perennial species is less than 1, even in exceptionally good years, and only a luxuriant growth of ephemerals in a good rain-year can lead to a reasonable production.

In the extreme desert near Cairo, the production of ephemeral vegetation has been measured after the upper 25 cm of the soil had been soaked by winter rain amounting to 23.4 mm (Abd El Rahman et al. 1958). Of this, 68% was lost by evaporation, and transpiration of the ephemerals during the winter months accounted for 7.3 mm, or 32%, which is the equivalent of 730 kg of water per 100 m^2. Over the same area, the ephemerals produced 9.384 kg fresh mass, or 0.518 kg dry substance. This gives a transpiration coefficient of 730:0.518 = 1409, which, compared with values for central European crops (400–600), is very high and is to be attributed to the low air humidity of the desert. Similar values were found at very low precipitation levels by Seeley (1978) for annual grasses in the Namib.

Zoomass in the desert is extremely low, and secondary production is thus almost nonexistent. Nevertheless, even in the desert, the food chains play a not insignificant regulatory role in the ecosystem (p. 146).

Finally, we will discuss the special investigations on the productivity of agaves and spherical cacti, such as have been carried out in the western part of the Sonoran Desert in California, with its summer drought period.

a) Nobel (1976) gives exact quantitative data (mean values) for *Agave deserti*, which also occurs in the eastern Sonoran Desert. The following data are for plants with an average of 29 leaves: length of leaves 30 cm, surface area 380 cm^2, fresh weight per leaf 348 g, dry weight 47 g, number of stomata per mm^2 30, number of roots per plant 88, root length 46 cm. The roots are spread out flat and radially in order to take optimal advantage of each rainfall.

The stomata open during the rainy season (November−May) for 154–175 days at a soil-water potential of −0.1 bar. Water uptake ceases when the potential decreases to −3 bar at the beginning of the drought period. The stomata open at night the first 8 nights of the drought period and then remain closed. A diurnal acid metabolism (CAM) does not take place until the following rainfall. The transpiration losses in 1975 were 20.3 kg per plant, corresponding to a rainfall on the densely rooted soil of 26.9 mm, or 35% of the annual precipitation. The transpiration coefficient, i.e., the ratio of the transpired water (in kg) to the amount of dry matter produced, was 25, which is quite low (in commonly cultivated plants with a C-3 photosynthesis, this value is 300–900). Per plant, 0.8 kg dry weight was produced annually. Growth is, therefore, very slow and only the older plants blossom once and then die, since in order to produce the large inflorescences all material and water reserves of the plant are exhausted. This is confirmed by the following observations (Nobel 1977a): coming into blossom, the old plant had 68 leaves which were 4.1 cm thick as the inflorescence was just becoming visible. After development of the inflorescence, the leaves shriveled up, faded in color and were only 1.4 cm thick. In all, the leaves lose 24.9 kg fresh weight and 1.84 kg dry weight during blossoming. Since water uptake is insufficient, 17.8 kg of water was taken from the leaves to supply the inflorescence. The dry weight of the inflorescence was 1.25 kg, and 0.59 kg were respired. A plant in blossom produces 65,000 seeds, 85% of which are consumed by animals. Not a single young agave could be located within a 400 m^2 area on which 300 agaves were found. Reproduction by means of seeds occurs only in years with ample rainfall, otherwise vegetative propagation by runners takes place. These values clearly indicate why agaves are hapaxanthic (monocarpic) species; i.e., plants which only produce blossoms and fruit once and then die.

b) A further detailed production analysis was carried out in the same area with the spherical cactus *Ferocactus acanthoides* (Nobel 1977b). Although this is also a species with a diurnal metabolism (CAM), the expenditure required for the blossoms is so low that it blossoms annually.

The plant studied was 34 cm tall, 26 cm in circumference and weighed 10.8 kg (with a water content of 8.9 kg). The loss by transpiration in 1 year was 14.8 kg, plus an additional 0.6 kg for the transpiration and growth of the generative organs. CO_2 assimilation resulted in the production of 1.6 kg in one year, one third of which was respired. The annual growth was measured at 9% and the transpiration coefficient was 70. Although still quite low, it was higher than that of the agave. The openning of the stomata corresponded to the conditions in the agave.

10. Orobiome III – Desert Mountains of the Subtropics

The air above extreme deserts contains so little water vapor that, even at considerable altitudes, rising air masses provide not precipitation. In the Tibesti Mountains (3415 m above sea level) in the central Sahara, an annual precipitation of only 9–190 mm (4 years) was recorded at an altitude of 2450 m, despite frequent clouds in the winter months. Arid conditions thus persist up to great altitudes. The occurrence of a number of mediterranean elements, however, indicates that conditions are somewhat more humid than they appear to be. *Erica arborea* has been found in gorges at 2500–3000 m above sea level, and *Olea laperrini,* a close relative of the olive tree, occurs in Hoggar at 2700 m as a relict form.

According to Ellenberg (1975), on the west slopes of the Andes, perarid desert extends up into the montane belt, followed by a dwarf shrub semidesert extending into the subalpine belt. Above 4500 m, there is tropical alpine grass semidesert, or "desert puna".

In the less arid Sonoran desert in southern Arizona the *Covillea,* or giant cactus, desert is succeeded by a belt with *Prosopis* grass savanna and many leaf succulents *(Agave, Dasylirion, Nolina).* Above this are several belts with evergreen *Quercus* species and *Arctostaphylos, Arbutus,* and *Juniperus* shrub layers. These are followed by coniferous forest belts: *Pinus ponderosa* ssp. *scopulorum* (further up with *Pinus strobiformis), Pseudotsuga menziesii* with *Abies concolor,* and, only in the northern slopes of the San Francisco Peak in northern Arizona, *Picea engelmannii* up to almost 3700 above sea level. Here, the annual precipitation increases substantially with increasing altitude.

11. Biome of the Namib Fog Desert

The Namib desert on the coast of southwest Africa has been selected as an example of a biome of ZB III and the subzonobiome of the fog desert since it differs vastly from all other deserts.

Although it is a subtropical desert and extreme in its lack of rain, the coastal strip is remarkable for the high air humidity (Fig. 79). There are about 200 foggy days annually and only small fluctuations in temperature, similar to a maritime region. The temperature is always cool, and there are only a few

Fig. 79. Climate diagram of Swakopmund in the Namib. The region is almost without rain but has 200 foggy days annually (precipitation not measurable) (Walter and Lieth 1967)

hot days each year. These peculiar conditions are due to the cold Benguela current (water temperature, 12–16 °C), above which a 600 m high cold air layer accumulates, with a bank of fog, so that, as a result of inversion, the warm easterly air streams cannot reach the ground (Logan 1960). Instead, a sea breeze arising daily in the west pushes the fog and cold air into the desert.

Thunderstorms with rain can only occur if the inversion layer is penetrated. In most years this is not the case, however, and rainfall is barely measurable. Heavy rains are rare and occur only once or twice in the course of a century, as for example when 140 mm of rain were recorded in 1934/1935 or over 100 mm in 1974/1975. The annual mean for Swakopmund is 15 mm, but this figure conveys little information (Fig. 79).

The soil is slightly moistened by fog or dew, the mean daily value being 0.2 mm, and the maximum 0.7 mm. The annual total of about 40 mm of fog precipitation is without effect because the moisture evaporates before it can enter the soil. The high air humidity is of benefit only to the poikilohydric lichens, which brighten every available stone and rock in the fog zone with their vivid colors, and to the "window algae" which grow on the under surfaces of the transparent quartz pebbles, where the moisture is retained longer. True fog plants such as the *Tillandsia* of the Peruvian desert (p. 133), which take up no water from the soil, are not found in the Namib. Only where wind-driven fog condenses against a rock face and the water seeps down into the crevices can plants (mostly succulents, Fig. 80) establish themselves. This happens on the isolated mountains rising from the flat platform of the Namib. The plain rises with a gradient of 1/100 from the coast toward the east and ends at the foot of the steep escarpment of the African highlands (100 km from the coast). The fog is noticeable up to 50 km inland and still contains drops of seawater from the spray, which accounts for the fact that the soils of the outer Namib are brackish.

In the Namib, perennial plants are only found in places where the soil contains water below a depth of 1 m, from the water supplies laid up in years with good rainfall.

After the 140 mm of rain in 1934/35, the desert turned green and was sprinkled with flowers, mainly ephemeral forms, including a particularly large number of succulent *Mesembryanthemums*. The latter were able to store so much water in their shoots that they flowered in the following year as well, although the roots and shoot base had already dried out. In rainy years, a large number of seedlings of perennial forms also develop, their roots rapidly penetrating to the deeper part of the soil, which stay moist longer. These plants can only survive the ensuing decades, however, where the soil contains large stores of water.

Fig. 80. In foreground flowering *Hoodia currorii* between white marble rocks (Witport Mountains, Namib); left background: *Aloë asperifolia;* right: *Arthraerua* in fruit. (Photo W. Giess)

After heavy rains, the water runs along deep, sand-filled gulleys in the direction of the sea, without reaching it. The water seeps down in depressions filled with alluvial soils and penetrates into the ground. The upper soil layers dry out down to a depth of about 1 m (less in sandy soils), but below this, the water can remain for decades and is available for plants with roots long enough to reach it. The salt washed out of the sand in the gulleys by the rainwater collects in the depressions. In this way, two types of habitat are formed, one consisting of nonhalophilic biogeocenes in both larger and smaller erosion gulleys (*Citrullus, Commiphora, Adenolobus,* and, where groundwater is more plentiful, shrubs such as *Euclea, Parkinsonia,* and *Acacia* spp.), and the other in large, shallow depressions with halophilic species (including chiefly the Amarantaceae *Arthraerua* or *Zygophyllum stapffii,* and *Salsola* a Chenopodiaceae). The plants grow out of the sand that drifts onto them, and this leads to the formation of small dune hillocks, which constitute the typical nebka landscape (Fig. 81). Presumably, all of the plants germinate in the same year with a good rain, since they are of about equal size, and will survive as long as the water supplies in the soil last. If a long period of time elapses before another good rain-year, the plants gradually die off and the dune sand is dispersed by the wind. If, however, they receive enough rain again within an appropriate time, they continue to grow. The year 1975 was another fairly good year, with over 100 mm of rain. The fog plays a substantial role in the survival of the plants since they can assimilate CO_2 in the water-saturated air without incurring transpiration losses and water consumption is thus low. Whether or not C_4 photosynthesis is involved has not yet been investigated; diurnal acid metabolism (CAM) is typical for succulents (Schulze et al. 1976).

Fig. 81. *Arthraerua leubnitziae* (Amaranthaceae) in a watercourse of the outer Namib. (Photo W. Giess)

Fig. 82. River bed (dry) of the Kuiseb near Gobabeb with *Acacia albida, A. giraffae, Tamarix usneoides,* and *Salvadora persica.* In the background the northernmost sand dunes of the dune Namib. (Photo W. Giess)

Apart from these three biogeocene complexes near the coast, i.e., the rock-face vegetation dependent upon fog condensation, the erosion gulleys with salt-free soil, and the brackish depressions, the oases of the large dry river valleys (wadis) in the central Namib also deserve mention: the Omaruru, Swakop, and Kuiseb. All arise in the highlands where there is a summer rainy season (mean annual rainfall of 300 mm), and are deeply incised into the Namib platform. The river bed is filled with sand into which the rainwater seeps, but water only reaches the sea in years with particularly good rains on the highlands. A continuous stream of groundwater is present at all times, however, so that water can be obtained from wells, although some of it is brackish due to inflow from the Namib. This groundwater allows gallery forests to develop (Fig. 82), with *Acacia albida, A. giraffae, Euclea*

Fig. 83. *Welwitschia mirabilis* on the edges of a wide, shallow watercourse in the Messum Mountains of the Namib desert. (Photo W. Giess)

pseudebenus, and *Salvadora persica* or, in more brackish spots, *Tamarix* and *Lycium* species. In places safe from floodwater, the forests can attain a great age. On the surrounding accumulations of sand, *Ricinus, Nicotiana glauca, Argemone,* and *Datura,* among others, may grow while the spiny, leafless *Acanthosicyos* (nara pumpkin) and *Eragrostis spinosa*, a woody thorny grass, thrive on the dunes. Where the groundwater forms ponds, *Phragmites, Diplachne, Sporobulus,* and *Juncellus* can be found.

All of these plants are adequately supplied with water and are capable of considerable production. There is also a diverse and abundant fauna in oases of this kind, including birds, rodents, reptiles, arthropods, etc. Elephants and rhinoceros could previously be encountered, but they have been exterminated by man, and only the pavians, dwelling in the rocky clefts, have survived.

The fauna of the nebkas in the desert is poor, consisting of a few rodents, reptiles, scorpions and saprophagous beetles. More species are encountered in the isolated mountains, particularly those farther inland, where summer rain is more frequent, so that water accumulates between the rocks, and shrubs grow in the cracks.

The foregoing description applies to the outer Namib. At a distance of 50 km from the sea, the inner Namib begins. This area receives sparse summer rain and is periodically covered by grass. Desert conditions are not so extreme, and the mobile game is able to find food and to take advantage of the isolated water holes. This part of the desert abounds in game, including large numbers of zebra, oryx antelope, springbok, hyaena, jackal, ostrich, and other birds. The central part of the Namib, which is uninhabited by man, has been declared a nature reserve and is being investigated from the Namib desert station in Gobabeb.

In the region between inner and outer Namib, the well-known *Welwitschia mirabilis* is found in large numbers (Fig. 83). It grows in wide, very shallow, erosion gulleys in which, owing to the barely perceptible gradient, the sparse summer rains accumulate and seep down into the deeper

layers of the ground. The roots of *Welwitschia,* which reach down to a depth
of 1.5 m, are able to exploit this water. Farther down is a hard chalk crust.
Welwitschia possesses only two ribbonlike leaves that continuously grow
are extremely sensitive to running water and to being covered by sand.
Welwitschia possesses only two ribbonlike leaves that continuously grow
from a meristem at the leaf basis on the turnip-shaped stem. They dry out at
the tip to an extent dependent upon the water supply. In rainy years, the
surviving portion is reasonably long, but in drier years, the leaves die off
almost down to the meristem, which greatly reduces the transpiration surface.
The leaves are highly xeromorphic, and the stomata are situated in pits. The
oldest specimen to be carbon-dated was approximately 2000 years old.

Certain ecosystems of the Namib are quite unique: (1) the bare dunes
south of the Kuiseb, (2) the guano islands, (3) the mating places of the seals,
and (4) the saltwater lagoons on the shore behind sand bars.

Organic detritus blown into the dune valleys by the wind consists of grass
remnants, protein-rich animal remains, and dead insects (butterflies). This
detritus is consumed by wingless psammophilic beetles (tenebrionids),
which are eaten by small predators (spiders, solifuges) or by the larger lizards,
sand-dwelling snakes, and golden moles (Kühnelt 1975).

Since the sand surface warms up to a temperature of 60 °C by day, most of
these animals avoid the heat by burying themselves in the cooler layers of
sand during the day and emerging at night. The dew which wets the sand in
the early morning provides them with water, for which they have developed
unique methods of uptake (Hamilton et al. 1976, Seely et al. 1976). The fauna
is rich in endemic species.

The guano islands are the nesting-places of the cormorants that feed on
the abundant fish in the cold seawater. The excrement of these birds
accumulates in the rainless climate and precludes any form of plant growth,
although it is harvested by humans for guano (phosphate fertilizer). A similar
situation prevails in the mating areas of the seals.

The lagoons are cut off from the sea by sand bars, over which the waves
only break during storms. Water lost by evaporation is replaced by water
seeping through the sand from the sea, so that these are aquatic habitats with
a very high salt concentration. We shall not consider them in any further
detail here. Like the Namib, every desert has its own peculiarities and must
be studied monographically. It would be beyond the scope of this book to go
into such detail (see Walter 1973, pp. 460–679; Walter and Breckle, Vol 2,
1984).

Zonoecotone III/IV – Semidesert

The boundary between true desert and semidesert, although not always clearly defined, is to be found in the zone where the increasing winter rains lead to the replacement of the "contracted vegetation" by a diffuse vegetation. About 25% of the total area in the semidesert is covered by vegetation, and the floristic composition of this plant cover varies just as greatly from one floristic realm to the other as is the case with the true deserts.

North of the Sahara, the malakophyllous *Artemisia herba-alba* and the sclerophyllous grasses *Stipa tenacissima* (halfa grass) and *Lygeum spartum* (esparto grass) are the most abundant species. Although *Artemisia* generally grows on heavy loess or loamy soil, it has been found growing in Tunis in places where secondary $CaCO_3$ deposits are present at a depth of 10 cm. At a depth of 5–10 cm, the soil is densely permeated by its roots, some of which even reach down 60 cm. *Stipa* prefers high ground with a stony covering. A soil profile revealed the following: 2–5 cm of stony pavement underlain by 30 cm of loamy soil with dense root growth, and, below this, a gravel layer with a hard upper crust, which appeared to present an obstacle to the roots but probably also represents a water reservoir. The numerous roots originating at the base of the grass tufts spread far on all sides at a depth of 10–20 cm, so that, although the tussocks themselves are 0.5–2 m apart, their root tips are, in fact, in contact with one another. Solitary individuals of *Arthrophytum* grow between the grasses. The soils are not saline. *Lygeum spartum,* on the other hand, is characteristic of soils containing gypsum (calcium sulphate) and even tolerates a certain amount of salt.

The halfa grass is cut and provides fibers for weaving, the production of coarse ropes, and paper manufacture. *Artemisia herba-alba* is found in the Near East and has in many places replaced the original grassland, after overgrazing, whereas *Stipa tenacissima* is to be found only from southeastern Spain and eastern Morocco as far as Homs in Libya.

With an increase in rainfall, isolated trees occur, such as *Pistacia atlantica* in the west and *P. mutica* in the east, or *Juniperus phoenicea*. It is possible that the latter was originally always present among the halfa grasses until it was eliminated by man. Thin stands of trees finally lead to the sclerophyllous woodlands.

In the transitional zone in California, *Artemisia californica* and the halfshrubs *Salvia* and *Eriogonum* spp. (Polygonaceae) are found.

The transitional zone in northern Chile is a dwarf-shrub semidesert with Compositae *(Haplopappus)*, columnar cacti, and *Pyua* (large Bromeliaceae). This is succeeded by savanna with *Acacia caven,* and the grass cover nowadays consists of annual European grasses.

The so-called "renoster" formation (*Elytropappus rhinocerotis,* Compositae) in South Africa can be regarded as typical of a winter-rain region with low rainfall. In Australia, where there are no true deserts, the transitional zone is occupied by the mallee scrub consisting of shrubby species of *Eucalyptus,* the branches of which originate on an underground tuberous stem (lignotuber). Thin stands of *Eucalyptus* trees with an undergrowth of *Meireana (Kochia) sedoides* sometimes occur.

IV Zonobiome of the Winter-Rain Region with an Arido-Humid Climate and Sclerophyllic Woodlands

1. General

As already stated, this zonobiome is best divided into five floristic biome groups, according to the various floristic realms into which it falls: (1) Mediterranean, (2) Californian, (3) Chilean (4) Capensic and (5) Australian. The largest of these groups is the Mediterranean, where the winter rains occur from the Atlantic Ocean to Afghanistan. The climate diagrams of the various groups are very similar in shape except with regard to summer drought, which varies in extent. The climate will therefore be generally termed mediterranean in type. The vegetation consists mainly of sclero-phyllous woody plants typical of winter-rain regions with only sporadic frost; cold periods of greater duration cannot be tolerated by the plants. The most favored time for growth is spring, when the soil is moist and the temperatures are rising, or autumn, after the first rain. The winter temperatures of 10 °C and lower are already too cool for growth.

The zonoecotones will be dealt with at the end of the discussion of the climatic subzonobiomes of the relevant biome groups. There are transitions from ZB IV to Zonobiomes V, VI, and VII.

The climate of ZB IV was not always the same as it is today. The widespread fossil soils, the developmental rhythm of the most common plants, and other factors (fossils) suggest that the climate was still tropical with summer rain in the Tertiary. The rain maximum was in all probability displaced to the winter months only shortly before the Pleistocene. The plants were obliged to adapt, and this meant a drastic process of selection. Only Tertiary plant species of drier habitats with small xeromorphic leaves could survive. The present-day reduction of activity in summer is imposed by drought and is not seen where the plants have sufficient water at their disposal. The ephemerals and ephemeroids confine their activity to the favorable months of spring or the wet autumn.

It is easier to comprehend the ecological behavior of the vegetation if the historical facts are borne in mind (Axelrod 1973, Specht 1973). Many taxa of ZB IV are closely related to those of Zonobiomes V and II. Examples are species of the genera *Olea, Eucalyptus,* etc.; *Quercus* aff. *ilex,* which grows in Afghanistan, with mainly summer rainfall; and the vegetation of the chaparrall of California, where rain falls only in winter, which corresponds to the encinal vegetation of the mountains of Arizona, with summer rain. In Volume 11 (643 pp., 1981) of "Ecosystems of the World", the aspects of

Zonobiome IV are discussed in 27 contributions by 32 author. Although a large quantity of valuable data is presented, a synthesis of the basic relationship is lacking. In this zonobiome, ecosystem research is only in its elementary stages (see di Castri et al. 1981).

2. Biome Group of the Mediterranean Region

The climatic conditions prevailing in this zone can be seen in the diagrams in Fig. 84. Cyclonic rains occur in winter, and the hot, dry summer is a result of the Azores high-pressure zone. Since some of our most ancient civilizations originated in the Mediterranean region, the zonal vegetation was long ago forced to give way to cultivation. The slopes have been deforested and used for grazing, with resultant soil erosion, so that nowadays only varying stages of degradation remain. There is no doubt, nevertheless, that the original zonal vegetation was evergreen sclerophyllous forest with *Quercus ilex*. Small remnants of this association have provided the following data concerning the original forests:

Quercetum ilicis

Tree layer: 15–18 m tall, closed canopy, composed exclusively of *Quercus ilex.*

Shrub layer: 3–5 m tall, *Buxus sempervirens, Viburnum tinus, Phillyrea media, Ph. angustifolia, Pistacia lentiscus, P. terebinthus, Rhamnus alaternus, Rosa sempervirens,* etc.; lianas–*Smilax, Lonicera,* and *Clematis.*

Herb layer: Approximately 50 cm tall, sparse, *Ruscus aculeatus, Rubia peregrina, Asparagus acutifolius, Asplenium adiantum-nigrum, Carex distachya,* etc.

Moss layer: Very sparse.

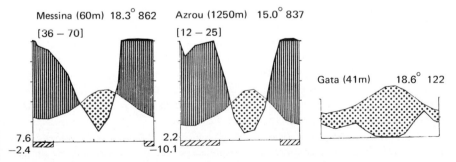

Fig. 84. Climate diagrams from Messina, Sicily; Azrou, in the montane belt of the Central Atlas, Morocco; and Cabo de Gata, in southeastern Spain (driest place in Europe). (From Walter and Lieth 1967)

A terra-rossa profile is usually found in the chalky regions beneath these low forests, consisting of a litter layer, a dark humus horizon, and, beneath this, a 1- to 2-m-deep, clay-containing, plastic, bright-red horizon. On cultivated land, the upper layers are missing due to erosion, so that the red color is visible at the surface. These are mostly fossil soils from a more tropical climatic period. Today, brown loamy soils are developing (Zinke 1973).

A change in the appearance of this region takes place in March, when many of the shrubs start to bloom. The height of their flowering season, as well as that of *Quercus ilex,* is in May, although *Rosa, Lonicera,* and *Clematis* are still blooming in June. A relatively dormant period then follows as a result of the coincidence of the hottest and driest seasons. Growth only recommences with the autumn rains, which may then even lead to an additional flowering of the sclerophyllous trees.

Quercus ilex extends from the western Mediterranean region to the Peloponnese and Euboea. Related species are found eastward in Afghanistan. *Quercus suber* grows in the west (not on limestone). These two species are replaced by *Quercus coccifera* in the eastern Mediterranean. The dominant species in the tree layer of the hot lower belts in Spain and northern Africa are the wild olive tree *(Olea oleaster), Ceratonia siliqua,* and *Pistacia lentiscus; Chamaerops humilis,* Europe's sole palm, also occurs here. Of special interest in Crete are Tertiary relict habitats of a wild form of date palm, which was mentioned by Theophrastus. A large stand grows by a small lagoon near Vai (on Cape Sideron, northeast corner of Crete), above ground water. In North Africa, from Morocco to Tunisia, the distribution of *Quercus ilex* is montane (see Fig. 85), above an intercalated coniferous belt consisting of *Tetraclinis (Callitris)* and *Pinus halepensis* (Aleppo pine). The southeast corner of Spain, with a rainfall of only 130 to 200 mm, is almost desert like (Fig. 84, Gata).

There remain only a few places, in the mountanous regions of northern Africa, where typical *Quercus ilex* forest still exists. Elsewhere, the trees are cut down every 20 years, while still young, and regenerate by means of shoots from the old stump. This leads to the formation of a maquis, consisting of bushes the height of a man. Maquis is also encountered on slopes where the soil is too shallow to support tall forest. Sclerophyllous species, usually shrublike in form, may develop into big trees in a suitable habitat and can achieve a considerable age. Imposing old specimens of *Quercus ilex* can be seen in gardens and parks. In places where the young woody plants are cut every 6 to 8 years and the areas regularly burned and grazed, higher woody plants are lacking and open societies called garigue are formed (*phrygana* in Greece, *tomillares* in Spain, *batha* in Palestine).

These areas are often dominated by a single species such as the low cushions of *Quercus coccifera* or *Juniperus oxycedrus* (*Sarcopoterium* bushes, too, in the east) or *Cistus, Rosmarinus, Lavandula,* and *Thymus.* On limestone in the south of France, the best grazing is provided by a *Brachypodium ramosum-Phlomis lynchnitis* community. In springtime numerous the-

Fig. 85. *Quercus ilex* forest above Azrou in the Central Atlas (Morocco). *Rosa siculum, Lonicera etrusca,* etc., in the undergrowth. (Photo H. Walter)

rophytes and geophytes such as *Iris,* orchids *(Serapias, Ophrys),* and species of *Asphodelus* put in an appearance on otherwise bare spots. An almost pure *Asphodelus* vegetation is all that finally remains in places seriously degraded by continuous fire and grazing. Although the garigue is a sea of flowers in spring, it presents a severely scorched aspect in late summer. If cultivation or grazing is stopped, then successions tending toward the true zonal vegetation take over, as shown in the scheme on p. 154 for the south of France by Braun-Blanquet (1925).

On sandstone or acid gravel, the successions take a course similar to that on limestone, except that the individual stages are of a different floristic composition. Characteristic species are, for example, *Arbutus* and *Erica arborea.*

In the continental Mediterranean region of southern Anatolia, *Pinus brutia* (related to *P. halepensis*) is widespread. It often constitutes the tree layer, below which sclerophyllous plants form a maquis. Since the pine is unable to regenerate in the maquis owing to lack of light, such woodlands can regenerate only after forest fires, which explains why the trees are all of much the same age. The natural habitat of the umbrella pine *(Pinus pinea),* which is often planted in Mediterranean regions, was probably the poor sandy areas on the coast.

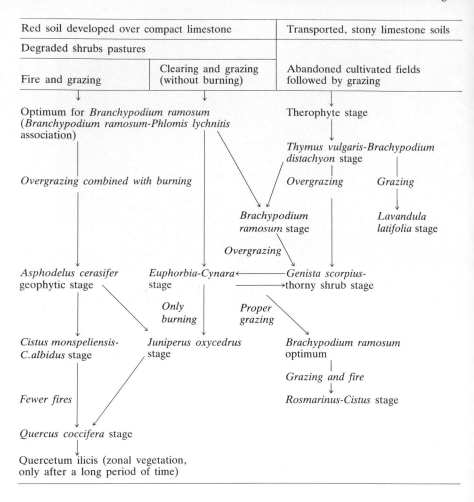

Red soil developed over compact limestone		Transported, stony limestone soils
Degraded shrubs pastures		
Fire and grazing	Clearing and grazing (without burning)	Abandoned cultivated fields followed by grazing

Optimum for *Branchypodium ramosum*
(*Branchypodium ramosum-Phlomis lychnitis*
association)

Therophyte stage

Thymus vulgaris-Brachypodium distachyon stage

Overgrazing combined with burning

Overgrazing   *Grazing*

Brachypodium ramosum stage

Lavandula latifolia stage

Overgrazing

Asphodelus cerasifer geophytic stage

Euphorbia-Cynara stage ←——— *Genista scorpius*-thorny shrub stage

Only burning   *Proper grazing*

Cistus monspeliensis-C.albidus stage

Juniperus oxycedrus stage

Brachypodium ramosum optimum

Grazing and fire

Rosmarinus-Cistus stage

Fewer fires

Quercus coccifera stage

Quercetum ilicis (zonal vegetation, only after a long period of time)

3. The Significance of Sclerophylly in Competition

In considering the ecophysiological conditions in the Mediterranean region, the question which immediately crops up is the degree to which the plants are affected by the long summer drought. Here a distinction must be made between sclerophyllous and malakophyllous plants, the latter being well represented (*Cistus, Rosmarinus, Lavandula, Thymus,* etc.). It should be borne in mind that the most favorable climatopes are nowadays occupied by vineyards or other cultures, the true Mediterranean species having been forced back into habitats with shallow soils, where they can be said to grow under relatively unfavorable conditions.

If the underlying rock is deeply fissured, the abundant winter rains can penetrate deeply and water is stored in the ground. This deep-lying water is available even in summer for plants capable of sending down roots through clefts in the rock to a considerable depth. In woody species, roots 5 to 10 m in length have been observed working their way down through the rock to the horizons which contain available water.

Observations on the cell-sap concentration of sclerophyllous plants over the entire course of the growing season revealed that the potential osmotic pressure increases only by about 4 to 5 atm (above 21 atm) during the dry season, which means that the water balance is not disturbed to any significant degree, and the hydrature of the protoplasm hardly falls. When the water supply is uncertain, such a balance can only be maintained by a partial closure of the stomata and resultant limitation of gaseous exchange. Measurements of transpiration confirm that water losses in summer are three to six times greater in wet than in dry habitats. On the other hand, the cell-sap concentration of the stunted individuals growing in extremely dry habitats reaches 30–50 atm. The climatopes which yield so much wine in autumn are far better provided with water, and a dormant summer period due to drought was certainly not a feature of the original sclerophyllous forests.

In contrast to the hydrostable sclerophyllous plants, the malakophyllous plants are highly labile in this respect. The cell-sap concentration of *Cistus*, *Thymus*, and *Viburnum tinus* can reach 40 atm in summer, with an accompanying drastic reduction of the transpiring surface, achieved by shedding the greater part of the leaves, in some cases leaving only the buds. Such species do not root deeply. The laurel *(Laurus nobilis)* is not sclerophyllous, and in the Mediterranean region, its natural biotope is invariably in the shade or on northern slopes; nowadays it can only be found as forest in the altitudinal cloud belt on the Canary Islands or in winter-rain regions with no pronounced summer drought, as for example in northern Anatolia.

The ecological significance of sclerophylly is thus to be seen not in the ability of sclerophyllous species to conduct active gaseous exchange (400–500 stomata/mm^2) in the presence of an adequate water supply but in their ability to cut transpiration down radically by shutting the stomata when water is scarce. This enables them to survive months of drought with neither alteration in plasma hydrature nor reduction of leaf area. In autumn, when rains recommence, the plants immediately take up production again. Sclerophyllous plants are therefore able to compete successfully in the winter-rain regions not only with nonsclerophyllous evergreen species which are sensitive to drought (e.g, *Prunus laurocerasus*) but also with deciduous trees.

The situation changes at once, however, in more humid winter-rain regions where the summer is not particularly dry or if the habitat, despite a typically Mediterranean climate, is itself perpetually wet, as is the case on northern slopes or in floodplain forests. On north slopes, the sclerophyllous species are replaced first by evergreen species like laurel and then by deciduous trees. The deciduous oak, *Quercus pubescens*, with its larger production of organic material, replaces *Quercus ilex*.

Deciduous trees such as *Populus* and species of *Alnus, Ulmus campestris,* and *Platanus orientalis* are found in the floodplain forest of the Mediterranean region, and in southwestern Anatolia, *Liquidambar orientalis,* a Tertiary relict, occurs. As soon as the point is reached at which the rivers dry up in the summer, however, deciduous woody species are no longer to be found. They are replaced by the evergreen sclerophyllous oleander *(Nerium oleander).*

The productivity of plants depends largely upon their assimilation economy (assimilat-haushalt) and is larger (a) the larger the proportion of the assimilated material which is used for increasing the productive leaf area, (b) the larger the ratio leaf area/leaf dry weight, i.e., the smaller the amount of material required to produce a given leaf area, (c) the greater the intensity of photosynthesis, and (d) the longer the time over which the leaves can assimilate CO_2.

Exact values for point a are not available, but it may be assumed that, in deciduous species, the contribution of leaf mass to the total phytomass is greater than in sclerophyllous species. As for the point b, the ratio is twice as large for the thin deciduous leaves as for the evergreen leaves. Measurements have shown that the intensity of photosynthesis per unit leaf area varies only slightly from deciduous to evergreen leaves. Regarding point d, evergreen leaves are of course at an advantage. This means that the deciduous species are superior in two respects and the evergreens in only one.

Exact calculations have revealed that in the humid, mild climate on Lake Garda in Italy, where *Quercus ilex* and *Q. pubescens* are found, the productivity in grams per gram dry branch weight was 22.9 for *Q. pubescens* as compared with 17.9 for *Q. ilex,* thus confirming the observation that deciduous species are able to compete successfully under these conditions of climate and habitat. On steep rock faces in the same climate, where a dry summer habitat results from runoff of the larger part of the rainwater, evergreen *Q. ilex* bushes grow. In such biotopes, *Q. pubescens* is unable to compete. In addition, *Quercus ilex* is protected from cold air pockets in winter on the steep cliff slopes. Its northern limit is determined by winter cold.

4. Mediterranean Orobiome

In the mountainous regions of the Mediterranean, a distinction must be made (Walter 1975b) between (a) the humid altitudinal belts in the mountains on the northern margins of the Mediterranean zone, in which, with increasing altitude, not only does the temperature decrease, but the dry season also disappears, and (b) the arid altitudinal belts with a summer drought, extending up to the alpine region.

In the humid altitudinal belts, the evergreen sclerophyllous forest is succeeded by a sub-Mediterranean deciduous forest with oak *(Quercus pubescens)* and chestnut *(Castanea).* Above this, at the summer cloud level,

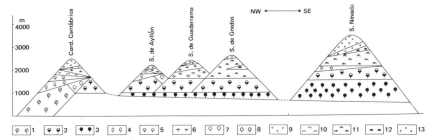

Fig. 86. Altitudinal belts in the high crystalline mountains of the Iberian peninsula shown in northwest–southeast profile: (1) deciduous oak forest *(Q. robur, Q. petraea);* (2) *Q. pyrenaica* forest; (3) *Q. ilex* forest; (4) beech forest *(Fagus sylvatica);* (5) birch forest *(Betula pendula);* (6) pine forest *(Pinus sylvestris);* (7) Mixed-deciduous forest *(Quercus, Tilia, Acer);* (8) high-altitude forest of the Sierra Nevada *(Sorbus, Prunus* etc.); (9) high-alpine grass and herbaceous vegetation; (10) dwarf-shrub heath *(Calluna, Vaccinium, Juniperus);* (11) broom heath *(Cytisus, Genista, Erica);* (12) thorn cushion belt; (13) *Festuca indigesta,* dry sward (H. Ern; from Walter 1964/68)

beech *(Fagus)* and fir *(Abies)* form a cloud forest. In the Apennines, the timberline is formed by beech, which also occurs on Mount Etna and in northern Greece. In the maritime Alps, the beech belt is succeeded by a spruce *(Picea)* belt and in the Pyrenees by a belt of *Pinus sylvestris* and *P. uncinata.*

In the arid altitudinal belts, there is no deciduous forest. The Mediterranean sclerophyllous forest is followed immediately by a series of various coniferous forest belts. For example, on the southern slopes of the Taurus in Anatolia, there is an upper Mediterranean belt with *Pinus brutia,* a weakly developed montane belt with *Pinus nigra* ssp. *pallasiana,* a highmontane belt with *Cedrus libanotica* and *Abies cilicica* (wetter) or species of *Juniperus* (drier), and a subalpine belt with *Juniperus excelsa* and *J. foetidissima.* But in the rainy northeastern corner of the Mediterranean in the Amanos Mountains, a cloud belt with *Fagus orientalis* is found. *Cedrus libanotica* occurs also on Cyprus, and a small relict is present in Lebanon at 1400–1800 m above sea level. On Cyprus and Crete, as well as in Cyrenaica, *Cupressus semper-virens* (cyprus) always occurs in the upper Mediterranean belt in its natural form with horizontal branches. The frequently planted columnar form is a mutation. In the Atlas Mountains, from the eastern High Atlas to the Tunisian border, the subalpine belt, at an altitude of more than 2300 m above sea level, consists of cedars *(Cedrus atlantica),* but the altitudinal belts vary greatly according to the course taken by the mountain ranges and the exposure of the slopes. Figure 86 shows the complicated order of the altitudinal belts in the Spanish mountain ranges.

A difference in the altitudinal belts in arid and humid regions is recognizable even above the timberline in the alpine region. Whereas the situation in the humid mountain climate is similar to that in the Alps, in the arid alpine regions the vegetation consists of thorny hemispherical cushion plants with many convergent species from different families. It is only possible to distinguish them when they are flowering or fruiting. This belt is followed by a dry, grassy belt, where hygrophilic plants (mostly endemic species related to arctic-alpine plants) are found on spots kept moist in summer by melting snow.

A survey of the Mediterranean altitudinal belts can be found in Ozenda (1975).

Interesting conditions are found in the orobiomes of Macaronesia, especially on the Canary Islands, which are affected by the northeasterly trade winds.

5. Climate and Vegetation of the Canary Islands

Macaronesia consists of the island groups of the Azores, Madeira, the Canary Islands and the Cape Verde Islands. The first three, with winter rain and summer drought, belong to Zonobiome IV, while the climate of the Cape Verde Islands is so dry that it must be assigned to Zonobiome III. The botanically most interesting and most thoroughly studied of these island groups is the Canary Islands, especially the islands of Teneriffe and Gran Canaria. Ever since Alexander von Humboldt interrupted his journey to Venezuela on Teneriffe in 1799, and distinguished five altitudinal belts after brief observations, numerous botanists have studied the flora of this island. This has led to a bibliography of 1030 titles (Sunding 1973). Recent plant sociological studies were carried out by Oberdorfer (1965) and Sunding (1972). Ecological investigations, however, are still in their initial stages (Voggenreiter 1974, Kunkel 1976). Here, only a brief description is presented.

The origin of these volcanic islands goes back to the Cretaceous. Gran Canaria rises to nearly 2000 m above sea level, and Teneriffe to a little over 3700 m. These very steep orobiomes differ from others of Zonobiome IV in that they rise directly out of the ocean and lie near 29° north latitude and are under the influence of the trade winds. This means that the wind-exposed northern slopes have climatic conditions differing from those on the wind-protected southern slopes.

Clouds from the trade winds are held back on the northern slopes, resulting in rainfall by the ascending clouds as well as fog, so that there is no summer drought. The warm, humid climate of the intermediate levels corresponds more to Zonobiome V with its evergreen laurel forests. In contrast, the southern slopes are especially dry in the lower levels and are more often swept by hot winds from the Sahara. Therefore, these islands possess local conditions such as are found in Zonobiomes III–V, which are increasingly subject to frost at higher levels. The Teide on Teneriffe is covered with an alpine rubble desert at altitudes over 3000 m above sea level, which is typical of tropical mountains.

The volcanic islands were colonized several times (mainly in the Tertiary) from the neighboring continent of Africa, which at the time was still inhabited by Tertiary evergreen forests. These tree species remained on the warm and humid northern slopes of the islands to the present day, presenting a sort of living museum after they had become extinct on the mainland. This has resulted in floristic relationships to distant elements on the southern tip of Africa *(Ocotea foetens)*, India *(Apollonias)*, various other tropical regions *(Persea, Visnea*, Theaceae, *Dracaena draco)* and to the humid Mediterranean *(Laurus azorica, Laurocerasus [Prunus] lusitanica, Phoenix canariensis)*. On the other hand, certain elements of arid regions have found appropriate niches in suitable lower levels and rocky locations *(Launaea, Zygophyllum*, succulent *Euphorbia* spp., *Kleinia* spp.). Many species are endemic, such as the numerous succulent Crassulaceans, which were previously assigned to *Sempervivum*, but are now considered endemic genuses *(Aeonium* with 33 spp., *Aichryson* with 10 spp., *Greenovia* with 4 spp., *Monanthes* with 15 spp.). In addition, eumediterranean elements also arrived, although probably not until the Pleistocence.

Since the colonization of the islands 500 years ago by Spain, further Mediterranean species, including goats, were introduced. The settlements with their cultivated land expanded increasingly, thereby endangering the original vegetation, especially the unique humid evergreen laurel forest with its many Tertiary relicts. This forest is cleared for its valuble wood; the litter layer and the humus are removed for the improvement of the agricultural soils, thus making regeneration of the forest in cleared areas virtually impossible. Less demanding woody plants, such as *Erica arborea* or *Myrica faya* grow in such areas, or the land is reforested with *Pinus* or even *Eucalyptus*. On Gran Canaria, only 2% of the original laurel forest still exists (Fig. 87) and it is still shrinking rapidly on Teneriffe. These beautiful islands have recently become even more endangered by profit-orientated mass tourism, as is the case for most impressive landscapes all over the world.

Visiting the islands after an absence of 40 years, one is taken aback to find only paved playgrounds and highways everywhere. Environmental protection usually does not become effective until there is almost nothing left to protect. Today's youth has never experienced the quietness and grandeur of undisturbed nature.

Kämmer (1974) studied the climatic conditions on Teneriffe in detail especially in regard to the significance of the fog precipitation filtered off by the trees in the cloud level. As a result of his measurements, carried out over a number of years, he came to the conclusion that the heavily increased ascending rains are more significant in the laurel forests than the relatively low additional precipitation provided by the fog. It is probably not correct to generalize the findings of Sunding in which a rain gauge placed in a clearing in the laurel forest registered an annual precipitation of 956 mm, while a second gauge situated beneath the trees to include water dripping from the leaves registered 3038 mm. Kämmer estimates the annual fog precipitation at

Fig. 87. Laurel forest on a northern slope on Teneriffe at 350 m above sea level. On the *left,* the treetops of *Laurus canariensis,* deformed by ascending trade winds, can be seen (Walter 1968)

approximately 300 mm. As we know from the tropics, epiphytes rely more on the frequency with which they become wet than on the actual amount of precipitation, and the epiphytic mosses depend on the low amount of evaporation. The short duration of sunshine and the resulting high humidity in the cloud level (especially in summer) are important factors for the laurel forest.

The climate diagrams in Fig. 88 provide general information on the character of the climate on Teneriffe. The climate on the seacoast at Santa Cruz is that of a semi-desert, while the annual precipitation on the southern coast, with only little over 100 mm, is characteristic of a desert climate. The climate of La Laguna, still below the cloud level, is typically Mediterranean and without frosts (with the exception of the year 1869). Izaña, at 2367 m above sea level and in the upper limit of the cloud level, obtains somewhat less precipitation. The amount of precipitation decreases with increasing altitude so that, similar to the situation in Mexico, the upper forest limit is also a moisture limit. Although Izaña does not have a cold season, frosts may occur between October and April.

The climate diagrams from Gran Canaria (Sunding 1972) indicate the same type of climate, with the moist arid station on the southeastern coast with 91 mm of rain annually, Las Palmas with 174 mm and the stations over 1500 m above sea level with over 900 mm. Here, the clouds often envelop the lower peaks.

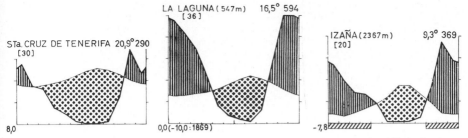

Fig. 88. Climate diagrams. Santa Cruz at sea level, La Laguna at the lower cloud limit, Izaña at the upper cloud limit (Walter 1968)

The vegetation map and the profile (A–B) in Figs. 89 and 90 illustrate the distribution on Teneriffe. A narrow desertlike area with Saharo-arabian elements such as *Launaea (Zollikoferia) aborescens*, *Zygophyllum fontanesii* (plus *Suaeda vermiculata* on Gran Canaria) is situated on the southern coast in the shadow of the trade winds. On the steep slopes, this is followed by a semi-desert with succulents, and is especially well-developed on the southern slopes. The montane forest belt is composed of remnants of laurel forests, above which *Pinus canariensis* forests are situated. On the dry southern slopes, the entire forest belt is made up of the latter. This three-needled pine species is related to *Pinus longifolia* of the Himalayan Mountains.

The peak of the Teide usually rises above the clouds. Above the forest limit, it is covered with shrub-like broom species *(Adenocarpus, Cytisus)*. Higher up, the alpine belt begins, in the lower part of which the white-blossomed broom *(Spartocytisus supranubium)* forms closed stands, which become less dense with increasing altitude. Such endemic species as *Sisymbrium bourgaeanum*, the violet-blossomed *Cheiranthus scoparius* and the several-meter-high *Echium bourgaeanum*, with its reddish inflorescences, also occur (color photograph in Walter 1968, Table II).

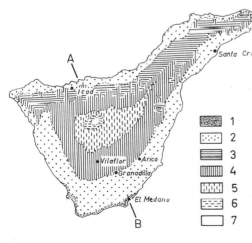

Fig. 89. Vegetation map of Teneriffe: (1) *Zygophyllum-Launea*-desert; (2) *Kleinia-Euphorbia* belt of the succulent semi-desert; (3) laurel forest and *Erica* belt in the north (trade wind exposed side); (4) pine forest-broom heath belt; (5) *Spartocytisus*-mountain semi-desert (temperate); (6) rubble belt with *Viola* and *Silene*; (7) mountain desert with cryptigamic species (cold). *A–B* profile cross-section illustrated in Fig. 90

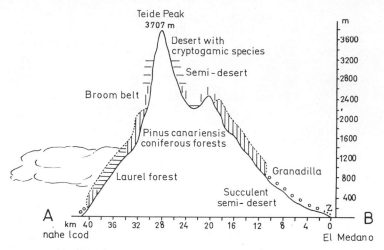

Fig. 90. NNW-SSE profile through the island of Teneriffe (see Fig. 89). The altitude is given on the right. z, *Zygophyllum-Launaea* desert on the sea coast at El Medano (Walter 1968)

The alpine rubble belt begins at over 2600 m above sea level. As a result of solifluction from alternating frosts, this layer is constantly in motion. Only individual rubble creepers, such as *Viola cheiranthifolia* and *Silene nucteolens* are able to survive in this habitat. At altitudes over 3100 ar 3200 m above sea level, only cryptogamic species are found: several blue-green algae *(Scytonema)*, mosses (*Weissia verticillata* and *Frullania nervosa*) and lichens (*Cladonia* spp.), etc.

The plant societies of Gran Canaria were studied in detail by Sunding (1972). The altitudinal belts are identical to those on Teneriffe, except that they do not rise over 2000 m above sea level, therefore barely over the upper forest limit. Two color maps in Sunding's publication are especially interesting. One illustrates the vegetation as it appears today, and one shows the potential vegetation and probably corresponds to the original situation as far as it can be reasonably reconstructed. Mankind has altered parts of the landscape irreversibly, such as through severe soil erosion on deforested areas, which are no longer capable of reforestation. These maps are presented in a smaller and simplified form in Fig. 91. The map of the potential vegetation indicates a very narrow desertlike zone situated mainly on the southern and eastern seacoasts. This is followed by a succulent semi-desert which occupies over half the entire surface below 400 m above sea level on the northern slope, and below 800 m above sea level on the southern slope. The remaining surface is covered by a *Pinus carnariensis* coniferous forest. The evergreen laurel forest in the broadest sense (including the drier form with *Myrica faya* and *Erica arborea*), was probably only found in the lower level of the forest belt on slopes exposed to the northeast. According to Sunding, the natural distribution of the broom belt was limited to the small area around the peak. Comparing this map with the present vegetation (not including the

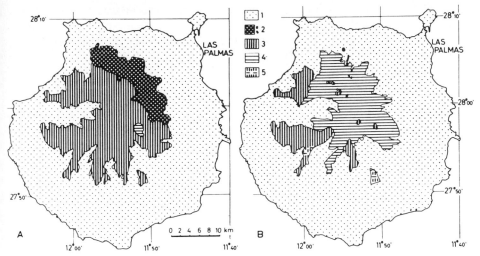

Fig. 91. Geographical distribution of vegetation on Gran Canaria as it was originally (A) and as it appears today (B). (1) succulent semi-desert (today most of the flat lowlands are under cultivation), (2) laurel forest including the drier myrico-ericetum form, (3) pine forest (today partly *Cistus* heath), (4) broom heath, (5) mixed stands of *Cistus* and broom. (Adapted from Sunding 1972)

settlements with their agricultural environs), an enormous change has taken place. The desert-like vegetation on the flat sea shore will soon be replaced by hotels or vacation homes with bathing beaches. The succulent semi-desert has expanded greatly at the cost of the forest belt, and now covers 78% of the island. Broom heaths have replaced the former forests in the upper ranges of the forest belt. The remaining forest is significantly smaller, and now consists almost only of pine. The former extensive evergreen laurel forest remains only in a few gorges as small remnants and only appears as black spots on the small map.

Natural vegetation, therefore, is only found on the steep, nearly inaccessible rocky cliffs of the succulent semi-desert. Ecologically, this is a highly heterogenous unit with an almost micromosaic structure of dry rock surfaces and shallow soils, creviced cliffs and rubble-covered slopes on which deeply rooting species are relatively well supplied with water, as well as ravines with exposed groundwater and dripping wet cliffs.

This creates a situation in which the most various ecological types may occur in neighboring niches, although under completely different conditions. At the one extreme are the *Euphorbia* with succulent trunks, which are able to survive long droughts, and of the other is the delicate Venus fern *(Adiantum capillus-veneris)*, which occurs in shady locations on constantly wet cliffs. Beneath the ferns are mats of moss covered with a calcareous crust, which remains after the water evaporates. The small amount of NaCl in the water may also become concentrated, making possible the occurrence of the halophytic species *Samolus valerandi* next to the ferns. Even if plant sociological surveys are limited to very small areas, random lists of completely heterogenous ecological types results, in which shallow and deep-rooting succulents and non-succulents are found, although they are dependent on entirely different niches. The presence of annual therophytes is of no significance since they develop in clearings, where they are free of competition during the short rainy season when all soils are moist. The occurrence of certain ecological types can only be explained with the aid of a detailed ecological analysis including information on the root and water distribution in the soil during the various seasons. An analysis of this type is quite tedious and requires very careful observations with well-aimed field experiments during all seasons and over a period of several years.

It is in this altitudinal belt of the succulent semi-desert that the palm *Phoenix canariensis* grew. Wild specimens of this palm no longer exist, although it may be found in the parks in the range of Zonobiome IV or V. It is more ornamental than the closely related date palm *(Phoenix dactilifera),* but its fruit is not edible. It was certainly originally situated in sunny locations with easily accessible groundwater, such as in the water-rich ravines.

The well-known dragon tree of the Canary Islands *(Dracaena draco)* probably also grew in similar habitats, yet today it is only found planted in gardens.

6. Arid Mediterranean Subzonobiome

Small arid regions occur in the Ebro basin of northeastern Spain (Walter 1973a) and, in an even more extreme form, in southeastern Spain (Freitag 1971). As an example of a larger arid region, we have chosen central Anatolia (Turkey).

Central Anatolia falls within the winter-rain region and is a basin, 900 m above sea level, completely surrounded by mountains. These mountains catch a large portion of the winter rains, and in May, the still wet but already warm ascending air masses lead to thunderstorms and a rain maximum (Fig. 92). The total annual rainfall amounts to less than 350 mm. There is a pronounced summer drought, and the months from December to March are cold (absolute minimum, −25 °C), with occasional intervening thaws. No forest is capable of developing under such conditions, and the pine forests of the encircling mountains (Mediterranean montane belt) are succeeded, via a shrub zone with *Juniperus, Quercus pubescens, Cistus laurifolius,* and *Pirus elaeagrifolia* and *Colutea, Crataegus,* and *Amygdalus* (dwarf almond) species, by steppe, which today is largely given over to arable land (winter wheat cultivation as "dry farming") or intensive grazing. This has resulted in degradation to an *Artemisia fragrans–Poa bulbosa* semidesert with many spring therophytes and geophytes. At greater altitudes, thorny cushions of *Astragalus* (Tragacantha) and *Acantholimon* (Plumbaginaceae), which are especially characteristic of the cold Armenian and Iranian highlands, are

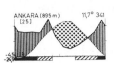
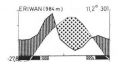
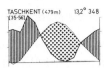

Fig. 92. Climate diagrams of Ankara (Turkey), arid Mediterranean climate. Homoclimes are Eriwan (Armenia) and Taschkent (central Asia, slightly lower and warmer). (From Walter and Lieth 1967)

encountered. Originally, central Anatolia was covered by herbaceous grass steppe *(Stipa-Bromus tomentellus-Festuca vallesiaca)* reminiscent of the east European steppes (p. 228), except for the Mediterranean floristic elements. The soil shows a typical chernozem profile (p. 224), although the A-horizon is not very rich in humus. The cold winter, and the summer drought, account for the brief growth period of only 4 months, whereby the occurrence of the rainfall maximum in May is of great significance.

The most favorable season here is the spring. The first geophytes are already flowering by February and March *(Crocus, Ornithogalum, Gagea,* etc.), followed, on overgrazed areas, by numerous small therophytes which, since they only root in the upper 20 cm of the soil, have disappeared again by June. The genuine perennial steppe species are fully developed by May and do not dry out until July. Their cell-sap concentration is low (10 to 15 atm), because the soil contains sufficient water in the spring, and rises only shortly before the plants die off. A whole series of species, including the thorny cushions, bloom during the main drought season. They are equipped with long taproots enabling them to take up water from the deeper soil horizons which are still moist in the summer: a root length of 7.65 m has been recorded in a 30-month-old specimen of *Alhagi.* Their cell-sap concentration, too, is below 15 atm.

The periphery of the Mediterranean steppe region was settled very early by man and may be looked upon as the cradle of human civilization. The Hittites of Anatolia were among these early settlers, as were the inhabitants of the fertile crescent formed by the mountainous slopes bordering Mesopotamia to the west, north, and east. The oldest traces of grain-growing have been found in the neighborhood of Jericho, Beidha, and Jarmo. Such steppe land provided suitable conditions not only for grain-growing but also for the support of cattle, and the surrounding forests offered both game and fuel. Inhabitants of the ancient settlements have completely ruined the natural vegetation in the intervening thousands of years, and in places which once were fertile country, there is now desert. Soil erosion has set in, and badlands with no sign of plant life are frequently encountered. The highly varied zonoecotones in the northern part of the Mediterranean region, with its wide west–east extension will be discussed later.

7. Biome Group of the Californian Region and Neighboring Country

This region, in western North America, is limited by mountain ranges (Cascades and Sierra Nevada) to a narrow strip on the Pacific Coast. The winter-rain region extends down the west coast from British Columbia to Lower (Baja) California, although, in the north, the rainfall is so high and the summer drought so brief that the forests are hygrophilic to mesophilic coniferous forests, rich in species, and can be regarded as Zonoectone IV/V (Barbour and Major 1977).

Central and southern California together form a sclerophyllous region, but Lower California is too arid for this type of vegetation (Fig. 93.) The north-south gradient of rainfall explains why only in the northern part of the sclerophyllous region do evergreen oak forests and sometimes even mixed deciduous species occur, whereas in the south a shrub formation known as *chaparral* predominates, corresponding to the Mediterranean maquis. Since the present-day flora of westernmost America is quite similar to that of the Pliocene, apparently little impoverishment took place in the Pleistocene, and the plant communities are therefore very rich in species. Such genera as *Quercus* and *Arbutus* are represented by a large number of species, and many other genera entirely absent in Europe are found, for example, the important shrub genera *Ceanothus* (Rhamnaceae), represented by 40 species, and *Arctostaphylos,* represented by 45. One of the main species is the Rosaceae *Adenostoma fasciculatum* ("chamise"), with needlelike leaves; its distribution fairly exactly reflects the extent of the sclerophyllous zone.

Detailed ecological studies were carried out by Mooney and Parsons (in Castri and Moony 1973) in an area of an *Adenostoma* chapparal in the mountains near San Diego (458–678 m above sea level), which has been under protection for 40 years. A station at 815 m yielded the following data: mean annual temperature 14.3 °C; absolute maximum 42.5 °C; absolute minimum −7.8 °C, frost may occur between October and May; mean annual rainfall 670 mm, mainly in December to March; evaporation 1625 mm/year, mainly during the 4 hot summer months. In years with very little precipitation, the soil may dry out to a depth of 1.2 m, below which, however, it is always moist.

Lightning-induced fires are common. The temperatures may reach 1100 °C in the flame, 650 °C on the ground surface and 180 °C–290 °C at a depth of 5 cm.

Even during the dry season, *Adenostoma* accomplishes over 50% of its new growth in the first 10 days after a fire, and develops 25-cm-long sprouts within 30 days. All plants of the species *Quercus agrifolia* and *Rhus laurina* develop sprouts. *Adenostoma* attains its densest cover 22–40 years after a fire and almost ceases growth after 60 years. The rejuvenation of such a stand occurs after the next fire. Approximately 50% of the shrub species rejuvenate by sprouting and the remainder by seed. After about 20 years, the stands are again closed. In the first years after a fire, steep slopes are subject to heavy erosion. The aboveground phytomass attains a value of 50 t/ha, while the

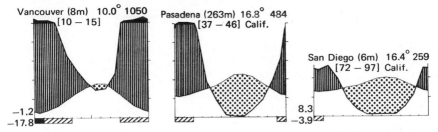

Fig. 93. Climate diagrams of stations on the Pacific coast of North America (from north to south) in the coniferous-forest region, sclerophyllous-forest region, and the region transitional to the desert. (From Walter and Lieth 1967)

underground phytomass is double that amount. The aboveground net production is approximately 1 t/ha/yr in the younger stands and decreases with age. The shrubs are normally photosynthetically active all year round. In the springtime, an abundant ephemeral vegetation develops, some species of which only germinate after a fire.

Adenostoma is predominant on southern slopes, while dense stands of *Quercus dumosa* grow on the northern slopes.

The strip of land immediately bordering the ocean north of latitude 36°N in California does not belong to the sclerophyllous zone. The fog resulting from the cool ocean currents renders the summers cool and wet, so that hygrophilic, northern tree species are found.

Unlike the maquis, the chaparral is the natural zonal vegetation corresponding to a relatively low winter rainfall of 500 mm. Fires are common in this region and were a natural factor even before the advent of man. The statistics of the US Forest Service reveal that fire caused by lightning is very common in the chaparral region, rendering constant fire-watching necessary during thunderstorms. If a fire occurs every 12 years, the character of the chaparral remains unchanged since the shrubs can sprout afresh. But if no fire occurs for a great length of time, then species such as *Prunus ilicifolia* and *Rhamnus crocea* infiltrate. If one fire follows another within 2 years, the seedlings of those shrubs which cannot resprout after a fire are killed off, and these woody species are eliminated.

The roots of sclerophyllous species reach far down into the ground because the upper soil layers are usually completely dried out in summer. By means of their roots, which may even penetrate from 4 to 8.5 m into the rock fissures, the plants are able to obtain a certain amount of water in the dry season. More details on the root system profiles may be found in Kummerow (1981).

Availability of water can be recognized from the observation that very soon after a fire in the height of summer, the shrubs begin to sprout again. After the loss of the transpiring surface, a small amount of water suffices for the buds to grow. The autumn rains have no direct effect, since it takes a month for the water to sink to a depth of 1 m. In the meantime, the

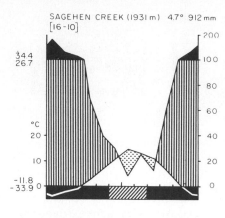

Fig. 94. Climate diagram of Sagehen Creek on a mountain pass of Sierra Nevada (1931 m above sea level) near Reno. The peak of the rain curve indicates summer thunderstorms in August. The absolute temperature maximum is 34,4 °C, and the minimum is −33.9 °C (Madroño 29, No 2, 1982)

temperature falls so much that growth stops. In April, when the water supply is good and the temperature rising, growth is at a maximum. The old evergreen leaves continue assimilation into the spring and are shed in June, by which time the new leaves can take over their function. All chaparral species possess mycorrhizae, and the *Ceanothus* species have nodules which assimilate atmospheric nitrogen. A detailed vegetation monography with much ecological data was published by Barbour and Major (1977).

Evergreen sclerophyllous oak forests are also found as a montane belt in North America, above the cactus desert in the mountains of southern and central Arizona, at an altitude of 1200–1900 m. This is known as the *encinal* belt, and on the basis of the distribution of the different species of *Quercus,* it can be subdivided into an upper and lower belt. The upper belt is succeeded by a *Pinus ponderosa* belt. The chaparral genera *(Arbutus, Arctostaphylos, Ceanothus)* form the shrub layer in the encinal forest. Although Arizona has two rainy seasons, the vegetation is strongly reminiscent of that of California, except that the sclerophyllous forests in the mountains are much better developed and are still original. Despite the fact that summer storms supplement the low winter rainfall, the summer drought is still very pronounced.

East of the Sierra Nevada, in the state of Nevada, the winter rains fall off to about 150–250 mm, and at an altitude of 1300 m, the cold season lasts 6 to 7 months. The climate diagram of Sagehen Creek (Fig. 94) illustrates this for a mountain pass with high precipitation and forest and bog vegetation. The climate diagram for Reno (Fig. 95) reflects the conditions on the lee side of the mountains with its *Artemisia tridentata* semi-desert—also designated as "sagebrush". Figure 96 illustrates the dependency of the precipitation on the relief in this region. The *Artemisia* semi-desert occupies enormous areas in Nevada, Utah, and the bordering states and is the cold-climate equivalent of the southern *Coleogyne* and *Larrea* semidesert. *Artemisia* prefers the heavy soils of the basins and gives way to *pinyon* on elevated stony ground. Pinyon consists of low scattered tree communities with *Pinus monophylla* or *P. edulis*

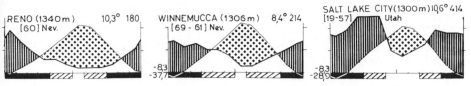

Fig. 95. Climate diagram of the sagebrush region (*Artemisia tridentata* semi-desert) around Reno, Winnemuca and Salt Lake City, Utah (transition grassland)

and *Juniperus* spp., including some cold-resistant chaparral species. True coniferous forests of *Pinus flexilis* and *P. albicaulis* commence at an altitude of about 2000 m in the mountains, whereas further east *Pinus ponderosa* is found. Higher up, these species are replaced by *Pseudotsuga* and *Abies concolor,* and the timberline, above 3000 m, is made up of *Picea engelmannii* and *Abies lasiocarpa.* The dry southern slopes are often devoid of trees, and *Artemisia* extends up to the alpine region. But the order of the altitudinal belts can vary greatly. The aspen *(Populus tremuloides)* is also frequently encountered.

Artemisia tridentata is a half-shrub 1.5 to 2 m high and attains an age of 25 to 50 or more years. Its taproot extends about 3 m into the ground and gives off horizontal lateral roots extending far on all sides. In spring, after the snow has melted and the water supply is good, the cell-sap concentration is very low (about 10 to 15 atm); it soon rises to about 20 to 35 atm and can even amount to 70 atm in summer if there is an acute water shortage. At this point, like all malakophyllus plants, it sheds its leaves. The sagebrush semidesert is confined to the arid brown, semidesert, salt-free type of soil, with *Artemisia tridentata* as the dominant species, frequently accompanied by the dwarf-shrub *Chrysothamnus* (Compositae). The region belongs to the arid ZB VIIa.

Depressions with no outflow are invariably brackish in such an arid climate. The saltpans and salt lakes in these regions are the relics of much

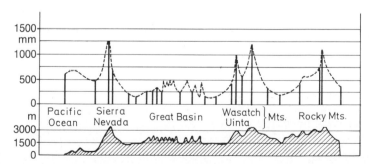

Fig. 96. The dependency of the precipitation *(above)* on the relief *(below),* illustrated in a west to east profile through the western part of North America at approximately 38 °C north latitude (Walter 1960)

larger Pleistocene lakes such as Lake Bonneville in Utah, the surface of which was 310 m above the Great Salt Lake and the extensive surrounding barren salt desert. Lake Bonneville covered an area of 32,000 km², with a maximum length of 586 km and a maximum width of 233 km. The present salt desert is more than 161 km long and 80 km wide. In 1906, the Great Salt Lake measured 120 km by 56 km when at its highest level. Its contours vary greatly, the average depth being a little above 5 m. Its salt content varies between 13.7% and 27.7%, when it is saturated. About 80% of the salt is NaCl; the remainder consists of $MgCl_2$, Na_2SO_4, K_2SO_4, etc. Halophytes occur around the saline areas, and the vegetation forms very distinct zones.

Hygrohalophytic *Allenrolfea* and *Salicornia* grow on the edge of the salt crust, followed by *Suaeda* and *Distichlis*. Further wide zones are occupied by *Sarcobatus,* which requires ground water, and the xerohalophytic species *Atriplex confertifolia*. Species of *Kochia* and *Ceratoides (Eurotia) lanata* then lead on to a nonhalophytic zone with *Artemisia tridentata*. With the exception of the salt-excreting grass *Distichlis,* all of the halophytes are Chenopodiaceae. The entire area is one large halobiome with biogeocene complexes.

The climate in Utah is very similar to that of Ankara (Anatolia). The marked predominance of *Artemisia* is the result of overgrazing; such grasses as *Agropyron, Stipa,* and *Festuca* were formerly widespread.

8. Biome Group of the Central Chilean Winter-Rain Region with Zonoecotones

Chile, lying at the western foot of the High Andes, is a long, narrow strip of land 200 km wide and 4800 km long, extending from 18° to 57° S. It exhibits every possible transition in vegetation, from the rainless subtropical desert in the north to, via a sclerophyllous region, the very wet, temperate and subarctic forests in the south. Winter rains are prevalent throughout (Fig. 97). The cold Humboldt current flowing along the entire coast modifies the summer drought so that temperatures are lower than those of California. The mean annual temperature of Pasadena, at 34° N, is 16.8 °C, whereas in Santiago, at 33° S, it is only 13.9 °C. The climates of these two regions have been compared by Castri (1973). Since Chile belongs to the Neotropical floristic realm, its flora is quite different from that of the Mediterranean region or California. Only the cultivated areas offer a similar appearance, since the same species are cultivated on the farms and in the gardens in all three regions. The sclerophyllous region occupies the central part of Chile and adjoins the arid regions in the north. It is represented only by remnants forming woodlands 10 to 15 m tall, with such xerophytic species as *Lithraea caustica* (Anacardiaceae), which causes rashes and fever if touched, *Quillaja saponaria* (Rosaceae), *Peumus boldus* (Monimiaceae), and the Lauraceae *Cryptocarya* and *Beilschmiedia,* which have a preference for wet ravines, in addition to a whole series of shrubby species. The endemic palm *Jubaea*

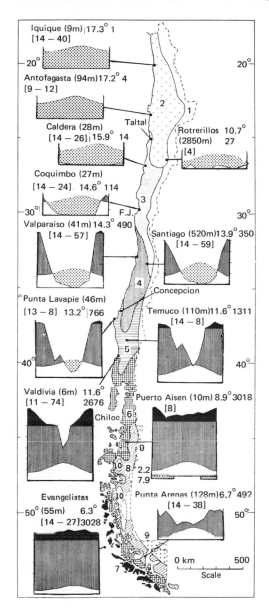

Fig. 97. Vegetation of Chile and climate diagrams. Northern Chile: (1) northern High Andes; (2) desert region; (3) dwarf-shrub and xerophytic-shrub region; (4) sclerophyllous region; (5) deciduous forest. Southern Chile: (6) evergreen rain forests of the temperate zone; (7) tundralike vegetation of the cold zone; (8) subantarctic deciduous forest; (9) Patagonian steppe; (10) southern Andes. (Modified from Schmithusen; Walter 1964/68)

chilensis grows in a narrowly limited area northeast of Valparaiso; columnar cacti (*Trichocereus*, Fig. 98), and the large *Puya* spp. (Bromeliaceae) as well as the thorny Rhamnaceae *Colletia* and *Prevoa* are found in dry, rocky habitats.

On the Chilean side, the Andes drop steeply: Aconcagua, at 7000 m, is only about 100 km from the coast. Talus communities predominate, and

Fig. 98. Landscape near Santiago (Chile). In the foreground, on rocky soil, flowering *Trichocereus*. In the valley, remnants of sclerophyllous vegetation. The grasses are adventitious annual Mediterranean species *(Avena,* etc.). (Photo E. Walter)

altitudinal belts are difficult to recognize. The sclerophyllous vegetation extends up to about 1400 m, and shrub communities lead on to the alpine region, with the occasional appearance of the conifer *Austrocedrus (Libocedrus) chilensis.* Alpine talus plants such as *Tropaeolum* spp. and *Schizanthus* (a Solanaceae with zygomorphic flowers), as well as Amaryllidaceae *(Alstroemeria, Hippeastrum)* and species of *Calceolaria,* are common. Flat, cushionlike plants are characteristic of the upper alpine belt *(Azorella* and other Umbelliferae).

The species occurring in the altitudinal belts of the orobiome, as well as to the south of the sclerophyllous zone, are already antarctic elements, as are the tree-form *Nothofagus* species. Immediately south of Concepción, with decreasing summer drought, *Nothofagus obliqua* forest is found (Zonoecotone IV/V). These trees lose their leaves in the cool winter months. Farther south, with a rainfall exceeding 2000 to 3000 mm, this forest is replaced by the evergreen Valdivian warm temperature rain forests of ZB V (Quintnilla 1974), which are scarcely less luxuriant than the tropical rain forests, while their timber mass is probably even greater. The woody species are partly neotropical in origin. Bamboos (Chusquea) are also common. Part are antarctic elements, as for example the evergreen *Nothofagus dombeyia.* A large number of ancient coniferous forms are found, especially in montane situations. Besides *Austrocedrus* and *Podocarpus* species, *Saxegothea, Fitzroya, Araucaria araucana (= A. imbricata)*, and *Pilgerodendron uviferum* also deserve mention. The climate is very wet and cool but frost-free, so that the evergreen forest gives way to the so-called Magellan forests which stretch almost to the tip of South America, gradually becoming poorer in species and decreasing in height untill finally they are only 6 to 8 m tall. The westerly islands are covered by bogs with cushionlike plants (*Sphagnum* is not important), a vegetation closely related floristically to that of the antarctic islands. Similar antarctic elements are found in New Zealand and in the mountains of Tasmania, indicating that these regions were formerly directly connected with each other via the Antarctic continent. The bogs can be considered to be antarctic tundra (ZB IX).

9. Biome Group of the South African Capeland

Although confined to the outermost southwestern tip of Africa the South African winter-rain region comprises an entire floristic realm, the Capensic. This small region is extraordinarily rich in species. In the Jonkershoek nature reserve alone, 2000 species have been recorded on 2000 ha and an equal number on the 50-km stretch from Table Mountain to the Cape of Good Hope. Included are 600 species of the genus *Erica*, 108 species of *Cliffortia* (Rosaceae), 115 species of *Muraltia* (Polygalaceae), 117 species of *Restio* (Restionaceae), and about 100 species of *Protea*. Particularly abundant among the sclerophyllous plants are the Proteaceae, a family which is otherwise well represented only in Australia, albeit by a different subfamily (a few genera also occur in South America). Many house plants of temperate latitudes originally came from the Cape [*Pelargonium, Zantedeschia (= Calla),. Amaryllis, Clivia*, etc.]. The climate diagrams for Cape Town and Tangiers are comparable, except that the annual rainfall of the former is 260 mm less, although the summer is slightly less dry (Fig. 99). The sclerophyllous vegetation of the Cape is known as *fynbos* and is a proteaceous scrub, 1 to 4 m high, very similar to the maquis. The only tree species, *Leucadendron argenteum* (silver tree), is of very limited distribution and is confined to the humid slopes of Table Mountain, at altitudes below 500 m. In wet ravines, forestlike stands can be found, but these are in fact the last outposts of of the wet-temperate forests of the southeast coast of Africa (ZB V).

The leaves of the Proteaceae are sometimes very large and, although they have very little mechanical tissue, are nevertheless sclerophyllous because of their thick cuticle. As in all sclerophyllous plants, the water balance of the Proteaceae shrubs is in a state of equilibrium, which means that the cell-sap concentration undergoes only very slight variations during the course of the year. The deeper layers of the soil, into which the roots penetrate, apparently contain water even in the summer. Cape soils are acid and poor in nutrients, which particularly suits both Proteaceae and Ericaceae. Fire is the most important ecological factor. In the year immediately after a fire, innumerable geophytes *(Gladiolus, Watsonia,* etc.), of which the Cape flora possesses about 350 species, make their appearance, followed by herbaceous species and dwarf shrubs. It takes about 7 years before the Proteaceae shrubs have grown up again, either from seedlings or as shoots from the old plant. Although these plants are capable of achieving considerable age, they lignify with time and flower less abundantly, so that periodic burning would appear to be advantageous. In this region, fire caused by lightning also seems to constitute a natural factor, although nowadays fires are deliberately set. It is interesting to note that bulb plants begin to flower only after a fire and otherwise grow vegetatively. The reason for this is probably the sudden removal of root competition by the bushes rather than a fertilizing effect resulting from the ash.

Rainfall increases with increasing altitude only on the southeastern slopes of the mountains, which are affected by the warm, humid air rising from the Indian Ocean. The rainfall recorded at the Table Mountain station is three times as high as that of Cape Town, 750 m lower in altitude (Fig. 99). The Capeland is mountainous, with basins scattered between the mountain ranges which often wear a "tablecloth," or covering of cloud, which is due to the warm, wet winds from the Indian Ocean and which rises up the southeastern slope and disperses again on the northwestern slope. This leads to a wet mist on the high plateaus of Table Mountain, and there is a tendency toward heathland *(Restio, Erica)* and even the formation of moors (mossy mats of *Drosera* and species of *Utricularia).* Succulents such as *Rochea coccinea* grow between dry boulders.

Moving inland, the rainfall decreases, particularly in the rain shadow of the mountain ranges (Fig. 99, right). For this reason, the sclerophyllous vegetation recedes farther up the seaward slopes. In the rain-shadow area, the dry form of Cape vegetation is met with, known as the renoster bush *(Elytropappus rhinocerotis,* Compositae). This is succeeded inland by the semidesert vegetation of the Karroo (p. 131).

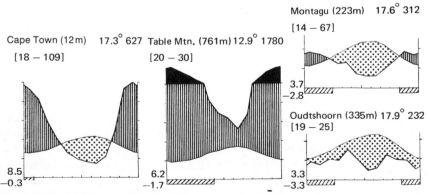

Fig. 99. Climate diagrams of South Africa: typical sclerophyllous region; damp montane climate (misty); transitional region; and typical Karroo climate. (From Walter and Lieth 1967)

Since the colonization of the Cape in 1400, the sclerophyllous vegetation, or fynbos, has spread extensively. Formerly, an evergreen-temperate forest with paleotropical elements stretched along the entire coast of southeast Africa beyond the southern tip of the continent (Cape Agulhas) (ZB V).

10. Biome Group of Southwestern and South Australia with Winter Rain

Perth, in southwestern Australia, lies at approximately the same latitude as Cape Town and has a very similar climate (Fig. 100). However, winter rain falls not only on the southwestern corner of this continent but also on the area around Adelaide in South Australia. On account of the peculiar floristic situation (p. 21), the sclerophyllous vegetation differs in character from that in other winter-rain regions of the earth. The tree form *Eucalyptus* spp.) dominates, and Proteaceae constitute a lower shrub layer or may even achieve predominance on the sandy heaths. The leaves of *Eucalyptus* are leathery rather than hard. Peculiar to soutwestern Australia are the "grass trees" *(Xanthorrhoea, Kingia),* the Cycadaceae *Macrozamia,* and species of *Casuarina.* Epacridaceae take the place of Ericaceae. Just as on the Cape in South Africa, the soils are acid and poor in nutrients, containing SiO_2 and iron concretions which are the lateritic crusts of an earlier geological age when the climate was tropical. The parent rocks are among the oldest geological formations of the earth. Indicative of the poverty of the soil is the fact that 47 species of *Drosera* (sundew) occur in the herbaceous layer of the forest around Perth. Wherever it is wet enough, bracken *(Pteridium)* is found.

Rainfall increases to the south of Perth (up to 1500 mm) but decreases to the north and inland. Each change in climate means that different species of *Eucalyptus* rise to dominance. The wetter the climate, the taller the trees and therefore the greater the leaf area per hectare. Because the leaves hang vertically, plenty of light can penetrate the space around the trunks, so that

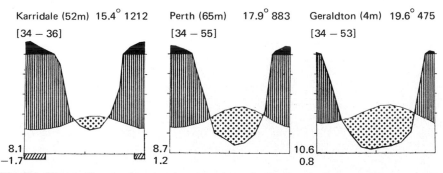

Fig. 100. Climate diagrams from southwestern Australia. Stations in the karri forest, the jarrah forest, and the shrub-heath (see also Fig. 104 Adelaide). (From Walter and Lieth 1967)

the shrub layer is usually well developed insofar as it has not been reduced by frequent fires.

"Jarrah" forest, which is completely dominated by *Eucalyptus marginata,* is characteristic of the climate corresponding to a Mediterranean type, with 650 to 1250 mm of rain and a summer drought. The trees can reach an age of 200 years and a height of 15 to 20 m (maximum, 40 m). In the wetter, southern regions, *Eucalyptus diversicolor* forms the "karri" forest, with trees reaching a height of 60 to 75 m (maximum, 85 m) (Zonoecotone IV/V). The canopy is closed to 65% coverage, and the undergrowth consists of a shrub layer and a herbaceous layer developed from the fronds of the bracken (Fig. 101).

In the drier "wandoo" zone, with a rainfall of 500 to 625 mm, *Eucalyptus redunca* dominates. The woodland is sparser and is now almost exclusively given over to sheep grazing. In the absence of suitable indigenous grasses, *Lolium rigidum* and the Mediterranean clover, *Trifolium subterraneum,* which buries its fruits underground and provides a source of nitrogen, are sown. Owing to the poverty of the soil, fertilization with superphosphate is essential before sowing; because of the large areas involved, both of these processes are carried out from the air.

In the zone receiving 300 to 500 mm of rain annually, many species of *Eucalyptus* are found scattered over the landscape (Zonoecotone III/IV).

Fig. 101. *Eucalyptus diversicolor* forest in Southwestern Australia. Undergrowth of *Acacia pulchella* and bracken *(Pteridium esculentum).* (Photo E. Walter)

Nowadays, however, this is the winter wheat zone, with farms of several hundred hectares, which are completely mechanized and run by two or three men. Rust diseases render wheat cultivation in wetter zones uneconomical.

Where the mean annual rainfall drops below 300 mm, *Eucalyptus* disappears, and the extensive grazing lands provided by the bush semidesert commence (p. 133).

In South Australia, there is no humid winter-rain region, but otherwise, the situation is similar to that in southwestern Australia, only rather more complicated because mixed forest communities occur, consisting of several species of *Eucalpytus*. Furthermore, the region is mountainous, which again leads to a marked differentiation of the vegetation.

Apart from the forests already described, there are vast areas of heathland with 0.5- to 1-m-tall Proteaceae. The bushes are capable of growing on the poorest of sands, and even the least demanding of the *Eucalyptus* species are unable to compete with them (Peinobiome). Such areas are not cultivated and are hardly used for grazing. It is all the more remarkable that so very many species grow on such an impoverished, sandy soil. On 100 m^2, 90 species have been recorded, including 63 small woody species, mainly Proteaceae or Myrtaceae. *Drosera* species and a tuberous *Utricularia* also occur.

Thorough ecophysiological investigations have been carried out on such a heath in South Australia at a rainfall of 450 mm annually and with 7 months of drought in summer. Soil temperatures at a depth of 15 cm varied between 4.1° and 36.0 °C and, at 30 cm, between 5.8° and 29 °C. The root systems of 91 species were dug up. The dominating sclerophyllous species are the shrublike *Eucalyptus bacteri,* nine Proteaceae, two *Casuarina* species, *Xanthorrhoea*, Leguminosae, etc.

The main growth season is the dry summer since the soil remains wet to a great depth. Smaller perennial species (42% of all species) root only in the upper 30 to 60 cm and grow in the spring. *Drosera* and orchids are ephemeroid species and root at a depth of only 5 to 7 cm. It is interesting to note that the water in sandy soil with a wilting point of 0.7 to 1% is very unevenly distributed because the larger species draw the rainwater toward their stems.

The composition of the heathland is conditioned by fire. Immediately after a fire, the grass tree *Xanthorrhoea* begins to sprout, and in fact this species blooms only after a fire. *Banksia* (Proteaceae) regenerates by means of seedlings, and its share in the aboveground phytomass rises to 50% by the 15th year after a fire. The larger part of the dry substances of 25-year-old specimens is accounted for by the large fruits which open only after a fire.

Banksia is one of the many *pyrophytes* very commonly encountered in Australia. *These plants are capable of regeneration only after fire* because the woody fruits do not otherwise open, which would suggest that fires caused by lightning have always played a natural role in Australia. Forest and heath are often burned down nowadays because the woody plants are of no commercial value and hinder grazing. In the farmer's opinion, "One blade of grass is worth more than two trees." Many Proteaceae and Myrtaceae, as well as the coniferous *Actinostrobus,* are pyrophytes, and even *Eucalyptus* species seed more prolifically after a fire.

The nutrients in a heath that has not been burned for a long time are bound up in the fruits of *Banksia,* old leaves of *Xanthorrhoea,* and the accumulating litter. A 50-year-old community degenerates on this account, and not until a fire has caused mineralization of the nutrients can a new succession be initiated.

Eucalyptus marginata woodland is ecophysiologically typical of a sclerophyllous vegetation. The roots penetrate the hard lateritic crust in places to a depth of more than 2 m. There is no summer dormant period, and the water balance is maintained by a partial closing down of the stomata from 10 a.m. to 3 p.m., with a resultant decrease in transpiration. The cell-sap concentration in winter was found to be 16.3 atm and is probably only slightly higher in summer.

Not only its flora, and thus the character of its vegetation, but also the fauna of Australia differs widely from that of other continents.

Monotremes, the most primitive mammals, occur only in Australia. The duckbill platypus, *Ornithorynchus anatinus,* belongs to this family. It lays one to three eggs which are hatched by the female. The spiny anteater *(Tachyglossus = Echidna),* in contrast, hatches only one egg in its brood sac; this animal represents a transition to the marsupials. With only a few exceptions, the latter are also confined to the Australian continent and include both herbivores and carnivores. The best known group are the kangaroos (Macropodiae), including the large kangaroo *Macropus.* As a grazing animal, the latter has had a considerable influence on the vegetation, although exact figures are not available.

11. The Historical Development of Zonobiome IV and Its Relationship to Zonobiome V

Various aspects of Zonobiome IV, including its historical development, are discussed in the volume by di Castri and Monney (1973). Zonobiome IV is closely related to Zonobiome V, since they may both be traced back to a common origin extending back to the tropical vegetation of the Tertiary, which reached into the higher latitudes at that time.

The further development of this vegetation to the present is summarized for California by Axelrod in the above-mentioned volume. The development in the Mediterranean region may by considered to be analogous.

Fossil evidence from the Eocene, at the beginning of the Tertiary, in today's temperate region of the Northern Hemisphere, indicates the former presence of tropical evergreen and deciduous species in a tropical climate with a pronounced summer rainy season. Studies on fossil saltwater molluscs indicate that the minimum temperature of the surface waters of the ocean was approximately 25 °C at that time, about 50 million years ago. The oceans gradually cooled during the Oligocene and Miocene, until the temperature minimum was only 15 °C by the Pliocene at the end of the Tertiary. The climate on the continents cooled correspondingly, and the flora became poorer in species requiring higher temperatures. Simultaneously, the distribution of the rainfall in California was altered, with a decrease in the summer maximum until it was no longer noticeable in the Miocene and had even become a small minimum by the Pliocene. During the Ice Ages of the Pleistocene, cold ocean currents developed on the west coasts of the continents, as well as climates with pronounced dry periods in summer and rain only in the winter months, as is typical of Zonobiome IV. During the Tertiary, the mountains in western North America and the Alpine chain in Europe rose continuously higher.

The result of these changes was the development of more arid climate regions and arid localities in unfavorable expositions of today's higher latitudes, which were the tropical zones of the Tertiary. Consequently, a selection took place among the evergreen species, favoring species with the typical leathery leaves found in the humid tropics (often referred to as the latitudes, which were the tropical zones of the Tertiary. Consequently, a withstanding long dry periods. The development of the summer drought climate (designated as "Mediterranean") on the west coasts of the continents during the Pleistocene led to the dominance of the sclerophyllic species, while the flora of the woody species declined. On the east coasts of the continents, which were subject to warm ocean currents, the humid climate with summer rain and somewhat lower mean annual temperatures than Zonobiome V remained unchanged. The transition from tropical humid to subtropical humid and to the temperate species-rich flora with evergreen leathery leaves is gradual on the humid east coasts of North and South America, Southeast Africa, Southeast Asia and eastern Australia.

Since the Tertiary species were already pre-adapted to dry locations, it was not necessary for the sclerophyllic vegetation of Zonobiome IV to develop by adaptation to the summer dry season. Only a limited development of new species took place, such as in California, for example, the genuses *Ceanothus* with 40 species and *Arctostaphylos* with 45 species, while others *(Adenostoma)* extended their distribution. *Arbutus* has more leathery leaves.

This developmental history explains why the same genuses are often found in Zonobiomes IV and V, although the species differ, as for example the sclerophyllic *Quercus* species of California and the evergreen *Quercus virginiana* with sclerophyllic leaves in southeastern North America (Zonobiome V). The sclerophyllus *Eucalyptus* species of Zonobiome IV in southwestern and southern Australia differ only sligthly from those in the summer rain region of Zonobiome V on the east coast.

Just as in the west, dry calcareous soils in the east are settled by a profuse Proteaceae vegetation, except that the species are not the same. The occurrence of the fossil "terra rossa" soils in the Mediterranean region may also be explained in a similar manner. This soil contains relicts of a tropical microscopic fauna which was able to avoid the effects of the summer drought at greater depths. The remaining fauna of Zonobiome IV confirms the observations made on the vegetation (see several contributions in di Castri and Mooney 1973). As concluded by Axelrod, fossil evidence in North Africa indicates a development similar to that of the Mediterranean vegetation. The conditions in Europe, however, are somewhat more complicated, since the climate of western Europe has been determined by the warm Gulf Stream since the post-glacial period.

The cold Canary Stream becomes effective south of the Canary Islands towards the coast of Senegal (Fog Coast). Zonobiome IV extends from the west, along the Mediterranean coast and further eastward. The last Ice Ages had an especially negative effect in Europe, practically exterminating the entire flora. During the post-glacial period the flora returned from a small number of refuges, but remained poor in species. A continual series of fossils from the Tertiary to the present, such as is found in California, is missing here. It is generally concluded, however, that the development of Zonobiome IV followed the same basic course wherever it now occurs, and that a type of climate corresponding to Zonobiome IV with zonal sclerophyllic vegetation had not yet existed in the Tertiary, although the sclerophyllic species certainly did exist in dry local habitats.

V Zonobiome of the Warm-Temperate Humid Climate

1. General

Zonobiome V cannot be sharply delineated since it is a transitional zone between the tropical-subtropical and the typical temperate regions, although too large to be considered an ecotone. Two subzonobiomes can be distinguished:

1. *A very humid subzonobiome with rainfall at all times of year or with a minimum in the cool season.* The principal vegetational period, which is invariably wet, is oppressive due to high temperatures. The regions involved lie on the eastern sides of the continents between latitudes 30° and 35° in both the Northern and Southern Hemisphere and are influenced by trade and monsoon winds. Temperatures drop quite severely in the cool season, and there may even be frost, but there is no cold season. Nevertheless, the vegetation spends the winter in a resting state.

2. *A subzonobiome with rainfall occurring principally in winter and no summer-drought season.* This region lies along the western seaboard of the continents, nearer to the poles than the first, adjoining the wet subzonobiome of ZB IV with winter rain.

In North America, the subzonobiome with winter rain stretches along the coastal regions from northern California to southern Canada. It is the zone of the *Sequoia sempervirens* forests, which can attain a height of 100 m. Farther north, this vegetation is succeeded by forests with *Tsuga heterophylla, Thuja plicata,* and *Pseudotsuga menziesii* (Fig. 102). *Prunus laurocerasus, Rhododendron ponticum,* and *Araucaria excelsa* flourish in gardens, thereby indicating mild winters. Still farther north, the temperature gradually drops, the climate becomes wetter, and diurnal and annual fluctuations in temperature are small. The maritime, frost-sensitive sitka spruce becomes predominant. The zone extends along a meridian up to the subarctic in Alaska, but regions corresponding to ZB VI or ZB VIII are barely recognizable. It is an extremely humid, maritime ecotone in which land cultivation is impracticable and the population is therefore sparse.

Studies based on the International Biological Program (IPB) have been carried out to examine these probably most productive coniferous forests in the world, especially the Douglas fir ecosystems (Pseudotsuga). A volume containing 11 contributions on the preliminary results obtained from studies in the years 1971−1978 has been published (Edmonds 1982), although the actual summary of these results has not yet appeared.

Fig. 102. Damp, oceanic coniferous forest with *Pseudotsuga menziesii, Tsuga heterophylla,* and *Thuja plicata* on the Hoh River (Olympic National Park). (Cf. climate diagram of Vancouver, Fig. 93). (Photo H. Walter)

An analagous situation is found in southern Chile. The subzonobiome with winter rainfall but no summer drought corresponds to the luxuriant Valdivian evergreen rain forests previously mentioned (p. 172). The Magellan forest, which continues to the south with both evergreen and deciduous *Nothofagus* species and well-developed moors, constitutes the transitional zone to the subantarctic Fireland and the islands.

The tall, frost-sensitive coniferous species of the Pacific coast of North America are wholly lacking in western Europe, where they died out during the glacial periods of the Pleistocene. The nearest counterparts to this subzonobiome in Europe are the heath formations on the coasts of northern Spain and southwestern France (Landes). The humid transitional zone is as split up as the coasts of Western Europe. It comprises Wales, western Scotland, and the island groups, including subarctic Iceland and the wettest parts of the Norwegian west coast with the Lofoten Islands, and extends to the arctic. Heather moors with birch and willow are the predominant form of vegetation at the present time (p. 190).

The southwest tip of Australia with winter rains and no summer drought (Karri forest, p. 176) also belongs to this subzonobiome. An extremely perhumid zone, however, consisting solely of Tasmania (with small *Eucalyptus* species and moors) and the southwest portion of the South Island of New Zealand, with Stewart Island, constitutes the transition to the subantarctic islands.

A completely isolated region belonging to this subzonobiome is found in northern Anatolia with Colchic forests in which *Rhododendron ponticum* and *Prunus laurocerasus* are native. These are offshoots of the luxuriant forests of the Colchic Triangle between the Caucasian Mountains and the Black Sea, with evenly distributed rainfall which may amount to as much as 4000 mm annually. Although the evergreen undergrowth has persisted, the tree layer with the Tertiary relict species *Zelkowa, Pterocaria,* and *Dolichos* and the lianas *(Vitis)* is deciduous. Despite isolated outbreaks of cold, citrus fruits are grown. A similar situation is encountered in the Hyrcanian relict forests on the south coast of the Caspian Sea, with *Parrotia* (Hamamelidaceae) and *Albizia julibrissin* (Mimosaceae) as relict species.

2. Humid Subzonobiomes on the East Coasts of the Continents

Because of the trade or monsoon winds on the eastern sides of the continents, there is an almost continuous sequence of wet subzonobiomes of ZB II to ZB V (via a wet subtropical Zonoecotone II/V) and to ZB VI (via Zonoecotone V/VI). But since the land masses of the Southern Hemisphere do not extend as far toward the pole as those of the Northern Hemisphere, and owing to the interruption of North and South America by the Caribbean Sea, the various zones are not easily distinguishable. The tropics are taken to end where frost occurs or where, even in the absence of frost, the mean annual temperature is below 18 °C and cultivation of tropical plants such as coconuts, pineapple, coffee etc., is no longer economically worthwhile and only tea, citrus fruits, and a few palms remain.

Frosts already occur within ZB V, but the mean daily minima of the coldest month are still above 0 °C, i.e., no cold season appears on the climate diagram. The annual means are slightly above or below 15 °C, at least for part of the time, and the forest trees are at least in part evergreen, whereas only a few shrub species remain evergreen in Zonoecotone V/VI. In ZB VI, in contrast, a cold season of 2–5 months is characteristic, and the woody species shed their leaves in autumn.

In eastern Asia, which is exposed to the east Asian monsoon and therefore has a ZB II, the humid subzonobiome of ZB V covers an unusually large region. Its northernmost limit at 35 °N just reaches the southern tip of the Korean peninsula and runs through the southern part of the main Japanese island of Hondo. The forests of this region consist of evergreen Fagaceae *[Cyclobalanopsis (Quercus)* and *Castanopsis]*, Myrsinaceae *(Ardisia)*, and Lauraceae *(Machilus)* and the shrubs commonly used as decorative plants in south central Europe *(Aucuba japonica, Euonymus japonica, Ligustrum japonicum)*, as well as the frostsensitive conifer *Cryptomeria japonica*. Farther north, deciduous tree species become predominant (Numata et al. 1972).

In China, the northern limit of the subzonobiome is slightly displaced inland toward the south, owing to the influence of outbreaks of cold winds in winter resulting from the high Siberian pressure. The southern limit, which borders the evergreen tropical–subtropical forests of southern China, is much less sharply defined. Canton itself is still within ZB II. Figure 103 shows the

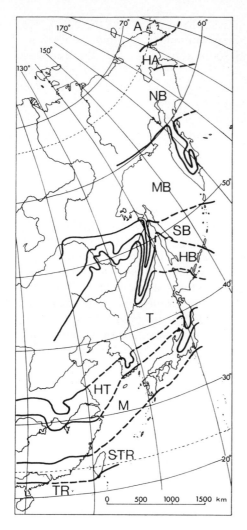

Fig. 103. Bioclimates of eastern Asia: TR, wet tropics; STR, wet subtropics; M, maritime warm-temperate ZB V; HT, ZE V/VNI; HB, hemiboreal mixed forest zone of ZE VI/VIII; SB, MB, and NB, southern, middle, and northern boreal zones of ZB VIII; HA and A, hemiarctic and arctic zones (ZE VIII/IX and ZB IX). (From Ahti and Konen 1974)

classification according to Ahti and Konen (1974), who also discuss the orobiome in Japan.

The southern tip of Florida, in the southeast of North America, is still subtropical, but even in Miami and Palm Beach, slight frosts may occur. The forests of evergreen oak *(Quercus virginiana)* stretch northward along the coast into North Carolina. The total area covered by ZB V is not large, since inland outbreaks of cold extend as far south as the Gulf of Mexico. Extensive sandy areas are occupied by psammobiomes with pine forests containing *Pinus clausa,* P. *taeda,* P. *australis,* and others, with an undergrowth of evergreens in places. In addition, there are the extensive *Taxodium-Nyssa* swamp forests (hydrobiomes) and two types of helobiome, the evergreen

Persea-Magnolia moor forests and the heather moors with Venus's-flytraps *(Dionaea muscipula)*. Large expanses of the coast itself are taken up by salt marshes (halobiomes).

In South America, the evergreen forests of eastern Brazil extend far to the south, from tropical to subtropical and even warm-temperate. The tropics end on the coast between Porto Alegre and Rio Grande. Even in Misiones and Corrientes in northern Argentina, the forests are subtropical, and along the great rivers, the Paraná and Uruguay, the forests penetrate the Pampa region as gallery forests. On the coast, ZB V ends near La Plata, and ZB VI is altogether lacking. On the high plateau more than 500 m above sea level in southern Brazil, the coniferous forest is composed of *Araucaria angustifolia* and has to be allocated to ZB V. Generally speaking, the forests in this region have been greatly reduced in area by clear-cutting.

The southeastern coast of Africa is also exposed to the southeast trade wind. The obstacle presented by the Drakensberg Mountains causes large amounts of precipitation, and tropical-subtropical forests are found in the coastal regions as far as East London. The region along the southern coast can be described as warm-temperate. In earlier times, the forests stretched without interruption up to the eastern slopes of Table Mountain near Cape Town. Nowadays, however, the larger part has been cleared or is secondarily occupied by the fynbos of ZB IV. The only large forest reserve, near Knyshna, contains tall, ancient specimens of *Podocarpus* trees and a large number of evergreen deciduous trees, including the "stinkboom" *(Ocotea foetans)*, which yields valuable timber.

The situation in Australia and New Zealand will be dealt with in the next section.

Scarcely any ecological studies have been made on ZB V, and therefore no details concerning its ecosystems can be given. Investigations would be particularly difficult since the majority of the forests contain a large diversity of species, and conditions are very favorable for growth. The decisive factor is undoubtedly competition, and this is not easy to estimate.

3. Biome of the *Eucalyptus-Nothofagus* Forests of Southeastern Australia and Tasmania

The wet tropical-subtropical evergreen rain forests on Australia's east coast have some Indomalayan elements foreign to the Australian realm. They extend as far as southern New South Wales on rich, usually volcanic soils. Only in southern Victoria and in Tasmania does the Australian element dominate with the genus *Eucalyptus,* combined, however, with some antarctic elements. Here, in this humid climate which lacks a cold season (Fig. 104), *Eucalyptus regnans* may reach a height of 100 m (earlier reports of 145 m can no longer be confirmed). *Eucalyptus gigantea* and *Eucalyptus oblique* attain similar heights. The most important of the antarctic elements are the evergreen *Nothofagus cunninghamii* and the tree fern *Dicksonia*

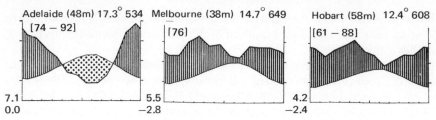

Fig. 104. Climate diagrams from the sclerophyllous region of South Australia and the warm-temperate regions of Victoria and Tasmania. (From Walter and Lieth 1967)

antarctica, as well as a series of other species in Tasmania. The vegetational types correspond to the occurrence of fire and are as follows:

1. In the wet parts of western Tasmania, where no fires occur, a tree stratum of *Nothofagus* and *Atherosperma moschata* (Monimiaceae) develops to a height of 40 m, and below it is a 3m-high stratum of the fern *Dicksonia* which is able to grow even when receiving only 1% of the total light. Hymenophyllaceae and mosses abound as epiphytes in the wet forests.

2. If forest fires occur every 200–350 years, mixed forest develops, consisting of three strata. Apart from the two strata mentioned above there is a further loftier stratum consisting of the three largest species of *Eucalyptus* (75 to 100 m). That the trees in this stratum are all of one age is an indication that germination of their seedlings took place simultaneously over an extensive area after a fire. Although the *Eucalyptus* and *Nothofagus* strata are destroyed in the fire, the fruits can still open, and the undamaged seeds are dispersed and germinate. The more rapidly growing *Eucalyptus* overtakes *Nothofagus,* with the result that two tree strata are formed. Although the tree ferns lose their leaves in a fire, they are able to develop new ones at the tip of the stem. Regeneration of *Eucalyptus* below *Nothofagus* is impossible owing to lack of light and can take place only after another fire.

3. If forest fires occur once or twice in a century, then *Nothofagus* is replaced by other more rapidly growing but lower-statured tree species such as *Pomaderris, Olearia, Acacia,* etc.

4. A pure, low *Eucalyptus* vegetation results where fire occurs every 10 to 20 years.

5. Still more frequent fires lead to a degradation of the forests. An open moor results, with "button grass" *Gymnoschoenus (Mesomelaena) sphaerocephala,* (Cyperaceae) and scattered Myrtaceae shrubs, together with *Drosera, Utricularia,* and Restionaceae.

4. Warm-Temperate Biome of New Zealand

New Zealand's forests warrant special mention. Although both islands are relatively near to the Australian continent and were probably directly connected with it in geological past, this connection must have been interrupted before the flora of the Australian realm was fully developed. There is not a single native species of *Eucalyptus* or *Acacia* in New Zealand, and the Proteaceae are represented by only two species.

In the north of North Island, there are still subtropical forests consisting of coniferous *Agathis australis* and palms, and along the coast there are even mangroves with low *Avicennia* bushes. The forest species are melanesic elements of the Paleotropic realm.

Forests of this type occur even on South Island, although its climate is definitely temperate, despite the absence of a cold winter season in the lower-lying country. The coniferous genera *Podocarpus* and *Dacrydium*, which are distributed throughout the entire Southern Hemisphere, are very common here. The antarctic element, represented by five evergreen *Nothofagus* species, plays an important role in the forests of both North and South Islands. These mutually exclusive forest species form a mosaic for which there is no satisfactory ecological or climatological explanation. The plant cover gives the impression of not being in a state of equilibrium with its present-day environment. Rather than being differentiated by soils or other ecological factors, it seems that the vegetation reflects mostly historical factors. Seventeen hundred years ago, North Island was covered by a thick layer of volcanic ash. The first pioneers were Podocarpaceae disseminated by birds; they are gradually being replaced by forests containing tropical elements as well as by *Nothofagus* forest in some of the mountainous regions. In the Pleistocene, South Island was covered by large glaciers, so that, since *Nothofagus* spreads very slowly, the process of recolonization is still going on.

In the extremely humid fjord county of southwestern New Zealand, where the rainfall exceeds 6000 mm, the *Nothofagus* forests are similar in nature to those of southern Chile. A peculiarity is, however, represented by the bare strips, 2 to 6 m wide, suggestive of avalanches, found in the middle of the forest on steep slopes. When the weight of the tree layer becomes too great for the rocky slopes, then the entire vegetation, inclusive of roots and soil, slides down owing to the force of gravity. The naked rock which remains is then recolonized by lichens, mosses, and ferns, followed by shrubs and, finally, trees, until the process repeats itself.

The imported European red deer presents a great danger to the forests of New Zealand, where originally the only mammals were bats. It is impossible to control the multiplication of the deer, and they hinder the regeneration of the *Nothofagus* forests, so that the danger of erosion and flooding is greatly increased. The Australian opossum (kuzu), also imported, is equally dangerous. It has confined itself to a tree species growing at the timberline in the high mountains. It strips the trees of their leaves completely, thus bringing about their death and increasing the danger of soil erosion on the steep slopes.

New Zealand provides an example of the extreme danger of disturbing the natural equilibrium by introducing new plants or animals. The damage done is often irreparable.

VI Zonobiome of the Temperate-Nemoral Climate

1. Leaf-Shedding as an Adaptation to the Cold Winter

A temperate climatic zone with a marked but not too prolonged cold season occurs only in the Northern Hemisphere; apart from certain mountainous districts in the southern Andes and in New Zealand, it is absent from the Southern Hemisphere. The phenomenon of facultative leaf-shedding has already been met within the tropics. There the leaves are shed only when the water balance is disturbed by a lenghty period of drought, and this leaf loss decreases water losses (p. 75). Leaf-shedding in the temperate zone, however, is an adaptation to the cold season. It is not facultative, but obligatory, occurring even if the trees grow in a greenhouse where they are protected from the cold of winter. The factor responsible for setting off the change in color of the leaves in autumn, even before the first frost occurs, is unknown. It may partly be due to the decrease in day length. A remarkable fact is that the various species of trees turn color within a very short period of time in central Europe, according to the phenological calendar, between the 10th and 20th of October, with no sharp distinction between west and east or between west and east or between lower or higher situations in the mountains. (Trees near street lamps remain green longer.) An evergreen broad leaf is neither resistant to cold nor to winter drought, that is to say, to sustained temperatures below freezing. In a central European climate, *Prunus laurocerasus* (cherry laurel) invariably freezes during severe winters, and even a light frost causes the leaves to excrete CO_2 by day, which means that although respiration continues, photosynthesis is blocked. *Ilex aquifolium* (holly) is atlantic in its distribution and *Hedera helix* (ivy) subatlantic, both species thus avoiding the eastern continental regions with their cold winters. The same holds true of the broom species *Ulex* and *Sarothamnus*. The alpine rose *(Rhododendron)* and red bilberry *(Vaccinium vitisidaea)* can survive the cold only beneath a covering of snow.

Loss of the thin deciduous leaves in winter and protection of the buds from water loss represents a saving of material as compared with the freezing of thick evergreen leaves. It is essential, however, that the leaves newly formed in spring have a sufficiently long, warm summer period of at least 4 months to produce enough organic material for the growth and maturation of the lignified axial organs and formation of reserves for the fruits and buds of the following year. Even in their bare winter condition, the twigs lose water, the extent varying from species to species. For this reason, the central

European beech is not found in the zone with a cold east European winter, although the oak extends as far as the Urals. In extreme continental Siberia, the only deciduous trees are small-leaved birch *(Betula)*, aspen *(Populus tremula)*, and mountain ash *(Sorbus aucuparia)*, which has small pinnate leaves. Where the summers are too cool and too short, evergreen conifers replace the deciduous species. The xeromorphic needles of the conifers are more resistant to cold in winter, and when warmer weather returns in spring, they are capable of starting photosynthesis immediately. In this way, the short vegetational season can better be exploited. Whereas deciduous trees require a vegetational season lasting at least 120 days with a mean daily temperature above 10 °C, the conifers manage with 30 days. However, the resistance of conifers varies from species to species. The yew *(Taxus)* does not extend farther east in Europe than the ivy *(Hedera)*. The Scots pine *(Pinus sylvestris)* and the spruce *(Picea abies)* are very resistant, and *Abies sibirica* and *Pinus sibirica (Pinus cembra)* survive in Siberia. But the deciduous, needle-leaved larch *(Larix dahurica)* extends farthest of all into the continental arctic region (up to 72°40′ N) where it exploits the short summers in a state of high productivity. Whether the species with deciduous or those with evergreen assimilatory organs are more successful in competition and rise to dominance thus depends upon external conditions and the ecophysiological characteristics of the species themselves.

2. Distribution of Zonobiome VI

The climate of ZB VI, with a warm vegetational season of 4–6 months with adequate rainfall and a mild winter lasting 3–4 months, is especially suitable for the deciduous tree species of the temperate climatic zone. Such trees avoid the extreme maritime as well as the extreme continental regions and favor what is termed the *nemoral zone.* In the Northern Hemisphere, a climate of this kind, with the rainfall maximum occurring in summer, is found in eastern North America and eastern Asia, between the warm-temperate and the cold- or arid-temperate zones. It is also found in western and central Europe north of the Mediterranean zone where, as a result of the influence of the Gulf Stream, winter rains are replaced by evenly distributed rainfall or by rainfall with a summer maximum and the cold season is relatively short.

The Mediterranean winter-rain region with sclerophyllous vegetation covers a very considerable distance from west to east and is replaced by different vegetational zones to the north. In the maritime region on the Atlantic coast to the northwest of Gibraltar, elements of evergreen warm-temperate laurel forests occur, with the evergreen species *Prunus lusitanica* (related to *P. laurocerasus)* and *Rhododendron ponticum* ssp. *baeticum.* Also to be found here are *Quercus lusitanica* ssp. *canariensis (Q. mirbeckii), Drosophyllum lusitanicum,* an interesting insectivorous plant, and the tiny *Utricularia lusitanica,* in addition to the epiphytic fern *Davallia canaricum.* This type of vegetation, however, is rapidly succeeded by Atlantic heaths, which extend in the coastal region as far as Scandinavia and are

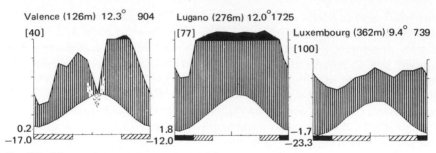

Fig. 105. Climate diagrams from the sub-Mediterranean zone (with no cold season), the warm and damp deciduous-forest zones, and the central European beech-forest zone. (From Walter and Lieth 1967)

replaced in the north by birch forests. Actual laurel forest are only found on the humid windward side of the Canary Islands (Teneriffe) or in the very similar Zonoecotone IV/V of northern Anatoloia. Farther to the east, a sub-Mediterranean zone is intercalated between the Mediterranean and nemoral zones. Although there are still winter rains, the summer drought is no longer pronounced, and frost occurs regularly in all months of the winter (Fig. 105, Valence).

Apart from one woody evergreen species *(Buxus),* all of the tree species in this zone are deciduous, e.g., *Quercus pubescens, Fraxinus ornus, Acer monspessulanum, Ostrya carpinifolia,* and the frequently cultivated chestnut *(Castanea sativa).* Consequently, this region cannot be termed a Mediterranean zone, but should preferably be considered as belonging to the deciduous-forest zone or, better still, as Zonoecotone IV/VI. To the northeast of this sub-Mediterranean zone is the steppe zone, which only farther north is replaced by various kinds of forest. In the Middle East, the Mediterranean sclerophyllous zone is succeeded by Mediterranean steppes and semideserts.

3. Atlantic Heath Regions

The Atlantic heaths represent stages in the degradation of deciduous forests, a process which has been going on since prehistoric times. Destruction is by now so complete that the heaths are often considered to be the true zonal vegetation. The fact that the soils are extremely poor and acid and capable of supporting only a weak kind of heath vegetation was formerly attributed to leaching, as a natural consequence of the humid climate. What has already been said with regard to tropical rain forests (p. 49) can also be applied here. As long as the natural forest vegetation remains untouched, leaching of nutrients from the biogeocenes does not occur, and the reserves of nutrients are mainly stored in the aboveground phytomass. But as soon as the forest is cleared and burned, the larger part of the mineralized nutrients is lost, and an impoverished soil results. If the ensuing heath vegetation is exploited or repeatedly burned, reforestation, in any case problematic, is rendered impossible. In uncolonized extreme oceanic regions on the Pacific coasts of northwestern North America, in the southwest of South America, and in Tasmania and New Zealand, areas which have similar temperatures and two to four times the rainfall of the Atlantic heath regions, we have seen forests growing in undisturbed luxuriance, with no sign whatever of degradation due to leaching of nutrients.

It is not easy to reconstruct the original composition of the west European forests. The more oceanic kinds of conifers were exterminated during the pleistocene and are absent from the European flora. In all probability, oak was the most abundant species (*Q. petraea* and *Q. robur,*) together with birch *(Betula)* in the north, as well as the evergreen species *Ilex acquifolium.* Heath, as an independent community, occurred only on shallow or peaty soils in the clearings and otherwise formed the forest undergrowth. Only after destruction of the forests did it take possession of the entire area.

In the southern part of the coastal zone, many species of broom dominate (*Ulex, Sarothamnus,* and *Genista* spp.), accompanied by various species of *Erica.* In the central districts, the broom species become scarce, *Ulex europaeus, Sarothamnus scoparius,* and *Genista anglica* being left as the most important representatives. At the same time, the dominance of Ericaceae greatly increases, especially *Calluna vulgaris, Erica cinerea,* and *E. tetralix. Empetrum, Vaccinium, Phyllodoce,* and *Cassiope* dominate in the north.

From $\frac{1}{4}$ to $\frac{1}{3}$ of Scotland is covered by *Calluna* heath, which is regularly burned. The iron podsol soils often have a cemented B-horizon forming hardpan ("ortstein"). *Calluna vulgaris* is absolutely dominant. It is a dwarf shrub achieving a height of about 50 cm which develops a dense web of roots in the upper 10 cm of soil, with a few roots going down 75–80 cm to the hardpan. Its very small leaves are sessile on short twigs, of which the larger portion is shed in autumn, which thus reduces the danger from winter drought in the cold season. The annual litter production of a dense community amounts to 421 kg/ha. If the heath is burned every 30 years, the development of the vegetation falls into the following three phases, each lasting 10 years:

1. The reconstructive phase of the dwarf shrub layer after the fire. Part of the nutrients is bound in the litter.
2. The phase of maturity, during which litter production increases and the rate of increase in phytomass drops.
3. The phase of degeneration, during which litter production remains constant but decomposition increases until a state of equilibrium is attained. After 35 years, the standing phytomass amounts to 24,000 kg/ha and the litter to 17,000 kg/ha.

As a rule, the heath is burned again after 8–15 years, without the phase of degeneration having been reached. In climates as humid as that of Scotland, fires are caused only by man. Natural fires caused by lightning very rarely occurred in the original forests, so that degradation has only taken place where man has intervened. All of the transitional forms from heath to moor are found. Four stages associated with increasing moisture are listed, the species being arranged in the order of decreasing abundance:

1. *Erica cinerea, Calluna vulgaris, Deschampsia flexuosa, Vaccinium myrtillus.*
2. *Calluna vulgaris, Erica tetralix, Juncus squarrosus.*
3. *Erica tetralix, Molinia coerulea, Nardus stricta, Calluna vulgaris, Narthecium ossifragum.*
4. *Erica tetralix, Trichophorum caespitosum, Eriophorum vaginatum, Myrica gale, Carex echinata.*

In Scotland, the heath is utilized for hunting and also for extensive sheep grazing, with 1.2–2.8 ha per animal. In Germany on the "Lüneburger Heide," which is also purely anthropogenic in origin, farming was formerly common (buckwheat cultivation). The upper 10 cm of the raw humus layer was dug out in squares, used as stable litter, and returned to the fields as manure. This process hindered reforestation. Nowadays the heath area is no longer cultivated, and forests have grown up from birch and pine seeds blown in by the wind. In some places, the heath is systematically reforested.

Apart from heaths, bogs are of frequent occurrence in extreme maritime regions, where the climate is oceanic, with small fluctuations of temperature. In Ireland, for example, the January temperatures are 3.5–3.7 °C, and those for July are 14–16 °C. Frosts may occur, but snow covers the ground only 3–10 days of the year. Rainfall amounts to 350–1000 mm annually and is evenly distributed, varying from year to year by 25% at the most. Owing to the prevalence of cloudiness, the amount of sunshine is only 31% of the possible maximum. Under such conditions, the danger of bog formation after deforestation is very great. Since a low, herbaceous vegetation loses less water owing to transpiration than does the tree stratum of the forest, a rise in groundwater level is noticeable after deforestation in humid regions. This, in turn, favors the growth of peat mosses, principally *Sphagnum* spp., although *Rhacomitrium lanuginosum,* too, is widespread. In regions where it rains on more than 235 days of the year, bogs may cover the entire area even in this rolling landscape. Such "blanket bogs" are found in West Ireland, Wales, and Scotland, where the largest encompasses 2,500 km^2.

In regions further removed from the Atlantic coast, heath formation presents no danger, since despite the fact that *Calluna* has very small leaves with a thick cuticle and stomata lying in hairy grooves, heather species are very sensitive to frost. The leaves of *Calluna* differ from true xeromorphic leaves in that the mesophyll has a very loose structure. Transpiration is relatively active in summer if the water supply is adequate, and in shady habitats, values can equal those of *Oxalis acetosella,* calculated on a fresh-weight basis. When water is scarce, however, transpiration is sharply decreased. These properties, nevertheless, do not suffice to prevent water losses during the long periods of frost. Even in the mild winters in Heidelberg, in southern Germany, *Calluna* dries out because it lacks a protective covering of snow. In the north, it is only found where there is a covering of snow each year.

Inland, in western Europe, heath is to be found in patches on the western slopes of the low mountains with an oceanic climate (Ardennes, High Venn, Eifel, Vosges, and even in the Black Forest on the Feldberg). It also extends as a narrow strip along the southern coast of the Baltic.

4. Deciduous Forests as Ecosystems (Biogeocenes)

A deciduous forest is a multilayered plant community, often consisting of one or two tree strata, a shrub stratum, and an herbaceous stratum. Numerous hemicryptophytes grow in the latter, as well as many geophytes which develop only in the spring. Illumination on the forest floor is too weak for the development of therophytes, i.e., annuals. A mossy ground cover is lacking since it would be covered by the falling leaves, and mosses are only found on rocks or tree stumps projecting above the ground. These groups of plants constitute synusiae (p. 15).

There are no virgin forests on euclimatopes in the western European deciduousforest zone, and the structure of the forests is determined by the type of management practiced. From the forestry point of view, it is the woody species that are of importance, and the herbaceous layer is influenced only indirectly. If forest grazing is practiced, on the other hand, it is the herbaceous layer which is changed by selective grazing of cattle, which also are a danger to the tree saplings. High forests (hochwald) run on a rational basis approach virgin forest, although they differ basically in the small number of species in the tree stratum, the fact of having trees all the same age, the lack of rotting wood on the forest floor, and their homogeneous structure (virgin forest is usually of a mosaiclike structure).

Managed beech forests are pure stands with only a herbaceous stratum in addition to the trees. Oak forests, on the other hand, are usually mixed stands of various deciduous tree species and possess a shrub stratum. Among the various types of deciduous-forest biogeocenes, a western mixed forest in Belgium (p. 9) and eastern oak forests on the forest-steppe margin have been investigated in detail. Further investigations have been carried out, but the results are not yet available [projects of the International Biological Program (IBP)].

The active layer in deciduous forests is the tree canopy, where both direct radiation from the sun and diffuse radiation are to a large extent transformed into heat. Of the incoming radiation, 17% is reflected by trees in leaf and about 11% by trees in the leafless condition, but in both cases the radiation reflected is less than that from meadows and cultivated land (25%). Only a minute portion of the daylight penetrates the forest vegetation. The following figures apply to eastern European oak forests (13- to 220-year-old stands):

In young forests in full leaf only 1.2% of the daylight penetrates halfway down, and 0.6% penetrates to the forest floor, whereas in very old forests, the figures are 20 and 2%, respectively. The mean daily temperature of the canopy in summer is 2 °C higher than that on the forest floor; the mean daily maximum, 11 °C higher; and the mean daily minimum, 3 °C lower. The mean humidity of the air is 98% on the ground and drops as low as 77% with increased height. The wind velocity in the forest is low, and since the forest floor is protected from direct radiation, the air in the forest remains cooler during the day than that in open stands.

The crowns of the trees intercept 11–12% of the rain, and the remainder either drips down or runs down the trunks. Where snow has accumulated on the forest floor in winter, it melts only in spring. The resulting water seeps almost entirely into the litter layer, whereas on open ground it would run off the surface of the still-frozen ground. In summer, transpiration of the tree stratum is so intense that the groundwater receives no additional input from forested areas. Water losses from the herbaceous stratum are 5–6 times smaller. A well-developed deciduous forest of the forest-steppe region uses nearly all of the incoming precipitation, whereas a beech forest in central Europe uses only 50–60%, although in summer months there is no surplus.

The productivity of a forest depends to a large extent upon its Leaf Area Index (LAI), that is to say, upon the ratio of total leaf area of the tree stand to the ground area covered by it. This ratio is limited to a maximum value, above which it may not rise, since otherwise the lower, overshadowed leaves would not be able to maintain a positive balance of CO_2 assimilation. This maximum, however, not only depends upon light intensity but also decreases in the face of inadequate supplies of water or nutrients. The LAI of a pure oak stand is 5–6 (higher in wet years), and in mixed stands with a good water supply, inclusive of all tree species and shrubs, it can exceed 8 m^2/m^2.

The productivity of the tree stratum has been studied in detail in a central European beech forest. The results are expressed in tons of dry substance per hectare and year for a 40-year-old beech stand in Denmark (Müller and Nielson 1965).

Gross production of the assimilating leaves = 23.5 t/ha
Respiratory losses (leaves 4.6 t/ha, branches 4.5 t/ha, roots 0.9 t/ha) = 10.0 t/ha
Annual production of leaves (2.7 t/ha), branches (1.0 t/ha), litter and roots (0.2 t/ha) = 3.9 t/ha
Wood production, aboveground (8.0 t/ha) and underground (1.6 t/ha) = 9.6 t/ha

On an average, 6 t/ha of the maximum of 8 t/ha of trunk wood is utilizable, which is equivalent to 11 m^3. The same weight of wood is produced by the spruce, but it occupies a mean volume of 17 m^3. Figure 106 shows how the production figures change with the age of the beech stand. Biomass figures for an oak forest in Belgium have already been quoted (p. 9).

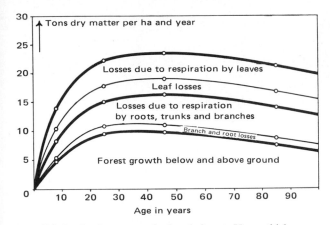

Fig. 106. Production curve of a beech forest. Upper thick curve: gross production; thin curve beneath it: primary production; middle thick curve: assimilates from leaves; thin curve below it: gross wood production; lowest thick curve: net wood production. (From Möller et al. 1954)

a) Example of a Primeval-Like Mixed Oak Forest Ecosystem

Of all the forests investigated so far, the most natural conditions are found in the mixed oak forest on the Vorskla, the left tributary of the middle Dnieper (IBP project of Leningrad University). The forest extends over 1000 ha, and 160 ha is protected and is an almost primeval forest consisting of 300-year-old trees. Here, at its southeastern limits of distribution, the deciduous forest receives ample sunshine and just sufficient rain. The climatic conditions are illustrated in Fig. 107. See p. 196 for the soil conditions. The forest possesses three ill-defined tree layers, and the shrub layer is almost nonexistent, but a herbaceous layer is present (Fig. 108).

According to Goryschina (1974), the phytomass (in tons per hectare) is as follows:

Aboveground	306.7 (leaves, 3.7; twigs and branches, 71.2; trunks, 230.8)
Subterranean	124.9
Total	431.6 (plus 10.7 for the herbaceous layer)

Losses due to dead wood are scarcely less than the increase in wood over the same period, i.e., net phytomass increase is practically zero, which is the situation in a primeval forest at its optimum phase.

Primary production is 8.9 t/ha, or 9.6 t/ha with the herbaceous layer. Subterranean production was not determined. As would be expected in a semiarid climate, primary production is lower than in western deciduous forests. The water budget will be dealt with later (p. 221).

If the chemical energy bound in the process of production is related to the incoming radiation energy for 1 ha of forest, the figures of 2% for gross

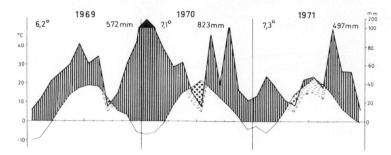

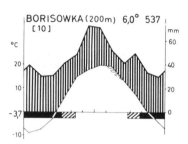

Fig. 107. Forest on the Vorskla. Climatogram for the years 1969–1971, showing that, in the steppe-forest ecotone, short drought periods occur at different seasons from year to year (the climate diagram of the next nearest station, Borisowka, does not show any such drought periods). The winter of 1968–69 was cold with little snow, whereas the following winter was warmer with more snow. The rain minimum in summer 1970 is anomalous. (According to Goryschina, from Walter 1976)

production and 1% for primary production are obtained. One-third of the incoming energy is used up in transpiration, and the remainder is transformed into heat. The leaf mass and leaf area formed annually increase rapidly during the first 20 years, but as soon as the canopy closes, leaf mass and LAI remain almost constant. The height of the canopy above the ground is all that changes with upward growth of the trees. The litter consists of the old leaves and twigs, and these, together with dead roots, constitute the total litter of the forest.

Only the new wood is accumulated, so that the standing phytomass of the forest steadily increases even with great age and may exceed 200 t/ha for a 50-year-old stand and 400 t/ha for a 200-year-old stand. The following figures for average yearly wood production, related to age, were found for eastern

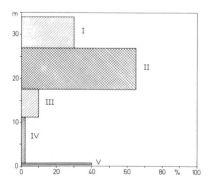

Fig. 108. Diagrammatic representation of the forest layers of the Tilieto-Quercetum aegopodiosum (300-year-old stand in the forest on the Vorskla): I, upper tree story of oldest oaks *(Quercus robur),* II, middle tree story of *Tilia cordata* (up to 200 years), *Acer platanoides,* and *Ulmus scabra (montana);* III, lower tree story chiefly with *Acer* and some *Ulmus* and *Quercus;* IV, shrub layer with tree saplings weakly developed; V, herb layer. There is no moss layer on the ground. Abscissa, horizontal projection of the various layers in percentage of total area; ordinate, height in meters. (According to figures given by Goryschina 1974)

European oak forests: 13 years, 3.8 t/ha; 22 years, 3.6 t/ha; 42 years, 4.3 t/ha; 56 years, 4.7 t/ha; 135 years, 0.4 t/ha; and 220 years, 0.0 t/ha.

Litter, too, accumulates in the forest, until a state of equilibrium is reached, at which time as much litter is mineralized annually as is produced in the course of 1 year. A portion of the most important minerals is bound in the litter (N, P, Ca, K), and thick layers of raw humus are therefore undesirable. Removal of litter by farmers for use in stables is particularly detrimental because the resulting loss of nutrients, particularly calcium, leads to impoverishment and acidity of the forest soils and a consequent decrease in wood production. Decomposition of litter also involves the mineralization of nitrogen compounds. The larger part of the nutrients in the lower, decomposing humus layer is available to the tree roots, which are therefore abundant in this soil horizon. Next to the water supply, the soil fauna is very important for forest productivity, whereas the part played by animal organisms aboveground is negligible, and only a few percent of the biomass is lost through insect feeding.

b) Ecophysiology of the Tree Stratum

The size of a tree renders it an unsuitable object for experimentation. Its form is to a large extent dependent upon its surroundings. The crown of a solitary tree is usually dome-shaped or spherical, whereas if the tree is part of a dense stand, the crown is usually very small. Since the leaves are arranged in several layers, the outer ones are exposed to the full daylight, and the inner ones grow in the shade. A distinction is therefore made between sun leaves and shade leaves, which differ in anatomical-morphological and ecophysiological properties and are linked by intermediate forms. Sun leaves are smaller and thicker, have a denser nervature, and possess more stomata per square millimeter on the undersurface. In other words, sun leaves are more xeromorphic than the large, thin shade leaves.

Structural differences between sun and shade leaves result from differences in the water balance at the time when the buds for the next spring are being formed. Twigs exposed to the sun transpire more actively, as indicated by an increase in cell-sap concentration. Cell-sap concentration of the sun leaves of a beech tree amounts to 16.3 atm, and that of the shade leaves is 11.6 atm (cf. also p. 135). The CO_2 assimilation also differs. In laboratory experiments, it has been established that, in darkness, shade leaves respire less actively than sun leaves: in beech, the shade leaves give off only 0.2 mg of $CO_2/dm^2/hr$ as compared with 1.0 mg for the sun leaves. This explains the finding that, in spring, the light compensation point (where respiration = gross photosynthesis) of shade leaves is 350 lux and that of sun leaves 1000 lux. Photosynthesis increases proportional to light intensity until a maximum is reached, which for shade leaves is 20% of maximum daylight and for sun leaves is about 40%. Thus shade leaves are better able to utilize lower light intensities and sun leaves higher intensities, although even the sun leaves do not appear to utilize available daylight to the full. The above figures apply to leaves oriented at right angles to the incoming light. Sun leaves near the apex of the tree, however, are nearly always rather steeply inclined. This protects the leaves from overheating and helps to reduce water losses and also means that more light can penetrate the outer canopy, to the advantage of the lower leaves. Leaves in the deep shade are always at right angles to the incoming light, and thus even with a LAI of 5 or more, a mean positive production is possible.

An exact production analysis was carried out by Schulze (1970), who directly measured the CO_2 assimilation of the beech *(Fagus)* in its habitat. It was shown that the annual production of sun leaves and shade leaves is the same because the shade leaves remain active longer in autumn.

If illumination is continuously below a certain minimum, respiration is no longer compensated for by photosynthesis, losses of material occur, and the leaves turn yellow and are shed. This minimum, expressed in percentage of full daylight, varies from one tree species to the next. A "shade" tree with a dense crown, such as beech, has a low light minimum (1.2%), and "light" trees, such as birch and aspen, with a thinner crown, have a higher light minimum (11%). The figures for species such as maple and oak lie somewhere between those mentioned above. This light minimum is valid for the crown of the tree and does not necessarily coincide with that light minimum which must be exceeded if the tree seedlings are to develop on the forest floor, although the values are correlated with each other. Beech seedlings require little light, whereas birch seedlings need at least 12–15% of total daylight.

Light conditions are of vital importance to trees in competition with one another. Light-demanding trees grow up within a few decades in a clearing, and under their canopy, shade-tolerating trees germinate and gradually grow higher, producing in turn a canopy so dense that the "light" trees are incapable of regeneration. In time, it is the species tolerating the most shade which achieves dominance, providing that the other local conditions are appropriate.

The zonal forests in central Europe consist of beech *(Fagus sylvatica)*. Only on very poor soils, places where groundwater is high, or in the drier biotopes is beech unable to compete successfully. In the western parts of eastern Europe where the climate is too continental for beech, it is replaced by another shade-loving species, the hornbeam *(Carpinus betulus)*. Still further east hornbeam is replaced by oak *(Quercus robur)*. On the eastern limits of distribution of deciduous forests in the southern Urals, the lime *(Tilia cordata)* becomes predominant.

c) Ecophysiology of the Herbaceous Layer (Synusiae)

The microclimate of the forest floor is vastly different from that of an open habitat. When the forest is in leaf, the light intensity on the forest floor is weaker, the temperature is more moderate, and the humidity of the air and upper soil layers is greater than that outside the forest. For these reasons, the herbaceous plants of the forest are shade-tolerating and hygrophytic, and their cell-sap concentration is very low.

On a clear day, light conditions on the forest floor can be very heterogeneous. Single rays of sunshine falling through the tree canopy cause sun flecks, and as the sun moves across the sky or the branches are moved to and fro by the wind, these flecks change their position and intensity.

If a leaf of a herbaceous plant is hit by a sun fleck, its illumination can rise 10- or even 30-fold, a factor of great significance for the plant's photosynthesis. It is therefore preferable, in determining the amount of light received by herbaceous plants in percentage of total daylight, to carry out the comparative measurements on a bright day with a more or less even cloud cover. Such measurements, however, provide us with no more than preliminary information. It would be better to measure the sum total of the daylight falling on a certain place on the forest floor with the aid of a selfregistering light meter.

Before the trees come into leaf, the herbaceous stratum is very adequately illuminated, but as the trees come into full leaf, the situation gradually deteriorates. The following values were obtained between March and June in an oak-hornbeam forest: March 12, 52%; April 15, 32%; May 10, 6.4%; and June 4, 3.7%. For a beech forest, the values were as follows: mid-May, 6%; one week later, 3%; and June 7, 1.5%.

The favorable light conditions prevailing before the trees come into leaf are exploited by the spring geophytes (*Galanthus, Leucojum, Scilla, Ficaria, Corydalis, Anemone,* etc.). They profit from the fact that, even in April, the litter layer in which they root is warmed up to 25–30 °C because of its uninterrupted exposure to the sun's rays. Geophytes benefit from the fact that heat capacity of the air-containing litter layer is small and, as a result, the temperature conductivity is very good. The trees come into leaf later because the deeper soil layers in which they root are slow to warm up.

In the short, early-spring season, geophytes flower and fruit and store up reserves in their underground storage organs for the coming year. When the trees are in leaf, the leaves of the geophytes turn yellow, and a dormant period begins for them. The death of the leaves is not, however, due to the deeper shade but is the expression of an endogenous rhythm; the leaves die even sooner in light. Apparently, geophytes are just the right plants to fill the vacant ecological niche in deciduous forests.

Spring geophytes of this kind are also termed ephemeroids. Although their vegetational period is just as brief as that of the annual ephemerals, they are perennial species with underground storage organs. The ecological behavior of the spring geophytes is very similar, and their developmental cycles are almost identical, so that they can be considered to form a functional unit, termed a synusia by ecologists. Synusiae have no independent material cycle and are thus merely constituent parts of the various ecosystems (see p. 15).

The synusiae of deciduous forests have been investigated in the abovementioned mixed oak forest on the Vorskla (Goryschina 1974; see also Walter 1976). The following five examples were studied in detail:

1. Ephemeroids: *Scilla sibirica, Ficaria verna, Corydalis solida, Anemone ranunculoides.*

2. Hemiephemeroids: *Dentaria bulbifera.*

3. Early-summer species: *Aegopodium podagraria, Pulmonaria obscura, Asperula (Galium) odorata, Stellaria holostea.*

4. Late-summer species: *Scrophularia nodosa, Stachys sylvatica, Campanula trachelium, Dactylis glomerata, Festuca gigantea.*

5. Evergreen species: *Asarum europaeum, Carex pilosa.*

The various synusiae grow at different light phases on the forest floor (Fig. 109) and are thus successively exposed to different conditions of transpiration, to which they adapt (as was discussed for dessert plants p. 135): *Aegopodium,* for example, first develops small light leaves but by summer has developed large shade leaves, which are replaced in autumn by very small, xeromorphic, cold-resistant leaves which persist throughout the winter (*Aegopodium* has no resting stage in winter). Exactly the same situation can be seen in *Stellaria* and *Asperula,* except that the different types of leaves develop successively on the same vertical shoot axis.

The assimilate economy, or rather the way in which assimilates are used, varies considerably from one synusia to the next. *Scilla* uses all of the assimilates stored in the bulb for formation of the flowering shoot and leaves, and only toward the end of the short vegetational period are the newly formed assimilates directed into the young bulb for use in the following year.

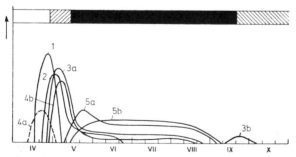

Fig. 109. Schematic comparison of the course of photosynthesis in the various synusiae of the herbaceous layer of a mixed-oak forest: (1) ephemeroids; (2) hemiephemeroids; (3) early-summer species (a, spring leaves; b, autumn leaves); (4) evergreen species (a, leaves from previous year; b, young leaves); (5) late-summer species (a and b, slightly different types). Ordinate: relative intensity of photosynthesis. Above: different illumination phases on the forest floor: white, light phase; hatched, two transitional phases; black, shade phase. (According to Goryschina, from Walter 1976)

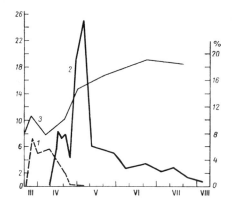

Fig. 110. *Asarum europaeum.* Photosynthesis curve of (1) overwintered leaves and (2) young leaves (in milligrams of CO_2 per square decimeter and hour) and (3) starch content of rhizomes (in percentage of dry weight). (According to Goryschina, from Walter 1976)

Dentaria, on the other hand, quite soon begins to lay up reserves in the rhizome and thus requires more time for flowering and fruit formation. *Aegopodium* exhausts its small reserves in order to produce the light leaves which then assimilate CO_2 so intensively that, by the beginning of May, they are able to lay up new reserves while at the same time delivering sufficient assimilates for formation of the shade leaves, after which they die off. The other species of the same synusia behave in a similar way.

In its large tuber, the late-summer plant *Scrophularia* stores much water but very few organic reserves for the formation of the first leaves. Assimilation of these leaves in the shade is so low that the shoot is not fully grown and in flower before autumn, and the fruits ripen late.

Figure 110 shows the assimilate economy of *Asarum.* Leaves from the previous year die off as soon as photosynthetic activity is fully established in the young leaves.

The phytomass of all the synusiae together is governed by the water budget of the upper layers of the soil, as can be seen from Fig. 111. Although

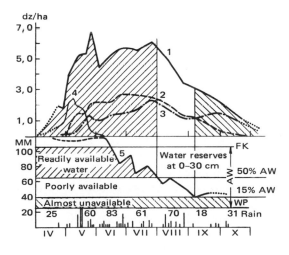

Fig. 111. Dependence of the phytomass of the herbaceous layer on the water stored in the upper 0–30 cm of the soil layer in mm (5), and the availability of the water: FK, field capacity; WP, wilting point, AW, available water. Phytomass: (1) herb layer, (2) *Aegopodium podagraria,* (3) *Carex pilosa,* (4) ephemeroids. (According to Goryshina, from Walter 1976)

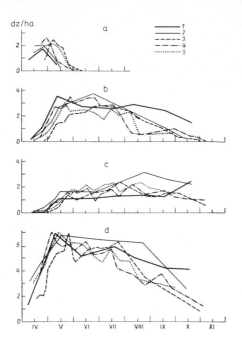

Fig. 112. Changes in phytomass of the various synusiae and of the total herbaceous layer from April to September in the years 1967–71 (air-dry mass in dz/ha, or 100 kg per hectare): (a) ephemeroids, (b) *Aegopodium podagraria*, (c) *Carex pilosa* (young shoots), (d) total herbaceous layer. (1) 1967; (2) 1968; (3) 1969; (4) 1970; (5) 1971. (According to Goryschina, from Walter 1976)

the total phytomass of the herbaceous plants is not large, its significance for the ecosystem lies in the fact that it is rapidly mineralized and thus contributes to a quick cycling. Leaf litter of trees, however, disintegrates slowly, and the nutrients contained in it are not available until the following year. The fluctuations in the phytomass of the herbaceous layer during several years is illustrated in Fig. 112.

The majority of species of the herbaceous layer are hemicryptophytes, which means that their regenerative buds form at the base of the shoot and spend the winter just at the ground surface, protected by a covering of autumn leaves and sometimes snow as well.

Illumination values have been determined for central European species. These plants possess an illumination maximum (L_{max}), since they are not found in full light, and minimum (L_{min}), since they avoid the deepest forest shade. The following limiting values, in percentage of total daylight, are given as examples (L_{max}–L_{min}): *Lamium maculatum*, 67–12%; *Lathyrus vernus*, 33–20%; *Geranium robertianum*, 74–4%; and *Prenanthes purpurea*, 10–5% (sterile to 3%).

The L_{max} value is dependent upon the water supply. Hygrophilic species require damp soil and cannot tolerate a high saturation deficit of the air such as would occur under conditions of full illumination, although in mountainous regions, where the humidity is higher, these plants are often found growing in meadows rather than in the forest. The fact that the species of the forest floor cannot compete with sun-loving plants in open habitats may also influence the L_{max} value.

A few species otherwise capable of tolerating full daylight can also be found in the forest, although there they are poorly developed and often sterile (the strawberry *Fragaria vesca,* for example). If the trees are felled, such plants develop luxuriantly in the clearings, flower, and bear abundant fruit. The L_{min} value represents the starvation limit, the light just sufficing for the production of the organic material indispensable for development. Plants which remain sterile manage with less light, as do the cryptogams (ferns and mosses). In general, the so-called "dead" shade of the forest commences at an illumination of 1%, below which point only the fruiting bodies of heterotrophic fungi or saprophytic angiosperms are to be found.

As the edge of the forest is approached and light conditions improve, the plants are better developed. Productivity of the herbaceous layer increases linearly with mean light intensity on the forest floor.

The illumination in percentage of daylight is not a good means of estimating photosynthesis because in the presence of heavy clouds the light may be very weak, whereas on a clear day it can be very much stronger. This is the reason why the production of the plants of the forest floor is negative in dull weather, or in other words, that the loss of carbohydrates by respiration in 24 hours is higher than the production by photosynthesis during such a day. On bright days, the balance is positive. It is only essential, however, for production to be positive over the entire season of growth. Certain plants of the forest floor (tree seedlings, *Oxalis, Asperula, Asarum, Viola,* and ferns) adapt themselves to the unfavorable light conditions prevailing after the trees are in leaf by drastically reducing their respiration when they reach the starvation point. This results in a lowering of the light compensation point and better production.

In other species with evergreen leaves, the point of light compensation remains unaltered. These plants live from their reserves in the summer and replenish them in autumn or spring, when the forest is bare (*Stellaria holostea, Hedera,* etc.).

Another factor of importance in the herbaceous layer is competition from the roots of the trees. Water plays a vital role in the dry forest regions bordering the forest-steppes. Trees, with their higher cell-sap concentration, are able to develop larger suction tensions in their absorbing roots, so that they are better able to obtain water from the soil than are the herbaceous plants. As a result, the floor of beech forests is bare (Fagetum nudum). If the roots of the trees are severed, thus excluding them from competition, herbaceous plants develop, which proves that water and not light is the limiting factor.

Trees also extract nutrients, especially nitrogen, on poor and very shallow soils, and herbaceous plants are obliged to make do with what is left. Only plants with low nutrient requirements, like *Luzula luzuloides, Deschampsia flexuosa, Potentilla sterilis, Vaccinium myrtillus, Calluna vulgaris,* etc., are found in such forests.

d) The Long Cycle (Consumers)

Food chains are of such diversity, and their connections of such complexity, that they have not yet been completely elucidated for any single ecosystem.

Plants are attacked by a variety of parasites, chiefly fungi, and by a large number of insect pests. The different plant organs serve as food for the various herbivores that constitute the food source of predators of the first order, both large (birds and mammals) and small (invertebrates). These predators, in turn, are consumed by predators of the second order, e.g., birds that catch predatory insects. Finally, there are predators of the third order, as for example the ticks and fleas on foxes, which feed on predatory birds. Only a very small part of the chemical energy in the food of animal organisms is transformed into secondary production, i.e., into animal body substance. The larger part is lost in excrement or used up in respiration.

So far, only isolated links in these complicated nutritional chains have been studied. Upon closer inspection, the leaves or other plant organs are often seen to be damaged. Twenty different insect species are known to live off the leaves, buds, bark, and wood of the oak alone, and the number of gall-forming insects specializing on oak trees is extremely large. The most important group of herbivores, the rodents, have been the subject of an exact quantitative investigation in the oak forest on the Vorskla. Numbers of individuals per hectare fluctuated within a single year from 70–90 to 400– 500. During the years 1967–1969, 4283 specimens of *Clethrinomys glareolus* and 864 specimens of *Apodemus flavicollis* were marked. Observation of their feeding habits revealed that they consume about 1% of the herbaceous layer phytomass (as well as fruits and fungi) in summer, but the damage done is even greater owing to the fact that many of the plants that have merely been nibbled by the rodents later wilt or die.

In addition to this, food hoards involving 20–45 kg dry mass are collected and not entirely used up. In 1968, 250 food nests of this kind were counted per hectare. In late autumn, the rodents collect the acorns that constitute their main source of food for 210 days of the year. A single animal can carry 10–13 kg of acorns into a storage chamber, and this food is only partially used up. Additional damage is caused by gnawing at tree seedlings, 8–46% of which die as a result.

The burrowing activities of rodents have an important and useful effect upon the environment. In 1967, 3350 entries to subterranean quarters were counted on 1 ha. The 2525 heaps thrown up as a result amounted to a volume of 2.84 m^3 and covered 1.18% of the total area. The total length of the surface passages was 2.9 km and that of the subterranean passages was 7.9 km. These passages occupied 3.4% of the whole area and accounted for 0.87% of the volume of a soil layer 10 cm thick. Both texture and chemistry of the soil are altered by burrowing activities. The earth thrown up is richer in nutrients and humus, its reaction is neutral, and aeration is improved, all of which benefit the forest vegetation.

e) Decomposers in Litter and Soil

The larger part of the annual waste of a deciduous forest forms a layer of dying leaves (litter) on the ground, which is immediately attacked and broken down by microorganisms, fungi and bacteria. The small animals such as insect larvae and other arthropods, but mainly earthworms, in whose excrements the bacteria are particularly active, feed on the litter as saprophages, decomposing it in the process and rendering it more accessible to the microorganisms. Exact quantitative studies on the activity of such animals in three deciduous-forest stands have been carried out by Edwards et al. (1970). Leaf fall begins gradually in autumn when the days begin to shorten and assimilation of the shade leaves is no longer adequate. Within a few days, all the leaves have fallen, and by about October 20th leaf fall is complete. Sugar, organic acids, and pigments are washed out of the leaf litter by rain, and the dead leaves turn brown.

The lower the C:N ratio in the litter, the more rapidly mineralization proceeds. By June, birch litter has lost about $\frac{4}{5}$ of its dry weight; lime litter, about $\frac{1}{2}$; and oak litter, which is not so readily broken down, only about one-quarter.

Mineralization of litter does not proceed to completion: humus matter is produced which, when saturated with calcium, forms the mull horizon containing large numbers of lumbricids (earthworms) or, under acid conditions, forms the moder horizon containing oribatids (mites) and collemboles (springtails). In extreme cases, if the reaction is very strongly acid, a raw humus layer accumulates, in which large quantities of fungal hyphae, but no animal organisms, are present.

Water conditions in the litter layer of the forest on the Vorskla, with its semiarid climate, present a special problem. In the litter, which consists mainly of oak leaves in various stages of decomposition (the leaves do not decompose easily), a lower layer (in an advanced state of decomposition) and an upper layer (less decomposed) can be distinguished. The maximum water content of the litter layer amounts to 1 mm, but because it often dries out regardless of the water in the upper soil layers, the litter itself contains no roots. The absence of capillary connections between litter layer and soil prevents evaporation of water from soil to air, but even if the litter dries out, the air within it remains water-saturated (hydrature about 100%) so that poikilohydric organisms (fungi, bacteria) remain continuously active.

Satchell (from Duvigneaud 1974) reported the activity (i.e., respiration) of the individual groups of soil organisms in an English oak forest on chalky soil as follows (values are kilocalories per square meter and year): diptera larvae, 9; collemboles (springtails), 2; oribatids (mites), 5; other arthropods, 24; molluscs, 3; enchytraeds, 167; lumbricids (earthworms), 54; nematodes (threadworms), 85; and protozoa, 12. This amounts to a total of 361 kcal/m^2 per year.

In addition, the activity of bacteria and actinomycetes amounts to 0.23 kcal/m^2 per year on oak litter, 1.68 kcal/m^2 on ash litter, 33 kcal/m^2 in humus, 17 kcal/m^2 in the A-horizon, and 25 kcal/m^2 in the B-horizon.

The greatest activity of all is developed by the fungi with 543 kcal/m² per year in the litter layer, 220 kcal/m² in humus, and 380 kcal/m² in the A- and B-horizons. This gives an annual total of 1143 kcal/m², in spite of the fact that the mass of these microorganisms is very small compared to that of the invertebrates.

Ninety percent of the dead wood dropping to the forest floor is destroyed by microorganisms, chiefly fungi. The soil itself also contains many complicated nutritional chains owing to the numerous predators among the soil invertebrates.

5. The Effect of the Cold Winter Period on Plants of the Nemoral Zone

Plant damage occurring in a cold winter can be due to one of two causes:

1. Direct damage due to freezing of tissue water. This is termed frost damage.
2. Drying-out of aerial organs, which even at low temperatures transpire to a certain degree. When the conducting vessels are blocked by ice, insufficient water reaches the aerial organs to meet these transpiration losses and the organs dry out. This is called *frost (or winter) drought* or *dessication damage.*

Plants are not equipped with any means of protection against the effects of low temperature, their own temperature always being that of the surrounding air. Their only possible adaptation is to become hardened. If the resistance of plant organs to cold is tested in summer by placing them in a refrigerator at various temperatures below 0 °C for 2 h, it can be shown that only a few degrees suffice to cause irreversible damage. The same plant organs tested in winter, however, tolerate much lower temperatures without undergoing damage because they have developed a so-called hardiness. This is a physiological process taking place in autumn with the onset of shorter days and the first cool nights. In spring, when the weather becomes warmer, the opposite process (dehardening) is initiated.

Hardening is connected with certain physicochemical changes in the protoplasm which are not as yet fully understood. The viscosity of the plasma increases and plasmolysis becomes concave rather than convex. It is accompanied by a sudden increase in cell-sap concentration of several atmospheres due to an increase in sugar concentration. Protoplasm in its hardened state is more or less inactive. The resistance to cold of the buds of European deciduous trees in their winter condition can increase from −5 °C in autumn to −25 °C or even −30 °C in January or February. A larger increase in resistance to cold is developed in a cold winter than in a mild one. Among related species of a genus, the resistance is greater the further the species advances into the continental region.

This hardening is a very complex process, involving several stages of development. The first stage is a result of the shorter days in Autumn and leads to a period of rest. Further hardening is achieved when the temperature decreases. It is more intensive in species subject to very low temperatures when the first severe frosts occur. If parts of hardened plants are suddenly exposed to extremely low temperatures, resulting in a "vitrification" of the protoplasm (without the formation of ice crystals), it is even possible to freeze them in liquid nitrogen ($-190\,°C$), or even at $-238\,°C$. They must, however, be warmed in serveral stages until they thaw, so that no plasma-damaging formation of ice crystals is able to occur. After such an experiment, plants of the cold climate zones, which are capable of hardening, remain alive. Near the coldness pole in eastern Siberia, the forest vegetation is normally subject to temperatures of $-60\,°C$ or below in the winter. Tropical species, or even those from Zonobiomes IV and V, cannot be hardened.

As a rule, "hardiness" suffices to prevent frost damage to the native trees in Europe in a cold winter, although exotic species imported from warmer climes, with no ability of "hardening," often suffer. If an early frost occurs before hardening has set in or if there is late frost after dehardening has commenced, frost damage is widespread. Late frost damage is most commonly found in young, newly opened leaves, and it can also cause cambium damage if the trees are already in sap and the protoplasm in an active state.

The eastern limit of the distribution area of the beech may possibly be conditioned by the frequent late frost damage of young leaves which reduces the powers of competition of the trees. An increase in resistance to cold by hardening can also be demonstrated in herbaceous forest plants even though they are not exposed to extremely low temperatures because of their covering of litter and snow. Resistance to cold in the evergreen leaves of, for example, *Anemone hepatica* reaches only to $-15\,°C$, that of the better protected flower buds to $-10\,°C$, and that of the rhizomes only to $-7.5\,°C$.

Damage due to frost drought is more difficult to detect. Shedding of the intensely transpiring leaves, bud protection by hard bud scales, and protection of the twigs by a layer of cork prevent the loss of large quantities of water by deciduous trees in winter. Nevertheless, a certain amount of transpiration can be measured in the bare twigs in winter. In experiments under identical frost conditions it is higher in deciduous trees than in evergreen conifers, and higher in deciduous species of southern origin than in those of northern. These transpiration losses become dangerous in spring, when the intensity of the incoming radiation increases and the air temperature rises while the ground is still frozen hard. Buds and twigs sometimes dry out as a result. Evergreen species such as holly *(Ilex)* and the broomlike shrubs such as *Sarothamnus* or *Ulex* are especially sensitive in this respect.

Frost damage usually occurs at the coldest time of the year, and frost drought is more common as spring approaches and on warm southern slopes. The latter should not be confused with late frost damage.

All examples so far quoted have been taken from European deciduous forests, which, with their very low number of deciduous woody plant species, are relatively simple systems. The tree stratum often consists of a single species. The geobotanical and ecological conditions of Central Europe are discussed in greater detail in UTB 284, 2nd ed., Walter (1979). The conditions described hold equally for the deciduous forests in eastern North America and Asia. The numerous species in the tree stratum of these forests, however, render ecophysiological investigations extremely difficult, and surveys of these regions are still lacking. The complicated floristic composition of such forests cannot be dealt with here.

6. Orobiome VI – The Northern Alps

This orobiome is well developed in the northern Alps. The mean annual temperature drops with increasing altitude in mountainous regions, and the vegetational period becomes shorter. Whereas direct insolation increases with altitude, diffuse insolation decreases, with the result that the temperature differences between northern and southern slopes become more pronounced. Owing to the ascending air masses, precipitation increases rapidly with altitude on the northern margins of the Alps, e.g., Munich at 569 m above sea level receives 866 mm, whereas Wendelstein at 1727 m receives 2869 mm.

Corresponding changes can be observed in the vegetation of the individual vegetational belts:

Belt	Vegetation
Nival	Cushionlike plants, mosses, lichens
Climatic snow line at 2400 m above sea level	
Alpine	Alpine meadows
Subalpine	Dwarf trees and shrubs
Timberline at 1800 m above sea level	
High-montane	Spruce forests
Montane	Beech-fir forest
Submontane	Beech forest
Colline	Mixed-oak forest

Since the Alps are interzonal, the succession of altitudinal belts on the southern margins is typical of a humid Orobiome IV, and the timberline is formed by beech *(Fagus)*. In the sheltered valleys, however, with their continental type of climate, the sequence is different in that there is no deciduous belt, and a pine *(Pinus sylvestris)* belt precedes the spruce *(Picea abies)* belt, above which a larch *(Larix)*—stone pine *(Pinus cembra)* belt extends up to the timberline. The timberline and the snow line are 400 – 600 m higher here as a result of less cloud cover and stronger irradiation.

The various parts of the Alps are distinguished as Helvetic (the northern Alps), Pennine (the central Alps), and Insubrian (the southern Alps). The corresponding altitudinal belts, in order of decreasing altitude, are as follows:

Helvetian	Pennine	Insubrian
	Alpine	
Alpine	*Larix-Pinus cembra* forests	Alpine
Picea abies forest	*Picea abies* forests	*Fagus sylvatica* forest
Fagus sylvatica forest	*Pinus sylvestris* forest	*Quercus pubescens* forest
Quercus robur forest	Steppe elements	Sclerophyllous forest (traces)
(Central European)	(Continental)	(Mediterranean)

In the central Alps, the uppermost forest belt consists of European larch *(Larix decidua)* and the stone pine *(Pinus cembra)*, which is related to the Siberian subspecies. The larch plays the part of the light-demanding pioneer species and is, in time, replaced by the five-needled cembra pine, which can better tolerate shade (p. 268). On paths left by avalanches, the larch may continue down to lower altitudes. Above the forest margin, the two-needled dwarf mountain pine *(Pinus montana)* occurs, but this is replaced by the shrublike alder *Alnus viridis* on wet habitats.

The ecology of the spruce biogeocene will be dealt with on p. 265ff.).

The northern margins of the Alps have been studied with regard to the factors responsible for setting the upper limit of the spruce forests. With increasing altitude, the period of vegetation is shortened, the summers become cooler, and the winters are both colder and longer. Although these climatic changes take place gradually, the timber-line in high mountains is, in contrast, very sharply drawn. The powers of growth of the trees seem to decrease quite suddenly, and only a very narrow zone consisting of stunted, low forms provides the transition from forest to treeless alpine belt. The question arises as to whether the short summer or the long winter is responsible for the cessation of tree growth, and it would appear that both factors are important. With a period of vegetation less than 3 months, the young needles are incapable of maturing properly, and their cuticle cannot attain the required final thickness. As a result, during the long winter with temperature inversion and much sunshine, and particularly in the strong sunlight of the spring when the ground is still frozen, large water losses occur, as indicated by rises in cell-sap concentration of 65 atm and more. Damage typical of frost drought is observable, and the needles drop off. Beneath a covering of snow, such events cannot take place, and this explains the ability of the stunted forms to survive some distance above the timber-line. It is apparently the combined effect of the two factors – the shortened period of vegetation and the increased danger of frost drought – that is responsible for the abruptness of the timber-line at a certain altitude.

Pinus montana, which grows above the limits of the spruce forests, manages with a shorter period of vegetation. But about 100 m further up, the phenomenon repeats itself. The needles are unable to mature, suffering damage from frost drought, and the upper *Pinus montana* limit is just as sharply defined as that of the forest.

The factors responsible for setting the limits of polar forests have not been investigated, but they are probably similar, apart from the fact that sunshine plays no part in the damage caused by frost drought. This factor is probably replaced by the drying effect of the strong, cold winds, as is borne out by the observation that, in the sheltered valleys, the timber-line pushes further to the north than on the watersheds. This case is the opposite in the Alps, since cold air is trapped in the valleys and the temperatures are lower than on the mountain ridges from which the cold air flows downward.

At its highest, the timber-line in the central Alps lies at 2000–2150 m and is, as already mentioned, formed not by spruce but by deciduous larch and evergreen cembra pine, which has relatively delicate needles. Continuous measurements of climatic factors and photosynthesis have been made here throughout the entire year, so that the productivity of larch and cembra pine can be compared. During the cold winter, photosynthesis is at a standstill, even in the evergreen cembra. In spring, the evergreen needles rapidly come into a state of activity, whereas, at these altitudes, the larch is not green until the middle of June and is beginning to turn yellow by the end of September. Not more than 107 days are at its disposal for production, as compared with the 181 days available to the cembra pine. However, the young larch has 3 to 6 times the mass of needles possessed by the young cembra pine, besides which, despite the brief period of vegetation, it assimilates 47% more CO_2 per gram of needle. The total production, therefore, of a 4-year-old larch is 4.5 times and that of a 12-year-old 8.5 times that of cembra of the same age. Only from the 25th year onward is the quantity of needles produced by the larch the smaller of the two, and larch begins to lag behind in its growth, particularly on raw humus soils. In time, the cembra succeeds in establishing itself as a shade-enduring species. The relationship of larch to cembra is similar to that of pine to spruce (p. 265).

Wood found in the subfossil humus deposits in the subalpine belt indicates that during the warm postglacial period, all belts were 400 m higher than they are today. The dwarf shrubs that are covered by snow in winter are thus partly relics of the former forests[2].

---

2 Recent studies have revealed that their positions varied only within 250 m of the present situation (Frenzel, pers. comm.)

Because of the thick covering of snow in the alpine belt, duration of the snow-free season ("aperzeit" from Lat. *apertus, open,* and German *Zeit, time*) is more important for the low alpine vegetation than air temperature. Snow cover depends to a large extent upon relief, wind direction, and exposure. Snow collects in the hollows and forms cornices on the lee side of ridges but is blown off the windward side. If the windward side is sunny as well, then the snow melts and the habitat is snowfree ("aper") all year round. The plants here (Loiseleurietum) are exposed to the same extremes of frost desiccation as those in mountainous tundras and are accompanied by exactly the same lichens (p. 274). A shady, windward slope, however, is not warmed up by irradiation. In the presence of large drifts of snow at the foot of a slope facing north, the snowfree season is reduced to a minimum on the so-called snow patches ("schneetälchen") or is completely lacking, the snow remaining throughout the summer. The snowfree season can, however, vary in length from year to year on the same habitat according to the snowfall. Its average length decreases with increasing altitude and is theoretically zero at the climatic snow line. In individual cases, however, such as on steep rock faces, the snow-free season can be very long, even above the snow line. This is why flowering plants are found in the Alps in the nival belt, above the climatic snow line.

In any case, the microclimate, even on sunny days, is propitious with regard to temperature. If leaves are exposed to direct sunlight, their temperature can be as much as 22 °C higher than that of the air. Every mountain climber is familiar with the warm niches which can be exploited by low, ground-hugging plants. In cloudy weather, the temperature differences tend to be equalized.

From what has already been said, it is clear that as far as the vegetation is concerned, there is no such thing as a standard climate in the steep alpine belt. Instead, this region is split up into very small climatic units which can differ vastly from one another within a very short distance, as for example on the sunny and shady sides of a boulder. The way in which the snow is distributed in winter is of primary importance and must be known in order to be able to estimate the length of the snow-free season and understand the pattern of vegetation.

Temperature inversions and cold air pools play an important role and can cause a reversal of the order of altitudinal belts (beech above spruce). Even in the height of summer, outgoing radiation in dolines can be accompanied by night frost, and trees are then unable to grow.

Another source of disruption of altitudinal belts is to be found in avalanches, in whose wake the alpine vegetation continues far down into the forest belt since here it is not subject to competition with the forest vegetation. Even on the shallow, impoverished dolomite soil, which undergoes little erosion, alpine exclaves can be found in the middle of the forest belt, and surviving habitats of alpine species in the bogs of the alpine foreland are familiar to the botanist. In habitats such as these, the unpretentious but slowly growing alpine species are less inhibited by the competition of other plants. The ecology of the Alps has been studied in detail.

The process of developing hardiness to frost in evergreen species takes an annual course similar to that observed in deciduous forest species. Hardiness is developed in late autumn, and the process is reversed in spring. Although spruce needles are killed by temperatures of $-7\,°C$ in summer, they survive $-40\,°C$ in winter. Despite the fact that they grow much higher up in the Alps than the spruce, the maximum degree of frost hardiness achieved by alpine species is usually lower (less than $-30\,°C$) since they are protected by a blanket of snow from the lowest winter temperatures. Only *Loiseleuria*, growing on windy habitats free of snow in winter, develops a greater degree of hardiness. Although the wind prevents the lowest temperatures from being reached, it enhances the danger of winter drought. Despite its xeromorphic structure, if *Loiseleuria* is suspended in the open in winter, it dries out within 15 days. In its snow-free natural habitat, however, it normally grows tightly pressed to the ground, and the sun thaws any snow held between its shoots, so that occasional water uptake is possible. The dwarf shrubs beneath their blanket of snow are not exposed to the dangers of winter drought.

The water budget is fairly well balanced in summer on account of frequent rainfall, and only when irradiation is intense or a strong wind blows are the plants subjected to a large degree of evaporation for a few hours. The effects of wind are lessened near the ground. Even in habitats that appear dry on the surface, such as talus or scree slopes and rock faces, the soil carries abundant water. In this kind of habitat, the plants develop an extensive root system or taproots capable of penetrating into the rocky crevices, even though normally their root systems are very shallow and confined to the upper soil layers. The propitious water balance is reflected in the low cell-sap concentration of 8 to 12 atm, and even in xeromorphic species such as *Dryas, Carex firma,* and *Androsace helvetica,* it never rises to 20 atm. Here again, it would probably be more correct to speak of a peinomorphosis resulting from nitrogen deficiency than of xeromorphosis, since the uptake of nitrogen is, in fact, more difficult at low soil temperatures. Luxuriant, hygromorphic herbaceous plants are found only on N-rich habitats such as the resting-places of livestock.

If the total water lost in one year by the plant cover of alpine meadow communities is calculated, the figure of 200 mm is obtained. Evaporation depends, above all, upon the wind and is for this reason influenced by the relief, although in the inverse direction to snow deposition.

The short growing season in the alpine belt gives rise to the question of adequate production just as in the Arctic. The days are shorter than in the Arctic, but the light intensity is stronger, and nocturnal temperatures are lower. Under favorable conditions of illumination, 100 to 300 mg of CO_2/dm^2 can be assimilated in 1 day. One month of good weather suffices for the accumulation of sufficient reserves for the coming year and for the ripening of seeds, but since the growing season lasts 3 months, adequate production is in any case ensured.

Primary production in plant communities depends largely upon the density of vegetation. The following values were found:

Closed mats	$50-276$ g/m^2
Dryadeto-Firmetum	91 g/m^2
Salicetum herbacaea	85 g/m^2
Oxyrietum	15 g/m^2
On limestone scree	1 g/m^2

Photosynthesis is less intense in dwarf shrubs than in herbaceous species, but since the total leaf area of the former is larger, and the period of vegetation longer in the lower alpine belt, dwarf shrubs achieve a higher primary production.

The most unfavorable conditions of all are met with on the snow patches ("schneetälchen") on the north-facing slopes in the siliceous rock regions. The snow melts very slowly from the edges of such patches, and the ground gradually becomes exposed. This means that, within a very small area, a zonation can be recognized in which the ground is free of snow ("aper") for ever-shorter periods. The soil of such habitats is rich in humus, slightly acid, and invariably well-provided with water from the melting snow, but for this reason also relatively cool. If the snow-free period lasts for 3 months, mats of *Carex curvula* develop. If the growth season is reduced to 2 months, the willow *Salix herbacea* predominates, a woody species of which only the tips of the shoots are aboveground so that its leaves form a compact sward. *(Salix herbacea* only bears fruit after a winter with particularly little snow, when the snow-free season amounts to 3 months.) There is also a scattering of very small plants such as *Gnaphalium supinum, Alchemilla pentaphylla, Arenaria biflora, Soldanella pusilla, Sibbaldia procumbens,* etc. Given an even shorter growth period, only mosses manage to grow because they require no time for the production of flowers and fruits. The most common of these is *Polytrichum sexangulare (P. norvegicum).* If the snow-free season is too short for such green mosses, then only a liverwort, *Anthelia juratzkana* (which looks like a moldering crust), is found. This moss grows in a symbiotic relationship with a fungus and is saprophytic to a certain extent. This latter zone does not become snow-free at all if there has been an unusually large amount of snow in the preceding winter.

On firn areas in the nival belt, the sole living organism is the alga *Chlamydomonas nivalis,* which gives the snow its rosy hue.

Since bare rock predominates in the alpine belt of the Alps, its chemical composition is of great significance for the vegetation because it governs the reaction of the soil. *The floristic differences between the calcareous Alps and the crystalline siliceous central Alps are very pronounced.* Accordingly, a distinction is made between limestone-demanding basophi species and limestone-avoiding acidophil species. Vicarious species like the alpine rose are often encountered: *Rhododendron hirsutum* on limestone and *R. ferrugineum* on siliceous rock or acid humus soil.

The ecosystems of the dwarf shrub heaths were studied in the years 1969–1976 as part of the International Biological Program (IBP) on the Patscherkofel near Innsbruck, Austria, above the present timber-line in the following three experimental areas (Larcher 1979):

1. *Vaccinium* heath (1980 m above sea level) in a basin protected from wind and winter snow: *Vaccinium myrtillus* 3, *V. uliginosum* 2, *V. vitis-idaea* 1, *Loiseleuria procumbens* 1, *Calluna vulgaris* 1, *Melampyrum alpestre* 1, mosses 1, lichens 1.
2. *Loiseleuria* heath (2000 m above sea level); dense stands, often snow-free in wind-exposed locations: *Loiseleuria* 5, *Vaccinium uliginosum* 1, *V. vitis-idaea* 1, others only +, lichens (*Cetraria islandica* 1, *Alectoria ochroleuca* 1, others only +).
3. Open trellis-like and lichen-rich *Loiseleuria* stands (2175 m above sea level) in extreme wind-exposed locations: *Loiseleuria* 3, stunted *Vaccinium uliginosum* 2, *V. vitis-idaea* 1, *Calluna* 1, others +, mosses +, lichens (*Cetraria islandica* 2, *C. cuculata* 1, *Alectoria ochroleuca* 1, *Cladonia rangiferina* 1, *C. pyxidata* 1, *Thamnolia vermicularis* and others +).

The climate is cold, with an annual temperature of little over 0 °C. Frosts may occur in any month (the absolute mininum is around −20 °C), although the daily maximum temperature may reach 20 °C in the summer months. In experimental area 1, snow lies approximately 6 months of the year; in experimental area 2 snow lies 4–5 months, and in area 3 it only lies in certain locations and only for short periods. The microclimate in 1 and 2 is somewhat

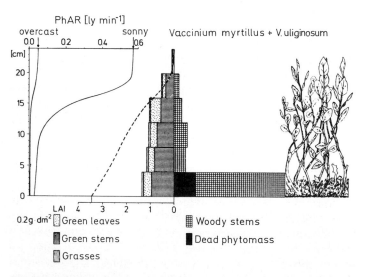

Fig. 113. *Center:* phytomass layering of the *Vaccinium* heath (*left:* assimilating portion; *right:* non-assimilating and dead portions). *Left:* cumulative leaf area index (LAI *dotted line*) and light decrease (PhAR) in the stand. (According to Cernusca 1976; from Larcher 1977)

warmer, while in 3, extreme temperature differences occur. The duration of CO_2 assimilation is approximately 100 days for the deciduous species and 140 days for the evergreen species. The composition of the stands is illustrated in Fig. 113, as well as the photosynthetically active radiation (PhAR) and the cumulative leaf area index (LAI). The wind is significantly decelerated in the dwarf shrub heath, even during strong storms, so that the humidity remains high. The annual precipitation in this area is approximately 900 mm, although it is only 100 mm in the summer months. On shale-like biotite gneiss, the soils are sandy, acidic iron podsols with thick raw humus layers and are only poorly developed in stand 3. They were formed from earlier stone pine forest stands. The humus is mineralized only slowly (nitrogen supply: approximately 3–4 kg/ha; in stand 3 only ⅓ as much).

Studies on the ecological production are illustrated in Fig. 113 and in the following table: live standing phytomass, dead portions and litter are given in g dry matter/m^2 of the dwarf shrub heath (1), the dense *Loiseleuria* heath (2) and the open *Loiseleuria* stand (3).

Experimental area	1	2	3
Live aboveground phytomass (max.)	983	1105	748
Adhering dead material	263	123	72
Live underground phytomass	2443	2200	803
Dead underground material	1549	608	56
Total live phytomass	3426	3305	1551
Total dead material	5238	4036	1679
Ground litter	819	1080	931

The ratio between the maximum aboveground live phytomass and the maximum underground live phytomass was 1 : 2.5 in 1, 1 : 2.0 in 2, and 1 : 1.1 in 3. The percent of assimilating mass to the total live phytomass was 55% for 1, 68% for 2 and was not calculated for 3. The determination of the net primary production is difficult and could only be estimated for the underground production. It is given as 485 g/m^2 per year for 1, 1317 g/m^2 per year for 2 and 108 g/m^2 per year for 3. Per hectare, that would amount to 4.8 t, 3.2 t or 1.1 t per year, respectively. The phytomass is probably constant, except for certain fluctuations, meaning that the stands are in an ecological equilibrium with their surroundings, whereby the increase in phytomass is inhibited by consumption (game animals, ptarmigans, arthropods) and by certain losses of material in winter (freezing and dessication of parts above the snow).

The photosynthesis capacity per unit surface area is similar for the leaves of deciduous and evergreen dwarf shrubs. In relation to the dry weight of the leaves, it is similar for deciduous dwarf shrubs and soft-leaved deciduous woody species, as well as for evergreen dwarf shrubs and conifers.

The narrow optimum temperature range for photosynthesis is between 10 °C and 30 °C for Ericaceae, and thereby corresponds to the normal temperatures on overcast and clear days in such stands; the temperature minimum for the assimilation of CO_2 in super-cooled leaves is −5 °C to −6 °C. The leaves are almost never subject to overheating or to the limitation of photosynthesis due to water shortage. Although the water supply is sufficient during the growth period, and the total amount of transpired water is in the range of 100−200 mm, restricted transpiration has been observed during periods of foehn (warm, dry south winds). Cuticulary transpiration is very low during the winter.

Heat damage in the summer only affects individual sprouts located above the loose stones or over raw humus soils void of vegetation. Damage from freezing in winter is possible only in snow-free conditions. Hardening protects the plants from frost damage, although late frosts, after the process of dehardening has begun, can be dangerous. Damage due to frost drought is difficult to recognize and usually the damage is the result of the combined effects of several factors. The arctic-alpine species *Loiseleuria procumbens* and *Vaccinium uliginosum* are completely frost-hardy.

During the main period of growth, respiration is significantly increased. At this time, the fat-storing *Loiseleuria* reduces its respiration coefficient to 0.8−0.9 until the phase of intensive growth is completed and then returns it to 1. During the growth period, the efficiency of the net primary production (in % of the photosynthetically active radiation) is 0.9% for the dwarf shrub heath, 0.7% for the dense *Loiseleuria* heath and 0.3% for open stands.

Although species of Ericaceae store much fat as well as starch, these substances are only partially mobilized. Most of this material remains in dead parts of the plant. On the first days of recurring frosts, the dwarf shrubs react immediately by transforming a large portion of their stored starch into sugar, whereby *Loiseleuria* turns red due to the presence of anthocyanin.

More detailed information is also available on the relatively small turnover of minerals by these plant communities with their low ash content.

Further investigations were carried out in the nival belt, i.e., above the climatic snow limit, on the Hohe Nebelkugel in the Stubaier Alps under very difficult conditions (Moser et al. 1977). One cabin had to be transported by helicopter, carefully isolated and well-grounded, since it was often subjected to heavy thunder and lightning storms.

This belt has no closed vegetational covering. In the 0.5 ha research area 3184 m above sea level, a flat ridge with seven flowering plants and several species of cryptogams, a northern slope with very sparse vegetation and a southern slope with 11 phanerogamic species on flat levels were chosen.

The climatic conditions do not at all correspond to those of the high arctic, and in the summer, they are more similar to those in the Páramos in the tropics. On clear days, the leaf temperature is often over 15 °C and sinks to below 0 °C at night without impairing the photosynthetic activity. The

Development of blossoms and phenophase
at 2600 - 3200 m above sea level

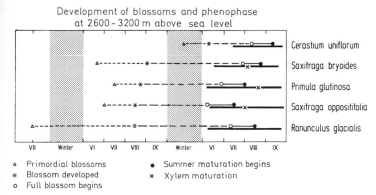

- Δ Primordial blossoms   • Summer maturation begins
- * Blossom developed   × Xylem maturation
- o Full blossom begins

Fig. 114. Phenology of nival plants. Vegetative parts = thick line, reproductive parts = thin line, parts green in winter = dashed line. (From Moser et al. 1977)

24-hour summer days of the arctic with the sun low on the horizon, however, are of relatively constant temperature.

Of the three chosen locations, the southern slope had the most advantageous light and temperature conditions. The phenology of the most important species is presented in Fig. 114. While *Saxifrage oppositifolia* blossoms early (the blossom organs are frost-resistant), *Cerastium uniflorum* blossoms late.

Primula spp. and *Ranunculus glacialis* store their assimilates in the form of starch, which is transformed to sugar in winter, while *Saxifraga* spp. stores fats. The location of the stored substances is illustrated in Fig. 115. It is interesting to note the provisional displacement from the leaves to the underground storage organs in *Ranunculus* even during intermittent periods of bad weather, which is reversed after the weather improves. This is important, since each snow cover in the summer could potentially last until the following spring. In general, the growth period on the southern slope lasts approximately 3 months, although due to the often poor weather conditions, only an average of 60–70 (15–100) days are suitable for production. At some locations, the plants may not become snow-free at all during some years.

Ranunculus glacialis achieves half of its production during the few light and warm days, and half on the many cool days with sparse lighting due to a low snow cover or fog. This species is capable of photosynthesis in the temperature range between −7° and +38 °C. Assimilation is greatest during full blossom and fruit formation, and may attain up to 0.056 g dry matter/dm^2 leaf surface/day in *Ranunculus glacialis* and 0.063 g in *Primula glutinosa* under optimum conditions; under favorable conditions, these plants assimilate between 0.015 and 0.020 g. In the course of the growth period, *Androsace alpina* increased its surface cover by 13.5%. The average rate of net assimilation was 0.058 g dry matter/dm^2 mat surface per day during this

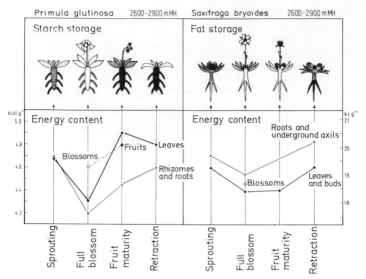

Fig. 115. Energy content of two nival species and the storage of reserves: *black* concentrated; *hatched* moderate; *dotted* traces; *white* no storage

period. Because of the poor cover protection, the primary production is extremely low at the nival level. Under optimum conditions and a cover of 10%, the production may be estimated at 0.66 g dry matter/m² per day.

No other mountain range on earth has been so thoroughly investigated as the complicated mountain system in the center of western Europe known as the Alps. The following fundamental works deal with this system: C. Schroeter, „Das Pflanzenleben der Alpen" 2nd ed., 1,288 pp., Zürich, 1926; J. Braun-Blanquet and H. Jenny, „Vegetationsgliederung und Bodenbildung in der Alpinen Stufe der Zentralalpen" (Denkschr. Schweiz. Naturf. Ges., 63, 183−349, 1926); H. Gams, „Von den Follatères zur Dent de Morcles" (Beitr. Geobot. Landesaufn. Schweiz, 15, 760 pp., 1927), E. Aichinger, "Vegetationskunde der Karawanken" (Pflanzensoz. vol. 2, 229 pp., Jena, 1933; and R. Scharfetter, „Das Pflanzenleben der Ostalpen" (Vienna, 419 pp., 1938).

There are also many maps which are important for understanding the various altitudinal belts in the Alps.

The network of weather stations in the Alps is more complete than in any other mountain range and includes a number of high altitude stations. H. Rehder (Flora B, 156, 78−93 pp., 1966) took advantage of these for the development of a climate diagram map of the Alps and its adjacent areas. Also of great ecological importance is the work of K. F. Schreiber „Wärmegliederung der Schweiz aufgrund von phänologischen Geländeaufnahmen in den Jahren 1969−1973" (4 maps, 1 : 200,000, Eidgen. Drucks. Zentr. Bern, 1977), which distinguishes 18 altitudinal temperature belts (three hot belts on the southern limits of the Alps, and three warm, three mild, three cool, three raw and three cold belts, followed by the alpine and the nival belts which are not further subdivided). A 1 : 500,000 map identifies the foehn areas of Switzerland in which the development of the vegetation may be up to 3 weeks ahead of other regions.

There are also a large number of vegetational maps available. Besides the many specialized maps, there are also general maps with the most important altitudinal belts. The following concern the eastern Alps: H. Mayer, „Karte der natürlichen Wälder des Ostalpenraums", (Cbl. Ges. Forstwesen 94, 147–153, Vienna, 1977) and his publication „Wälder des Ostalpenraums" (G. Fischer, Stuttgart, 344 pp., 1974); H. Wagner, „Karte (1 : 1,000,000) der natürlichen Vegetation" in the Österreich Atlas (1971); P. Seibert, „Übersichtskarte der natürlichen Vegetationsgebiete von Bayern 1 : 500,000 mit Erläuterungen" (Schriftenreihe f. Vegetationsk. H. 13, Bad Godesberg, 1968), which includes the northern limits of the Alps, and thereby leads to the next section of that mountain range.

The Central Alps are presented and discussed in the vegetational maps of E. Schmid (in 4 parts, 1 : 200,000), Geobot. Landesaufnahme Schweiz 39, 52 pp., 1961). The following belts are distinguished (which also correspond to altitudinal belts): (1) *Quercus pubescens* belt (on limestone) and *Quercus robur-Calluna* belt with the chestnut (on acidic formations) in the hot altitudinal belts. (2) *Quercus-Tilia-Acer* mixed deciduous forest belt in the warm and mild temperature belts. (3) *Fagus-Abies* belt in the cool temperature belt. (4) *Picea* conifer forest belt in the raw and lower cold temperature belt. (5) *Vaccinium uliginosum-Loiseleuria* belt, including the entire alpine or cold belt. In addition, there is the *Pulsatilla* steppe belt in the continental inner Alpine valleys with *Pinus sylvestris* in the lower range, followed by a *Picea* belt and a *Larix-Pinus cembra* belt above this, which extends to the much elevated timber-line. The dry foehn valleys are also characterized by the presence of *Pinus sylvestris* with *Erica carnea*.

An especially large number of ecological vegetation maps with detailed information in a scale of 1 : 100,000 (to 1 : 10,000) for the western Alps, and to some degree for other regions, are published regularly by P. Ozenda in the "Document de Cartographie Écologique" (Grenoble). This series also includes the altitudinal belt ascending from the Mediterranean coast in the south (therefore, within Orobiome IV). This large and unique cartographic work is especially worthy of mention. The detailed color vegetational maps provide exact information on the altitudinal belts and on their dependence on exposition and geological formations.

In „Vegetation Südosteuropas" by I. Horvat, V. Glavač and H. Ellenberg (Stuttgart, 768 pp., 1974), the vegetation of the Dinarian Alps and the adjacent mountains of the Balkan Penninsula is discussed.

Zonoecotone VI/VII – Forest-Steppe

The deciduous forests of the temperate zone are confined to climatic regions of an oceanic nature, where there are no sharp extremes of temperature and rainfall is more or less evenly distributed throughout the year, usually with a summer maximum. Steppe and desert occupy the continental regions, which are much more extensive in the Northern Hemisphere. In a continental climate, the temperature amplitude is greater and the summers are hotter, but the winters are much colder, so that the annual mean temperature is lower than that in oceanic regions of same latitude. This is accompanied by a decrease in annual rainfall and more arid summers.

The zonoecotone between deciduous forest and grassy steppe in eastern Europe is the forest-steppe. It is not a homogeneous vegetational formation like the tropical savanna, but rather a macromosaic of deciduous-forest stands and meadow-steppe. At first, the former predominate, the steppes forming scattered islands. But the more arid the climate becomes, the more the situation tends to be reversed, until finally, small islands of forest are left in a sea of steppe. Relief and soil texture (Fig. 116) determine the predominating vegetation. Forests are found on well-drained habitats, slightly raised ground, the sides of the river valleys, and porous soils, whereas meadow-steppes occupy badly drained, flat sites with a relatively heavy soil. Grasses and the tree seedlings compete with one another, and if (as happens in the course of reforestation experiments) the young tree plants are protected for the first couple of years from competition with the roots of the grasses, they are able to grow on the steppe, although they are incapable of regeneration. In previous times, fires caused by lightning and the grazing of big-game herds encouraged the growth of the steppe, which nowadays, however, is entirely given over to farming, except few protected areas.

The forest zone, forest-steppe zone, and steppe zone of eastern Europe are readily distinguishable from one another on a climatic basis. Climate diagrams for the forest zone reveal the absence of a period of drought, whereas diagrams for the steppe zone always indicate the presence of such a period. Although no drought is to be detected on the diagrams for the forest-steppe zone, a dry period is always recognizable, which is not the case for the forest zone (Figs. 117 and 118).

During the postglacial period, the boundary between forest and steppe shifted. In the soil profile beneath the present-day forest stands, it is possible to find "krotovinas," the deserted burrows of steppe rodents *(suslik,* or European ground squirrel, *Citellus citellus)* (Fig. 121). Since these animals

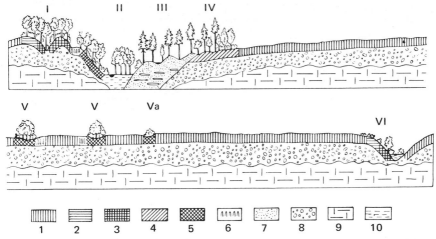

Fig. 116. Relation between vegetation, soil, and relief in the forest-steppe. Soils: (1) deep, poorly-drained chernozem with meadow-steppe; (2) degraded chernozem and (3) dark-gray forest soil (both well drained); (4) porous sandy-loamy forest soil; (5) light-gray forest soil; (6) solonetz on flat terraces or around depressions with no outflow and with soda accumulation; (7) fluvioglacial sands; (8) moraine deposits or loesslike loam; (9) preglacial strata; (10) alluvium in river valleys. I, Oak forest on well-drained elevations or on slopes; II, floodplain forests (oak, etc.); III, pine forests on poor sands with *Sphagnum* bog in wet hollows; IV, pine-oak forests on loamy soils; V, aspen groves in small hollows (pods), in spring containing water that seeps away only slowly (soil in central portion is leached); Va, the same, but with willows; VI, ravine-oak forest, with steppe-shrubs at upper margins. (Modified from Tanfiliev and Morosov; Walter 1964/68)

never inhabit forests, it must be assumed that before the forest-steppe was inhabited by man the forests were in the process of advancing, on account of the climate's becoming more humid after a warm optimum had prevailed. Later shifts in the boundaries cannot be detected because of the large degree of human interference.

The replacement of the forest zone by steppe in continental regions is governed by the supply of water. In order to understand this, it is necessary to return to the forest on the Vorskla (p. 195) and consider its water economy. In place of the brown forest soils of central Europe, dark gray, lightly podsolized soils cover a thick layer of loess (Fig. 121). The climate is similar to that of Uman (Fig. 117), with a moderately dry period in August and an annual precipitation of 537 mm, although as the climate diagram in Figure 107 shows, the weather may vary considerably from year to year. Water turnover is almost exclusively confined to the upper 2 m of the soil, and no water percolates to the deep-lying groundwater.

Figure 118a shows the isopleths of soil-water content in a normal hydrological year. The soil is only moistened to any depth when the snow melts in April, especially if it has fallen in large quantities (Fig. 118b). A winter with little snow has an unfavorable effect on the water content of the soil even if followed by a summer with plentiful rain (Fig. 118c). In any case,

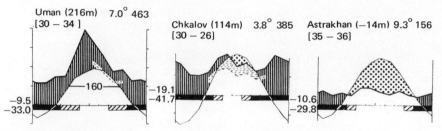

Fig. 117. Climate diagrams from the forest-steppe zone (with dry season), the steppe zone (with drought and a longer dry season), and the semidesert (with long summer drought). (See also Fig. 9, Odessa). (From Walter and Lieth 1967)

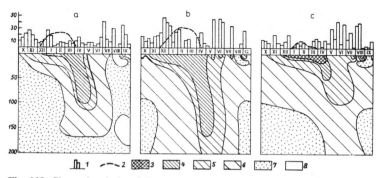

Fig. 118. Chronoisopleths of the water content of the forest floor for a forest on the Vorskla (on the edge of the forest-steppe): (a) hydrological year 1958–59, with precipitation at almost normal values; (b) 1966–67, a wet year with even distribution of precipitation; (c) 1968–69 a wet year, but with a snow-free winter and a summer precipitation maximum. (1) Precipitation (in millimeters); (2) depth of snow covering (in centimeters). Degree of soil soaking: (3) above field capacity; (4) field capacity. Percentage of water available: (5) 100–50%; (6) 50–15%; (7) 15–0% (wilting point); (8) no available water. (According to Goryschina, from Walter 1976)

it is clear that the forest uses up all of its water and the deeper soil is always dry. Such is the situation in the euclimatopes, but on the southern slopes with drainage and high evaporation, steppe develops because the soil water is insufficient to support forest.

In August and September, even the grass steppe dries out, water supplies being insufficient to cover losses due to transpiration. This is not harmful to the grasses, although the trees suffer damage if they lose their leaves too early or if entire branches die off.

Precipitation decreases and temperature increases toward the southeast in the forest-steppe, the patches of forest becoming smaller and smaller and increasingly confined to the northern slopes, until finally, on the southern-most limits of the forest-steppe, only oak-sloe *(Quercus* and *Prunus spinosa)* bushes are left in the gorges.

In the forest-steppe, grasses and tree seedlings compete with one another. Clements et al. (1929) demonstrated that on what was in the 1920s still original tallgrass prairie corresponding to the meadow-steppe, planted tree seedlings could only hold their own if all the grass roots were removed from the vicinity.

The amount of water required by a forest increases with its age. Experiments in reforestation have revealed that young forest plantations grow relatively well, but that, with time, the tips of the older shoots dry off and fresh shoots are then put out from below. The trees, therefore, develop abnormally as a result of the water shortage. If groundwater is available, however, healthy stands develop. Savannalike communities are missing in the forest-steppe because individual deciduous species are unable to compete successfully with the grasses. Only low shrubs such as *Spiraea, Caragana,* and *Amygdalus* are common, although they are generally to be met with on stony ground less suited to the dense root systems of the steppe grass (p. 80).

The steppe component of the forest-steppe, the meadow-steppe, will be considered in the next section, which deals with Zonobiome VII.

VII Zonobiome of the Arid-Temperate Climate

1. Climate

In Eurasia, this continental zonobiome stretches from the mouth of the Danube across eastern Europe and Asia, almost to the Yellow Sea. In North America, it occupies the entire Midwest, from southern Canada to the Gulf of Mexico. The degree of aridity varies considerably, and four subzonobiomes can be distinguished: (1) a semiarid subzonobiome having a short period of drought, with steppe and prairie vegetation (Fig. 117, Chkalov); (2) a very arid subzonobiome with a type VII climate (rIII), i.e., with as little rain (falling in winter) as the subtropical desert climate; (3) a subzonobiome similar to 2, but with summer rain; and (4) deserts of the cold mountainous plateaus (Tibet and Pamir). Thus Subzonobiomes 2 and 3 are genuine deserts, although the winters are cold. For practical purposes, it is helpful to distinguish an arid semidesert ecotone with a type VIIa climate between Subzonobiomes 1 and 2 and 1 and 3 (Fig. 117, Astrakhan). The semideserts (the sagebrush region in North America) are also more arid than the steppes but less arid than the deserts, and the vegetation is of a transitional type, although there is a well-developed drought lasting about 4–6 months.

2. Soils of the East European Steppe Zone

The east European steppes are the cradle of the science of soil types, the foundations of which were laid by Dokutchayer (1846–1903) and Glinka (1867–1927). There is no other region of comparable area where the parallel zonation of climate, soil type, and vegetation can be seen so clearly. It must be added, however, that very little remains of the original vegetation. The conditions responsible for the clear zonation are the extreme uniformity of relief and the fact that the parent rock is to a large extent homogeneous (loess). The climate changes steadily from northwest to southeast, the summer temperatures and potential evaporation rising while the rainfall decreases, so that the aridity becomes more and more pronounced. The boundary between forest zone and forest-steppe zone coincides with the boundary between humid and arid regions. This means that to the north of this demarcation line, the annual rainfall exceeds the potential evaporation, whereas to the south, the latter is the higher of the two (Fig. 119). In depressions with no outflow, saline soils form.

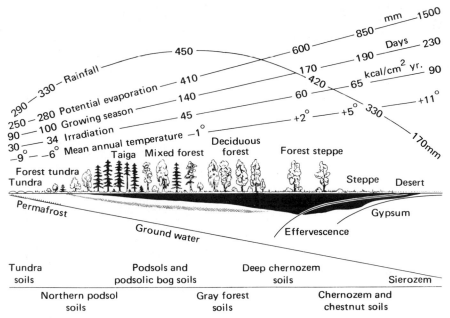

Fig. 119. Schematic climate, vegetation, and soil profile of the east European lowlands from northwest to southeast (black: humus horizon; diagonal shading: illuvial B-horizon). Vegetational season in the tundra corresponds to the number of days with a mean temperature above 0 °C, elsewhere to the number of days with a mean temperature above 10 °C. (From Schennikov; Walter 1964/68)

The distribution of the various soil types is depicted in a simplified form in Fig. 120.

Humid regions have a typical podsol soil and slightly podsolized gray forest soil, whereas arid regions have soils ranging from chernozem to chestnut and arid brown (burozem). The soil types are recognizable from their soil profiles, which are shown in Fig. 121.

The chernozems are A–C soils, or pedocals, without a clayey illuvial horizon (B). The zones are subdivided as follows: northern, thick, normal, and southern chernozems. The humic A-horizon consists of a black A_1 layer, a slightly lighter A_2 layer, and a loess layer slightly colored by humus A_3. Below this is the C layer, which is original, unchanged prismatic loess. In the thick chernozem, the humus layer goes down 170 cm, its thickness decreasing both to the north and south. Normal chernozem has the highest humus content, 7–8% (in the eastern steppe regions, the humus content is even higher). There is no translocation of clay in the chernozems, but in spring the downward flow of water from the melting snows carries with it calcium carbonates dissolved out from the upper horizons. If HCl is applied to soil from these leached upper horizons, no effervescence occurs–only with soil from deeper layers is a positive reaction obtained. The more arid the climate, the nearer the effervescence level to the surface. Somewhat below this

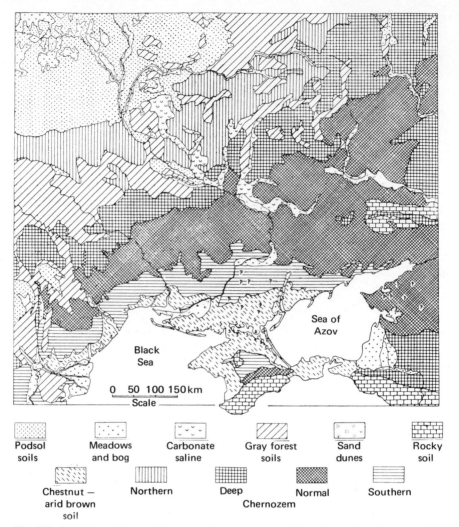

Fig. 120. Soil map of the east European steppe region and the adjacent forest regions. P, Pod (hollows with no outflow, in the steppe); S, saline soils (solonchak). (From Walter 1960)

so-called effervescence horizon, the dissolved carbonates precipitate, usually in the form of very fine $CaCO_3$ threads reminiscent of mold (pseudomycelia). Further south, these carbonates also precipitate as small white nodules (bjeloglaski). Besides $CaCO_3$, the humus-filled cross sections of the abandoned burrows of ground squirrels (krotovinas) are recognizable in the soil profile.

The changes in soil profile take place gradually from north to south, in conformity with the changes in climate, and reflect the increasing aridity.

Beneath the forest of the forest-steppe zone, the upper soil layers remain wetter. The A_0 horizon is made up of litter which mixes only slightly with the

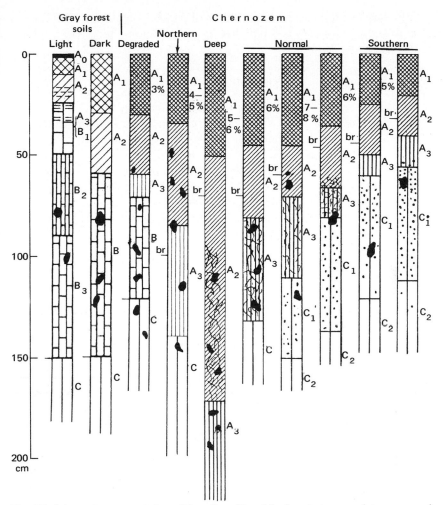

Fig. 121. Schematic representation of the soil profiles of the forest-steppe and steppe zones (west of the Dnieper), from north to south. Percentages, humus content of the A_1-horizon; br, effervescence horizon; wavy lines, pseudomycelia ($CaCO_3$); small dots, $CaCO_3$ nodules; large black spots, krotovinas (abandoned ground squirrel burrows); horizontal dashes, laminated structure in forest soil. (From Walter 1960)

mineral soil, so that the humus horizon under the moist hornbeam forests *(Carpinus)* is a light gray, and that under the dry oak forests dark gray. The good friable structure is lost, and the soil becomes laminated. Beneath the humus layer, there are mealy, bleached sand grains and below this a compact B-horizon, indicating the beginning of podsolization. There is hardly a trace of this, however, in the degraded chernozem beneath the shrubby oaks which constitute the last outposts of the forest. Below the most humid parts of the meadow-steppes, the soil is typical northern chernozem, with a very deep effervescence level and no $CaCO_3$ precipitations.

On the basis of the surviving remnants of natural vegetation, it has been possible to show that every soil type has its corresponding plant community, as in the following summary:

Soil type	Vegetational unit
Gray forest soil	Oak-hornbeam and oak-forest
Degraded chernozem	Oak-blackthorn (sloe) bush *(Prunus spinosa)*
Northern chernozem	Damp meadow-steppes with abundant herbaceous plants
Thick chernozem	Typical meadow-steppe
Normal chernozem	Feather grass *(Stipa)* steppe with abundant herbaceous plants
Southern chernozem	Dry *Stipa* steppe, few herbaceous plants

This scheme makes it possible, with the aid of a soil map, to reconstruct the original vegetation.

3. Meadow-Steppes on Thick Chernozem and the Feather Grass Steppes

The word steppe comes from the Russian "step", and its use should be confined to those grass steppes of the temperate zone, such as the prairie and the Pampa, that are comparable with the east European steppes. No steppes of this type occur in the tropics, where it is more appropriate to refer to tropical grassland. The word steppe often conjures up a picture of dreary, poor vegetation, although the very opposite holds true for the northern variant of the east European steppe. Nowadays these are the most fertile parts of Europe, with the best chernozem soils. In their natural condition, they excel even the lushest European meadows in the abundance of their colorful blossoms. Only in autumn do they give the impression of dryness.

It has already been said that the forest-steppe is a macromosaic of deciduous forest and meadow-steppe. Since the deciduous forests have already been dealt with, the meadow-steppes will now be considered in more detail.

The seasonal course of events is as follows (see Figs. 122 and 123). When the snow melts, the steppe soil is thoroughly wet, the temperature rises, and a profusion of spring flowers develops. At the end of April, the mauve blossoms of *Pulsatilla patens* appear, *Carex humilis* begins to shed its pollen, and at the beginning of May, they are joined by the large golden stars of *Adonis vernalis* and the pale blue inflorescences of *Hyacinthus leucophaeus*. By mid-May, the steppe is verdant, and *Lathyrus pannonicus, Iris aphylla,* and *Anemone sylvestris* are in flower among the sprouting grasses. The most colorful stage is reached at the beginning of June, when innumerable *Myosotis sylvatica, Senecio campestris,* and *Ranunculus polyanthemus* are in

Fig. 122. Meadow-steppe in spring. Vertical projection, quadrats in decimeters. Top: beginning of April, brown aspect with patches of mauve *Pulsatilla patens, Carex humilis* in pollen. Middle: end of April, yellow aspect due to *Adonis vernalis,* pale blue *Hyacinthus leucophaeus.* Bottom: end of May, blue aspect due to *Myosotis sylvatica,* white *Anemone sylvestris,* yellow *Senecio campestris,* a few *Stipa* in bloom. (According to Pokrovskaja; from Walter 1968)

bloom. At this point, the first plumes of *Stipa ioannis* appear, and by early summer, the long feathery awns of the various *Stipa* species, interspersed with the panicles of *Bromus riparius* (closely related to *B. erectus),* are swaying in wavelike motion in the wind. Intermingled with these plants are the blossoms of *Salvia pratensis* and *Tragopogon pratensis.* Toward the end of June, the flowers of *Trifolium montanum, Chrysanthemum leucanthemum,* and *Filipendula hexapetala* whiten the steppe, a colorful contrast being provided by *Campanula sibirica, C. persicifolia, Knautia arvensis,* and *Echium rubrum.* At the beginning of July, when *Onobrychis arenaria* and *Galium verum* come into flower, the glorious colors begin to near their end.

From mid-July onward, the plants begin to wither. The dark-blue panicles of *Delphinium litwinowi* and, later, the red-brown candles of *Veratrum*

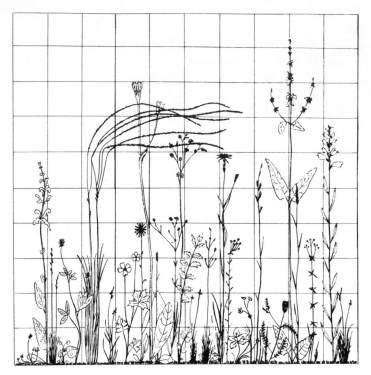

Fig. 123. Early summer aspect of the meadow-steppe. Many herbs in flower among the flowering feather grass *Stipa ioannis* (those above 40 cm in height are *Salvia pratensis, Hypochoeris maculata, Filipendula hexapetala, Scorzonera purpurea, Phlomis tuberosa,* and *Echium rubrum*). (From Walter 1968)

nigrum now make their appearance. From August onward, the steppes look dry and remain in this state until they are covered with snow.

This description shows that the dry meadows and steppe-heath of central Europe merely represent poor extrazonal outposts, on dry shallow habitats, of the meadow-steppes of humid climates. In floristic composition, the two are very similar, except that in central Europe, sub-Mediterranean elements (such as orchids) not found on the steppes are also present.

Further to the south of the meadow-steppes of the forest-steppe zone is the feather grass steppe, on normal and southern chernozems. Various species of *Stipa* predominate, and faced with increasing dryness, the less drought-resistant herbaceous species are incapable of successful competition and gradually recede. The density of the plant cover decreases to such an extent that the ground is covered in spring with the moss *Tortula (Syntrichia) ruralis* and the alga *Nostoc*. In spring, geophytes such as *Iris, Gagea,* and *Tulipa* and some winter annuals *(Draba verna, Holosteum umbellatum)* are abundant. *Paeonia tenuifolia* is especially striking. Other herbaceous species make their appearance in summer *(Salvia nutans, S. nemorosa, Serratula,*

Jurinea, Phlomis, etc.) and are joined later by Umbelliferae *(Peucedanum, Ferula, Seseli, Falcaria)* and Compositae *(Linosyris).*

Further to the south, the density of the vegetation decreases still more. Apart from the feather grasses, *Stipa capillata* and *Festuca sulcata* and herbaceous plants with very long taproots *(Eryngium campestre, Phlomis pungens, Centaurea, Limonium, Onosma)* are common.

On the chestnut soils, sagebrush *(Artemisia)* species become more abundant and initiate the transition to sagebrush semidesert.

4. North American Prairie

Although the conditions prevailing on the prairies and on the steppes are very similar, the situation in the former is more complicated. Whereas the steppes stretch at a latitude of about 50° from the outposts of the Carpathians far beyond the borders of Europe to the east, the prairies, although beginning south of a latitude of 55° in Canada, extend in a north-southerly direction beyond a latitude of 30° and are succeeded by *Prosopis* savanna. Furthermore, the extensive plains of North America rise gradually to 1500m above sea level, and precipitation decreases from east to west, but temperature rises from north to south. This means that there is no clear-cut soil zonation, but rather a checkerboard arrangement of soil types (Fig. 124).

The individual vegetational zones, such as tallgrass prairie, mixed prairie, and shortgrass prairie, succeed one another from east to west with increasing aridity, but within each zone there is a floristic gradient from north to south. *Andropogon* species, that is to say, grasses of southern origin, are more common in the prairie than *Stipa.*

In North America, too, there is a transitional zone of forest-steppe, in which the sides of the valleys and light soils are forested and the flat watersheds with heavy soils support grassland.

Tallgrass prairie corresponds to the northern meadow-steppe on thick chernozem, but the prairie soils are wetter, the chalk is completely leached, and there is no effervescence level. The question as to why no trees grow on the prairie has been settled experimentally by planting tree seedlings with and without the competition of grass roots. The results showed that if such competition is excluded, trees are, in fact, able to grow. Wherever prairie fires do not occur, and where human interference is excluded, the forest slowly encroaches upon the prairie, with a bush zone in the vanguard, at a rate of about 1 m every 3 to 5 years. Statistics show that, for 1965, an average of one fire caused by lightning occurred per 5000 ha of prairie, and it is clear that such fires are a natural environmental factor favoring the grasses. It must also be borne in mind that, in earlier times, the prairie vegetation was much favored by the large herds of grazing bison. A natural experiment was provided by the catastrophic drought of 1934–41, the effects of which on the

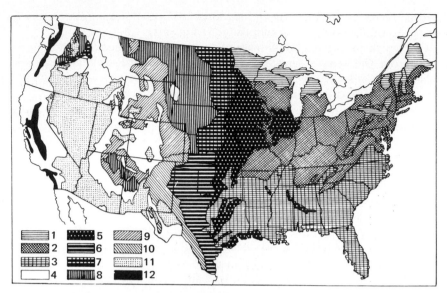

Fig. 124. Soil map of the United States: (1) podsol soils; (2) gray-brown forest soils; (3) yellow and red forest soils; (4) mountain soils (general); (5) prairie soils; (6) southern chernozem; (7) northern chernozem; (8) chestnut arid brown soils; (9) northern arid brown soils; (10) southern arid brown soils; (11) gray desert soils (sierozems); (12) Pacific valley soils. These soil types correspond to: (1) coniferous-forest zone; (2 and 3) mixed-forest and deciduous-forest zone; (5) tallgrass prairie; (6–10) mixed and shortgrass prairie; (11) sagebrush semidesert (in the north) and other desert types (in the south). (Based on U.S. Dept. of Agriculture map; Walter 1964/68)

prairie vegetation were still evident in 1953. Such recurrent periods of drought every century are undoubtedly partially responsible for the absence of trees on the prairie.

Tallgrass prairie is just as abundant in herbaceous plants as are the meadow-steppes and is floristically even richer. At the height of the flowering season in June, 70 species bloom simultaneously. The majority of the grasses *(Andropogon scoparius* and *A. gerardi),* being southern elements with a C-4 photosynthesis, do not flower until late summer, and in normal years they are not troubled by lack of water, since the prairie soils are moist to a great depth. The grasses themselves are 40–100 cm tall, even 1–2 m with their inflorescenses. In the mixed-prairie zone, apart from the tallgrasses *(Andropogon scoparius, Stipa comata),* there are many shortgrasses *(Bouteloua gracilis, Buchloë dactyloides).* In the shortgrass prairie, the latter assume the dominant role and herbaceous plants disappear. *Opuntia polyacantha,* however, is abundant, particularly in overgrazed areas (Küchler, 1974). Grazing tends to alter the appearance of the prairie slightly in the direction of an apparently greater degree of aridity, the tallgrass prairie turning into mixed prairie, and this, later, into shortgrass prairie. The $CaCO_3$ deposits in the soil indicate an increasing aridity toward the west. In the shortgrass prairie, $CaCO_3$ nodules are found at a depth of only 25 cm, the humus horizon is very shallow, and the plant roots are shorter since they only have to penetrate a short distance into the horizon with the chalk deposits, these being an indication of the mean depth to which the soil contains moisture. As part of the US/IBP, French (1979) published ten contributions in a volume containing ecological studies on production, consumers and grazing problems. The net production is approximately 2 t/ha in the shortgrass prairie, 3 t/ha in the mixed prairie and 5 t/ha in the tallgrass prairie.

5. Ecophysiology of the Steppe and Prairie Species

The cold winter, on the one hand, and the drought of late summer, on the other, limit the vegetational season of the steppe plants. Only about 4 months of favorable growth conditions occur, in spring and early summer. Most of the species are hemicryptophytes, and during this brief period they are obliged to build up a large productive leaf area at the smallest possible cost of material. Exact determinations of leaf area indices have not been made, but on the meadow-steppes, the values are probably similar to those for deciduous forests. Nevertheless, the total leaf area varies greatly from year to year according to the rainfall. The figures for the aboveground phytomass of the feathergrass steppe, which is poor in herbaceous plants, are 4530–6250 kg/ha in wet years and 710–2700 kg/ha in dry years. This means that an insufficient supply of water is countered by a reduction in transpiring surface, and a consequently smaller productivity. The underground phytomass remains unchanged and is much larger than that aboveground:

Meadow-steppe: Phytomass, 23.7 t/ha (84% underground)
                           Annual production, 10.4 t/h
Feather grass steppe: Phytomass, 20.0 t/ha (91% underground)
                           Annual production, 8.7 t/ha

The aerial parts dying off each year form a litter layer on the ground (steppe felt), amounting to 8–10 t/ha on the meadow-steppe and 3 t/ha on the dry steppe. When the underground parts die, they are converted into humus by the soil organisms. In spring and summer, the litter layer undergoes intense decomposition, with a minimum at the commencement of the drought period and a maximum at the beginning of winter. The seasonal changes of the environmental conditions and biotic parameter for the largest steppe reservation can be seen in Fig. 125. In protected areas, the accumulation of too much litter is detrimental to the regeneration of the grasses. As a consequence, the plant cover becomes patchy, and weeds such as *Artemisia, Centaurea,* and thistles become established. If the steppe vegetation is to be maintained in its original form, a certain amount of grazing is therefore indispensable. This was provided in earlier times by gazelles, saiga antelope,

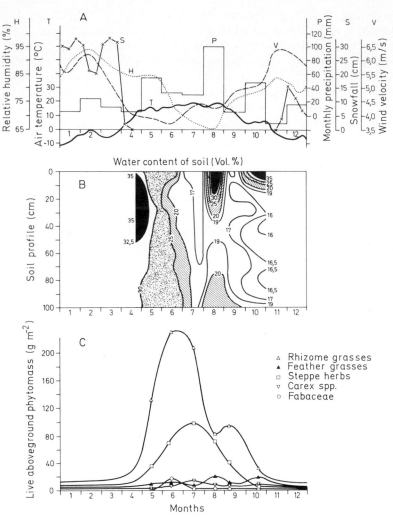

Fig. 125. Seasonal changes in environmental conditions (A, B) and biotic parameters (C–H) of a meadow-steppe ecosystem in the largest central chernozem steppe reservation during 1957. (A) Meteorological factors, (B) water content of soil, (C) aboveground phytomass, (D) phenology of the vegetation. (E) dead aboveground phytomass, (F) invertebrate zoomass, (G) humus content, (H) number of rodents (predominant vertebrate group)

wild horses and donkeys, and, above all, by the innumerable steppe rodents (ground squirrels, etc.) and locusts. Earthworms and burrowing rodents contribute substantially to the mixing of the humus with the mineral soil. Occasional naturally occurring steppe fires led to the destruction of the accumulated litter. Nowadays, in the steppe reservations, the grass is mowed every 3 years for hay in order to reduce such accumulation.

A similar state of ecological equilibrium exists between the grasses and herbaceous plants in the steppe as between woody plants and grasses in the

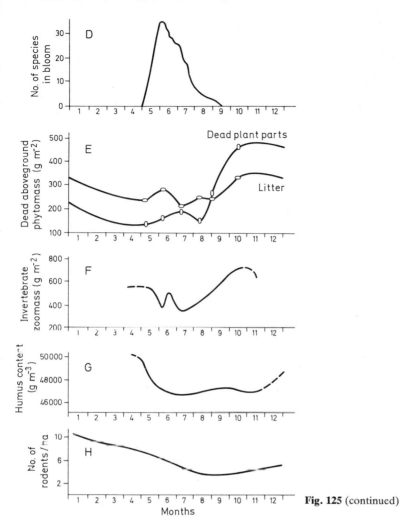

Fig. 125 (continued)

savanna (p. 80). Grasses possess a very intensive, finely branched root system, whereas that of the herbaceous plants is extensive, often with very long taproots. The nature of the water economy of the steppe herbaceous plants places them among the group of malakophyllous xerophytes. Their cell-sap concentration is very low in spring. Temporary periods of dryness lead to wilting, accompanied by a sharp rise in cell-sap concentration. Late-flowering species that bloom when the drought commences cut down their transpiration losses by allowing their leaves to wither. The flowers and fruits require little water and receive the necessary building materials from the withering plant organs.

Typical of the open spaces of the steppes are the so-called tumbleweeds (*Eryngium, Falcaria, Seseli, Phlomis, Centaurea, etc.*). The rigid stem supporting the spherical dried-out inflorescence breaks off at the root collar and is rolled across the steppe in the wind, scattering its seeds as it goes. The rolling plants often get entangled with each other and form enormous masses which are driven at great speeds across the steppes by the wind.

Stipa species regulate their transpiration by rolling up their leaves as well as closing their stomata, in this way also reducing photosynthesis. The distribution of the various species is determined by their adaptation to specific habitat conditions.

The water economy of the steppe-heaths of central Europe has been the subject of many investigations. This type of vegetation, an extrazonal relict of a xerothermic period of the postglacial era, is confined to warm loess or calcareous slopes or to sandy soils and is made up of very hydrolabile, malakophyllous steppe species. The dryness of such biotopes in central Europe is due to the small field capacity of the soils and the high potential evaporation on southern slopes rather than to the climate itself. Instead of a long drought in the late autumn, there are frequent but brief dry periods.

6. Asiatic Steppes

After the interruption presented by the southern Urals, the east European steppe continues into the more continental climate of Asia, although east of Lake Baikal it is interrupted by numerous mountain ranges and is confined to the basins and wide valleys. Only in Outer Mongolia and Manchuria is the steppe once more recognizable as a distinct zone (Fig. 126). The west Siberian steppe is similar in character to that of Europe, with certain floristic differences. *Lilium martagon* ssp. and *Hemerocallis* spp. are frequently encountered in both, but since the steppe climate east of Lake Baikal is extremely continental, with very little snow in winter and a dry spring, the spring flora is missing. *Filifolium (Tanacetum) sibiricum* is well represented; its leaves turn a brilliant red in autumn. A further consequence of the extreme continental climate is seen in the large numbers of what, for us, are alpine elements (species of the genera *Leontopodium, Androsace, Arenaria, Kobresia,* etc.) (Walter 1975a).

Innumerable small lakes dot the very flat northern Siberian–Kasakhstan steppe; paradoxically, these lakes result from the semiarid climate. In a wet climate where every hollow overflows with water, a river system can develop, whereas under semiarid conditions the situation is different, and each small hollow has its own catchment area. The hollows form wherever puddles accumulate after rain, and the water seeps slowly into the ground. As a result, the soil particles become more compactly distributed, and the soil volume decreases. This leads to settling of the soil surface and a deepening of the hollows. Similar lake flats have been observed by the author in the semiarid Pampa of Argentina, in western Australia, and from the air in North Dakota (United States). If such small lakes have subterranean drainage, however

Fig. 126. Grass steppe of *Poa botryoides, Koeleria gracilis, Festuca lenensis,* and others in highlands of Mongolia west of Ulan-Bator. (Photo P. Hanelt; from Walter 1974)

little, the water remains fresh, but if the water is only lost by evaporation, they become brackish (soda in slightly arid regions, otherwise chloride–sulphate).

In the eastern parts of the European steppe, and in North Dakota, the edges of the small and usually round lakes are lined with aspen *(Populus tremula* and *P. tremuloides),* and the steppe appears to be dotted with small groves. Aspen groves of this kind (often birch in Siberia) constitute the forest element of the forest-steppe where the nemoral zone peters out in the continental climate and the steppe meets up with the boreal coniferous-forest zone (i.e., in Zonoecotone VII/VIII, as in western Siberia and the Canadian steppe region).

7. Animal Life of the Steppe

Just like the American prairie, the steppe was originally big-game territory. Even in the eighteenth century, the tarpan was still extant *(Equus gmelini),* but in 1866, the last existing specimen was presented to the zoological garden in Moscow. The artiodactyls were well represented, and it is assumed that the aurochs *(Bos premigenius)* originally inhabited the steppe, gradually withdrawing to the forests in the face of man's encroachment. The antelope *(Saiga tatarica)* survived longer and is even seen today in a few places (reserves near Astrakhan and elsewhere). Both red and roe deer were previously found in the forest-steppe, and the wild pig *(Sus scrofa)* visited the watering places and lived in the reed stands. Light grazing is essential if the steppe vegetation is to be preserved, but man has completely exterminated both big game and predators, leaving only the myriads of rodents.

Today the fauna can only be observed in a few reserves in the Siberian and Mongolian steppe, where nomads still graze their herds.

A very important role in the formation of the chernozems is played by the soil organisms, above all the earthworms. The larger of these *(Dendrobaena)* work through the soil in all directions and to a considerable depth. As many as 525 passages/m^2 have been counted in the uppermost meter, and 110/m^2 at a depth of 8 m. The smaller forms *(Allophora* spp.) tend to be confined to the upper soil layers. The soil is so thoroughly mixed by the earthworms that the lower layers are enriched with organic material from the surface; in addition, the earthworm passages make root penetration easier. Next in importance are the ants, which also stir up the soil, followed by the rodents with underground burrows. The activities of the latter animals are well revealed by the "krotovinas," which are sections through tunnels that have been filled with humus soil from above and therefore show up in the profile as black circles in the loess soil. The rodents throw up earth from the deeper layers, and 175 such heaps have been counted per hectare, occupying 0.5–2% of the area. In time, this activity leads to the formation of a microrelief with small successions of vegetation. Steppe shrubs *(Caragana frutex, Amygdalus nana,* etc.) frequently colonize the heaps because they are safe from the competition of grass roots. Rodents, too, help to loosen the soil.

Up to about 200 years ago, the steppes were only lightly grazed since the nomads were constantly on the move with their herds, and the vegetation remained almost unaltered. But during the past two centuries, the change-over to cultivation and the introduction of intensive grazing has destroyed the steppe ecosystem, and as a result, the fauna has suffered considerably. Only a few animal organisms have adapted to the changed conditions. Rodents are a pest to the farmers, and insects that were harmless on the steppe now attack sugar beet and grain in large numbers and must be combatted by chemical means.

Cattle grazing has degraded the plant cover, and large areas have become potential breeding-grounds for locusts. The regeneration of steppe vegetation is a very slow process. It is impossible, however, to go into the question of the secondary successions here.

8. Grass Steppes of the Southern Hemisphere

Compared to the area of the grass steppes of the Northern Hemisphere, the area occupied by those of the Southern Hemisphere is relatively small. The largest continuous area is the Pampa in the eastern Argentinian province of Buenos Aires and parts of the neighboring provinces. It may be considered a semi-arid variant of Zonobiome V with relativly high precipitation in the summer.

The Pampa lies between 32° and 38° S, extends over about 0.5 million km^2 and borders directly on the Atlantic coast. It is thus situated in a warm-temperate region and corresponds to the southernmost parts of the prairies of Oklahoma and Texas. Rainfall reaches 1000 mm in the northeast of the Pampa and diminishes to 500 mm in the southwest, at its dry limit.

Although these values appear high at first sight, it has to be remembered that temperatures, and thus potential evaporation, are also very high (Buenos Aires: mean annual temperature, 16.1 °C). Despite this, the climate of the Pampa has always been considered to be humid. The question has continually arisen as to why the Pampa is bare of trees. The simplest explanation, as in all cases where no other explanation is available, is that the vegetation is of anthropogenic origin, having arisen from an earlier forest vegetation as a result of fires set by man.

This assessment of the climate has, however, proved to be incorrect. Even in the wettest parts of the extremely flat Pampa, many shallow lakes with no outflow (locally known as lagoons) are present, besides innumerable small pans which, although they contain water in spring, are dried out in summer. The water in the lagoons contains soda and is strongly alkaline. The soils surrounding the pans are alkaline (solonetz) and support the grass typical of saline soils *(Distichlis)*. All of this points to a semiarid climate such as has already been encountered in the forest-steppe. Measurements of potential evaporation have shown that, in the coastal regions of La Plata, evaporation and rainfall are equal but that, in the Pampa, evaporation exceeds rainfall. In the more humid parts of the Pampa, the negative water balance amounts to 100 mm, and in its arid parts reaches 700 mm. In January and February, rainfall is at a minimum, and the potential evaporation is particularly high because daytime radiation is very intense and only in the evenings and at night do severe thunderstorms occur. Although wellsupplied with water in the spring, the vegetation is severely scorched by January.

On well-drained ground in a forest-steppe climate, a woodland vegetation is to be expected, and in fact, in the vicinity of the coast, small groves of *Celtis spinosa* (tala) do occur on slight elevations with a porous limestone or sandy soil. On poorly drained ground, there is a grassy vegetation. Bushes grow on stony hillocks (Frangi 1975). Almost nothing remains, however, of the original vegetation of the Pampa. On the grazed areas, European grasses have been introduced. They are softer than the Pampa grasses and are preferred by the European breeds of cattle. The many planted exotic trees, whose roots are protected from competition by the intensive root system of the grasses, grow well everywhere. Judging from small remnants occurring on ungrazed patches, it can be concluded that, on the humid northeastern portion of the Pampa, *Stipa–Bothriochloa laguroides* steppe composed of about 23 graminids and 46 herbaceous species originally prevailed (Lewis and Collantes 1975). The soil profile beneath such a steppe has a thick humus horizon (1.5 m) and is reminiscent of thick chernozems or prairie soils. There are signs, however, of alternate high and low water content, and the soils are a transitional form leading on to the subtropical grassland soils of southern Brazil. There is no indication that forests existed here previously. Where the groundwater table is high, there are stands of the dense tussocks of *Paspalum quadrifarium,* which, at a very high groundwater table, are replaced by *Distichlis* on alkaline soils (pH 8–9).

Fig. 127. Southern tussock Pampa with *Stipa brachychaeta* (central province of Buenos Aires). (Photo E. Walter)

The dry southwestern Pampa was previously tussock grassland with *Stipa brachyaeta* and *S. trichotoma* and was almost entirely lacking in herbaceous plants. "Tussock" indicates a growth form wich, although completely lacking in the Northern Hemisphere, is widespread in the Southern Hemisphere, with its mild winters. A tussock consists of bunchlike tufts sometimes more than a meter high, in which the hard, old, withered leaves are intermingled with the fresh, young, green leaves, thus providing the tussock grassland with its perpetual yellowish color (Fig. 127). These grasses are of little grazing value, and for this reason they are often ploughed under to give the European grasses a better chance to establish themselves.

Toward the west, where the rainfall has decreased to 500 mm annually, falling mainly in the summer, and the loess soil is replaced by light sandy soil, the Pampa is replaced by xerophytic *Prosopis caldenia* woodland. As the rainfall decreases even further, the woodlands are succeeded by *Prosopis* savanna (Fig. 128), which strongly recalls the *Acacia* savanna of Southwest Africa. At the same time, large stretches of saline soil are found, bearing a halophytic vegetation. At a rainfall of less than 200 mm annually, there is a *Larrea* semidesert on stony ground, with many broomlike bushes belonging to various families (Caesalpinaceae, Scrophulariaceae, Capparidaceae, Compositae). With such a small transpiring surface, and by drastically cutting down transpiration during the 6 months of drought, the semidesert vegetation is able to survive with the meager amount of soil water at its disposal. This amounts to 50–80 mm annually on flat ground, 25–55 mm on sloping ground, and more than 140 mm in small valleys into which water drains.

Fig. 128. Tree savanna with *Prosopis caldenia* and a grass cover of *Stipa tenuissima* and *S. gynerioides* between Santa Rosa and Victoria (Argentina). (Photo E. Walter)

The *Larrea* semidesert extends along the eastern foot of the Andes to the northern part of Patagonia, where, south of 40° S, strong west winds blow continuously across the Andes, which at this point are rather low (pass altitude, 1000 m). The wind is, however, of a foehn character, descending and dry. The eastern margins of the mountains have a rainfall of 4000 mm annually and support *Nothofagus* forests and are succeeded to the east by dry *Austrocedrus* forest and, following on these, a bushland of beautiful red-blossomed Proteaceae *Embotrium coccineum*. The woody plants then disappear, and the Patagonian steppe commences. Only 100 km from the Andes, the annual rainfall is 300 mm and, even farther away, diminishes to 160 mm. Apart from the true steppeland on the westernmost margins of Patagonia, where low tussock grasses are predominant *(Stipa* and *Festuca)*, it is more correct to speak of Patagonian semidesert, which is characterized by xerophytic cushion plants belonging to completely different families (Compositae, Umbelliferae, Verbenaceae, Rubiaceae, etc.) (Fig. 129). The ground is in many places 60–70% bare. The cushionlike form appears to be an adaptation to the constant strong wind (mean velocity, 4–5 m/sec); within the cushions, a propitious microclimate can be achieved, protected from the effects of the wind.

The Patagonian tussock grassland has much in common with that of Otago on the South Island of New Zealand, situated in the lee of the New Zealand Alps, with a rainfall of 300 mm. Both lie south of a latitude of 40° S. Low tussock grasses predominate *(Festuca nova-zelandiae, Poa caespitosa)*, but at an altitude of 750–2000 m, where snow remains for 2–3 months of the

Fig. 129. Patagonian semidesert with cushions of *Chuquiraga aurea* near Manuel Choique (province of Rio Negro). (Photo E. Walter)

year, they are replaced by taller tussock grasses, 1.5–2 m high *(Chionochloa, = Danthonia)*. Fire and grazing are partially responsible for the fact that tussock grassland has spread widely in places at the expense of the original *Nothofagus* forests. So far, no ecophysiological investigations of these grasslands have been undertaken.

9. Subzonoecotone of the Semidesert

Semidesert is distinguishable from true desert by its diffuse vegetation, although the ground is only covered to about 25%. In the true desert, the density of the vegetation is still lower, and at the same time a change from a diffuse to a contracted vegetation takes place. The plant cover of the semideserts differs greatly. In the frost-free subtropics and in the tropics the plant cover consists mainly of woody plants and succulents, and in the temperate zone with cold winters, mainly of half-shrubs, especially the genus *Artemisia*. This holds true for the semideserts of Eurasia as well as North America. The characteristic cushion plants of windy Patagonia have already been mentioned.

Saline soils are widespread, as would be expected from the greater aridity of the semidesert. This is particularly marked in eastern Europe, where the broad expanses of the "sivash" to the north of the Crimea dry out in summer and are covered by a salt crust. The salt dust is blown north by the wind and deposited in the southern chernozem and chestnut soil zones, and causes solonization of the soil. In spring, the salt is washed out of the upper soil

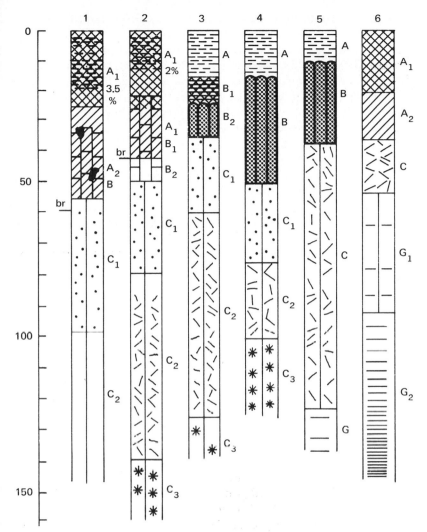

Fig. 130. Soil profiles in eastern Europe, weakly to strongly saline: (1) slightly solonized southern chernozem, some compaction (A_2B); (2) dark-chestnut soil with B-horizon; (3) light-chestnut soil, strongly solonized (A, poor in humus and laminated; B, columnar and very compact); (4) typical columnar-solonetz soil; (5) solonetz changed by rising groundwater; (6) typical solonchak with high groundwater and A_1 rich in humus. C_1, $CaCO_3$ nodules; C_2 (in 2–4) and C (in 5–6), gypsum tubules; C_3, gypsum druses; G, G_1, and G_2, gley horizon (groundwater). (From Walter 1960)

layers by water from the melting snows, and the sodium-humus sol thus formed carries the sesquioxides (Fe_2O_3, Al_2O_3) along with it into the deeper soil layers. Here precipitation occurs, and a compact B-horizon of strongly alkaline reaction (Na_2CO_3) is formed (Fig. 130). The amount of salt deposited increases steadily toward the south.

Humic material is entirely leached out of the A-horizon, and the strongly alkaline B-horizon becomes harder and harder and, owing to the alternate swelling in the humid season and shrinkage in summer, assumes a columnar structure. This so-called columnar solonetz resembles the podsols in certain respects, although the latter are strongly acid in reaction, their peptization being effected by H-ions. Beneath the B-horizon of the solonetz soils, the very slightly soluble $CaCO_3$ precipitates as chalk nodules, followed by gypsum as tubular or druselike deposits, and the readily soluble salts are washed down into the groundwater.

If the groundwater rises, as is happening on the slowly sinking north coast of the Black Sea, a wet saline soil known as solonchak is formed. The groundwater is drawn to the surface by capillary forces and evaporates. This results in a horizon containing gypsum tubules above the gley horizon which is followed by the humus horizon, which bears a white salt crust in the dry season. Humus soles are not formed since the humus is precipitated in the presence of such a high salt concentration.

The steppe grasses recede on solonetz soils, to be replaced by *Artemisia maritima* ssp. *salina* and *A. pauciflora,* as well as species of the genera *Camphorosma, Limonium, Kochia, Petrosimonia,* etc. Ground lichens *(Aspicilia)* and species of *Riccia* and *Nostoc* are also found (in North America, *Ceratoides lanata, Atriplex confertifolia, Kochia* spp., etc.)

On slightly elevated, nonsaline ground, a semidesert arid brown soil, or burozem, is formed. Its upper horizons contain only 2–3% humus and are brown in color. The effervescence level is at a depth of 25 cm, and plant cover amounts to less than 50%. The vegetation consists of *Festuca sulcata* and the low half-shrubs *Pyrethrum achilleifolium, Kochia prostrata* and *Artemisia maritima* ssp. *incana,* which avoids saline soils. Only solitary individuals of *Stipa* species are seen, but in spring many ephemerals appear.

In the Caspian lowlands, the two communities often form a mosaic on the burozem and solonetz soils because of the nature of the microrelief. *Salicornia* and *Halocnemum* predominate on very wet solonchak, and *Suaeda, Obione, Petrosimonia, Limonium caspica, Atriplex verrucifera,* etc., where it is less wet (see also Levina 1964, Walter and Box 1983).

After the Caspian Sea receded from the delta region of the Volga-Ural river system, the southern part of the Caspian lowlands was left covered with alluvial sand, upon which *Artemisia maritima* ssp. *incana, Agropyron cristatum, Festuca sulcata, Koeleria glauca,* etc., grew. But the vegetation was destroyed by grazing, the sand became mobile, and large bare wandering dunes, or barchanes, were formed. Whenever the sand becomes more stationary, a pioneer vegetation consisting of *Elymus giganteus* and *Agriophyllum arenarium,* a chenopod, can gain a foothold, followed by species of *Salsola* and *Corispermum.* In the dune valleys, *Aristida pennata* and *Artemisia scoparia,* among others, make their appearance, and gradually the zonal vegetation is restored.

Sand dunes, particularly those devoid of vegetation, store water. Groundwater is always present beneath the dunes, and this leads to the formation of small freshwater ponds in the dune valleys, around which *Elaeagnus angustifolia,* willows, and poplar grow. Attempts to get willow *(Salix acuminata)* and poplar to grow on the sandy areas were initially successful. The plants developed well for the first 4 years at the expense of the water stored in the ground, but when this was exhausted, they died.

In Kazakhstan, there are large areas of semidesen between the southern Siberian steppe in the north and the desert in the south. It is comparable to the sagebrush zone of North America, with *Artemisia tridentata* (see p. 169).

10. Subzonobiome of the Middle Asiatic Deserts[3]

This region lies to the north of the limit of date cultivation. In the Russian literature, a distinction is made between the Middle Asiatic and the central Asiatic desert (Fig. 131). The former comprises the Irano-Turanian desert region occupying the southern portion of the Aralo-Caspian desert and the southern part of Kazakhstan, including Dsungaria. The central Asiatic desert comprises part of Dsungaria, the Gobi desert, the western part of Ordos on the great bend of the Hwang-Ho, Ala-Shan, Bei-Shan, and the Tarim basin

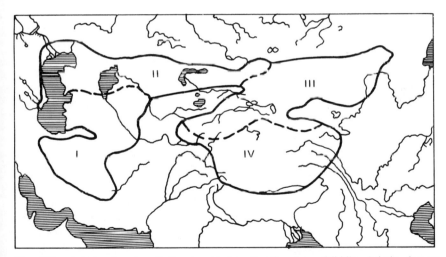

Fig. 131. Asiatic deserts of the temperate climatic zone. Middle Asiatic deserts: I, Irano-Turanian (in parts almost subtropical); II, Kazakhstan-Dsungarian. Central Asiatic deserts: III, in strict sense (hot summer); IV, Tibetan cold high-mountainous desert. (From Petrov; Walter 1964/68)

---

3 Twenty-five photographs of the vegetation of the Middle Asiatic region are to be found in Walter (1974, 1976)

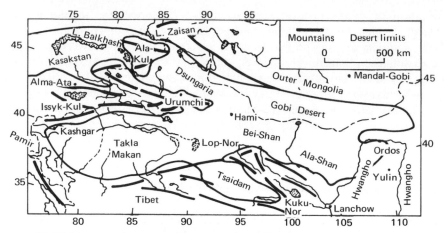

Fig. 132. Divisions of the central Asiatic desert regions. Dsungaria is a transitional region to Middle Asia. (From Petrov; Walter 1964/68)

(Kaschgaria), together with the Takla-Makan desert and the more elevated Tsaidam basin (Fig. 132).

The Tsaidam basin is succeeded by the high mountain desert of Tibet together with the Pamir in the extreme west.

In Middle Asia, cyclonic rain is still received from the Atlantic Ocean, falling in the winter in the southern parts, and mostly in the spring and summer in the north; in any case, the soil here is always wet in spring after the snows have melted. Rainfall diminishes from west to east. Floristically, the Irano-Turanian element is strongly represented. In contrast to the situation in Middle Asia, the source of the moisture in central Asia is to be found in the extensions of the east Asiatic summer monsoons. Winter and spring are extremely dry, and the aberrant rainfall distribution accounts for the predominance of east Chinese elements in the flora (see Fig. 133, Denkoi).

In these regions, the most detailed ecological investigations have been carried out on the vegetation of the Middle Asiatic desert in the Aralo-Caspian lowlands (formerly Turkestan). The rainfall in the entire region amounts to less than 250 mm. Since the winters are cold, evaporation at this time of year is very low. This is the reason why the annual evaporation at the Bay of Bogaz is only 1100 mm. The various types of vegetation are determined by the soils (biogeocene complexes):

1. *Ephemeral desert.* This is found on loesslike, salt-free soils that are very wet in spring but dry from May onward. During the brief period of vegetation, lasting from the beginning of March until May, annual species and geophytes develop, the most common of which are *Carex hostii (C. stenophylla)* and *Poa bulbosa.* Here and there, the 2-m-high *Ferula foetida* is encountered. The 40–50 annual species manage to produce ripe seeds

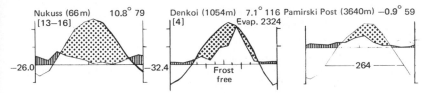

Fig. 133. Climate diagrams of Nukuss in Middle Asia with winter rain, Denkoi in central Asia with summer rain, and Pamirski Post in the cold desert (only 264 days with a mean temperature above −10 °C). (From Walter and Lieth 1967)

Fig. 134. Dry takyr areas, with the ground split up into polygons. A *Tamarix* bush has established itself on drifted sand. (Photo P. Hanelt)

within 30–45 days. In years with a good rainfall, the desert presents the appearance of a meadow, producing a dry mass of 0.5 to 2.5 t/ha. It provides grazing for 3 months but is completely lifeless for the rest of the year.

2. *Gypsum desert.* This is stony desert (hamada) on the high plateaus of table mountains (mesas). The soil contains up to 50% gypsum, which has the property of storing water. The situation is similar to that in the Sahara. Therophytes develop in the spring, but otherwise gypsum plants provide a ground cover (0.1%, except in the erosion gulleys, where plants are more abundant). There are also a few halophytes.

3. *Halophyte desert.* This type of desert is found more extensively, on soils with groundwater close to the surface, in the lower reaches of rivers, in hollows (shory), or around salt lakes. Most of the plants are hygrohalophytes (*Salicornia, Halocnemum, Haloxylon, Seidlitzia,* etc.).

4. *Takyrs.* These are seemingly bare, clayey, flat expanses which are flooded in spring by surface water running down from the mountains, but which soon dry out again (Fig. 134). The shallow pools left behind warm up rapidly and harbor 92 species of Cyanophyta, 38 of Chlorophyta and other

algae, producing in all 0.5 dry matter t/ha with an N content of 4.5% (N fixed from the air by Cyanophyta). Lichens *(Diploschistes,* etc.) colonize slightly higher areas, but flowering plants are rare.

5. *Sand deserts.* These are particularly widespread: Karakum (black sand) between the Caspian and the Amu-Darya River, and Kysylkum (red sand) between the Amu-Darya and Syr-Darya rivers. The sandy soil favors the growth of a denser vegetation (p. 116).

Extensive ecological investigations carried out at the desert station Repetek since 1912 provide a basis for the following discussion of the region.

11. Biome of the Karakum Desert

The Karakum desert covers an area of 350,000 km^2 and occupies the southern part of the Turanian depression between the Caspian Sea in the west and the Amu-Darya River in the east (Fig. 135). The latter arises at 5000 m above sea level in the mountains of Pamir and flows for 2600 km. This sand desert is a geographically well-delineated biome of the temperate-desert subzonobiome of Zonobiome VII. It occupies a large basin that has been in the process of being filled with loose alluvial rocks by the Amu-Darya since the Tertiary. In the course of redeposition by wind, the dust particles have been deposited as loess on the Kopetdag slopes in the south, and the sand has formed a dune region (Fig. 136).

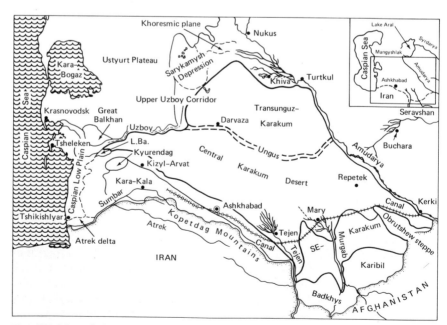

Fig. 135. Map of the Karakum desert (thickly outlined)

Fig. 136. Karakum desert with slightly raised dune relief and sparse growth of shrubs (*Haloxylon persicum,* etc.) Ground only covered in spring with ephemerals and ephemeroids. (Photo Petrov)

The Amu-Darya originally emptied into the Caspian Sea, but was displaced to the east by the delta deposits of the Murgab and Tedzhen rivers coming from the south, so that it now empties into the Aral Sea. But the river is still important to the region since its water infiltrates to feed a groundwater lake underlying the entire central Karakum. In the few places where this groundwater comes to the surface, saltpans form.

The water budget for the biome as a whole is given below. Inflow to the subterranean groundwater lake is as follows:

Groundwater infiltration from the Amu-Darya (mean)	150 m³/sec
Rainwater seepage in the barchane region	30 m³/sec
Infiltration from Murgab and Tedzhen	21 m³/sec
Subterranean inflow from Kopetdag (from the south)	20 m³/sec
Seepage from elevations and takyrs	1 m³/sec
Total inflow	222 m³/sec

Approximate losses are as follows:

Evaporation from saltpans above high groundwater table (average)	165 m³/sec
Groundwater losses due to phreatophytes (i.e., plants dependent on groundwater)	57 m³/sec
Total losses	222 m³/sec

Although the groundwater moves slowly from east to west, it has been proved that none of it flows into the Caspian Sea. The water is slightly brackish, although it is covered by a lens of fresh water beneath the bare dunes, owing to rainwater seepage, even in years when as little as 100 mm of rainfall is recorded. Wells sunk in such places yield good drinking water.

The climate of the Karakum can be seen on the climate diagram for Nukuss (Fig. 133). Fifty to seventy percent of the precipitation falls in spring. The winter is cold, but no lasting snow cover forms. The summer is hot, with a period of extreme drought. Potential evaporation amounts to 1500–2000 mm, or 10–15 times the precipitation.

Differences in vegetation and soil provide the basis for a distinction between the following biogeocene complexes: (1) psammophyte complex, covering 80% of the area, (2) takyr complex, and (3) halophyte complex.

The ecological studies that have been carried out for many years in the protected area in the vicinity of the desert station Repetek will be very briefly summarized. The 34,000 ha consists of 14,000 ha of bare barchanes, 18,000 ha of plant-covered dunes, and 2000 ha of dune valleys. The treeshrubs *Haloxylon persicum* (white saksaul) and *H. ammodendron* (= *aphyllum*) (black saksaul) are characteristic of the sandy desert. Within the psammo-phyte complex, several distinct biogeocenes can be recognized. The first is the biogeocene of Ammodendretum conollyi aristidosum on light, shifting sand, with a pioneer synusia of *Aristida karelinii* on the dune ridges and shrubs (*Ammodendron conollyi, Calligonum arborescens, Eremosparton,* and others) on the upper slopes. The second biogeocene is Haloxyletum persici caricosum on stationary sand of the lower slopes and on sandy areas, with synusiae of spring and summer ephemerals and ephemeroids (143 species, 24 of which are common). The third biogeocene is that of the deep dune valleys with groundwater at a depth of 5–8 m, which can be reached by the roots of the salt-tolerant *Haloxylon ammodendron,* a tree that attains a height of 5–9 m and forms small woods. Here, too, a number of synusiae can be recognized in the undergrowth (ephemerals, halophytes). The most widespread biogeocene is Haloxyletum persici caricosum (Fig. 136) with an open shrub layer 3–5 m tall (100–300 plants per hectare), where aphyllic *Caligonum* spp. (Polygonaceae) are found. The age composition of the shrubs is as follows:

Age in years	1	2	3–5	6–10	11–15	16–20	> 20	Dead
Percentage	8	3	1	14	20	41	11	2

Ecological studies have revealed that the shrubs remain active throughout the year since the upper 2 m of sand always contains available water. The osmotic potential drops slightly during the drought, but water deficits are never very large; the otherwise very intensive transpiration is reduced to $\frac{1}{3}$ or $\frac{1}{2}$ in the drought period. Photosynthesis continues uninterrupted throughout the summer and is particularly intense in the ephemerals, which have only a short vegetational season.

In the Ammodendretum biogeocene on shifting sand, 83 mm of water was found in the upper 2 m of sand, although at the end of the vegetational season only 34 mm was left. This is a difference of 49 mm, of which 37 mm was used by the plants for transpiration, and apparently only 12 mm evaporated.

Conditions on the fixed sand with denser vegetation are less favorable. It was found that 30 mm of water was transpired by the shrubs (H. *persicum* alone, 16 mm) and 17 mm by the dense undergrowth of *Carex physodes,* which gives a total of 47 mm. Water in the soil amounted to 62 mm in spring and 8 mm in autumn, which means that, in all, 54 mm was lost (thus 7 mm evaporated).

In wooded deep dune valleys situated above groundwater, transpiration amounted to a total of 149 mm (all shrubs, 138 mm; thin carpet of *Carex physodes,* 11 mm), although the soil contained only 76 mm. Assuming that *H. ammodendron* takes the 108 mm that it requires from the groundwater, the 41 mm needed by the rest of the plants can be supplied by the water in the upper 2 m. In the case of *H. ammodendron,* 14% of the precipitation runs down its trunk, which makes it easier for the taproots to penetrate the soil to greater depths in the rainy years.

Thus, despite the very high intensity of transpiration of Karakum shrubs per gram of fresh weight, the total transpiration per unit area is low owing to the small leaf area. The main adaptation of the plants is their aphylly and their ability, during drought, to shed any small leaves that may have formed in spring.

The aboveground phytomass was as follows: on shifting sand, 80 kg/ha (25% *Calligonum,* 12% *Aristida*), and on sand covered with vegetation, 2.4 t/ha (85% *Haloxylon persicum.* 10% *Carex physodes*). The subterranean phytomass is much larger but was only determined in the dune valleys with *Haloxylon ammodendron;* aboveground, 6.4 t/ha (82% *Haloxylon)* and subterranean, 19.4 t/ha (49% *Haloxylon*). The annual primary production for *Haloxylon ammodendron* amounted to 1.17 t/ha aboveground and 2.11 t/ha belowground. These are high production figures for a desert and are only possible because of the groundwater.

The takyr biogeocene complex is found where water of episodic rivers flooding large plains has deposited a layer of clay, above which the water stands until it evaporates. Algal masses, chiefly cyanophytes (N-fixing as well), develop on the water, and an algal skin is left when the water evaporates. The polygonic cracks formed on the dry ground rapidly close again when rainwater swells the soil.

Lichens develop in places that are wet but not flooded, and further up the slopes ephemerals can be found. The phytomass (= primary production) in the three biogeocenes is as follows: algae, 0.1 t/ha; lichens, 0.3 t/ha: and ephemerals, 1.2–1.6 t/ha.

The halophyte biogeocene complex is found where saltpans have formed as a result of a high groundwater table. The biogeocenes, often consisting of only one species, are arranged concentrically around the central portion, which has a salt crust. For the pioneer species *Halocnemum strobilaceum,* a phytomass of 1.76 t/ha (roots accounted for 1.04 t/ha) and an annual production of 0.5–0.7 t/ha were found.

The delta region of the Amu-Darya supports luxuriant *Populus-Halimodendron* forests rich in lianas, as well as large expanses of reeds *(Phragmites).* The following figures were obtained for aboveground phytomass and primary production: floodplain forest, 77.8 t/ha aboveground phytomass and 11.4 t/ha primary production; reeds, 35 t/ha phytomass and the extremely high figure of 18 t/ha for primary production.

Animals play a very important role. The herds of antelopes, donkeys, and wild horses have been replaced by millions of karakul sheep, which graze all the year round on the sandy desert and yield the Persian lambskins. The animals prevent the sandy expanses from becoming overgrown with *Carex physodes* and the moss *Tortula desertorum*. At the same time, the sheep tread the seeds from the shrubs into the soft ground, which facilitates germination. The soil is churned up by the innumerable rodents and the giant tortoises (100 per hectare) that feed on the ephemerals for 2.5–4 months and spend the rest of the year asleep in the ground.

However, the zoomass of all three biogeocenes is, as usual, not large: mammals, 0.3–1.4 kg/ha; birds, 0.02–0.07 kg/ha; reptiles, 0.21–0.7 kg/ha (excluding tortoises); invertebrates, maximum 15 kg/ha (above- and belowground).

Cycling of mineral elements was also studied, but the decomposers were only dealt with summarily. A detailed description of the Karakum desert may be found in Walter (1976).

12. Orobiome VII(rIII) in Middle Asia

The altitudinal belts of Orobiome VII are particularly interesting in Middle Asia, where the orobiome falls within Zonobiome VII(rIII). The mountains, which rise to over 7000 m above sea level, are part of the Pamiro-Alai and Tyanshan systems. Almost every sequence of belts here exhibits peculiarities, depending upon the nature of the local ascending air masses. Nevertheless, two main types of sequences can be distinguished: (1) an arid sequence with no forest belt and (2) a more humid sequence with 1–2 forest belts (Stanjukovitsch 1973).

In the most extreme cases in central Tyanshan, the semidesert belt is followed by a mountain steppe belt from 2000 to 2900 m above sea level, with an admixture of alpine elements such as *Leontopodium alpinum, Polygonum viviparum, Thalictrum alpinum,* etc., from 2600 m upward in the subalpine belt. Steppe elements also extend into the lower alpine belt (*Kobresia* swards), up to 3500 m, above which they disappear. The transition from high-montane to subalpine and alpine belts is a gradual one.

The explanation for this remarkable mixture of steppe and alpine elements is that the steppe plants require 4 months of favorable conditions for vegetation and, on account of the intense radiation, these are found up to 3500 m above sea level in the arid mountainous climate. The remaining 8 months of the year may be arid or cold. The alpine elements, however, can manage with a shorter period of vegetation but are able to go on growing as long as conditions are moist enough for them to compete successfully with the steppe plants (Walter 1975a).

In less extreme situations, on the northern slopes in subalpine belts, stands of *Juniperus* trees occur. These take on an espalier form on the ground in the alpine belt proper.

Of special interest are humid belts in which semidesert, with xerophilic trees *(Pistacia, Crataegus)*, and mountainous steppe is succeeded by a deciduous-tree belt with wild fruit trees such as *Amygdalus communis, Juglans regia, Malus sieversii*, and species of *Prunus* and *Pyrus*. The capital of Kazakhstan is Alma-Ata, (Kazakh = apple father.) When the fruit is ripe in the *Malus* belt, the entire population gathers around camp fires to cook apples. The deciduous belt is followed by a coniferous belt consisting of *Picea schrenkiana* and, in places, *Abies semonovii*, which is, in turn, succeeded by the alpine belt.

Afghanistan, with the Hindu Kush Mountains (Freitag 1971), provides another example of similar vegetation.

The succession of altitudinal belts in the Front Range of the Rocky Mountains near Colorado Springs provides an example of an orobiome in the semiarid climate region VII of North America. The shortgrass prairie at the foot of the mountains is succeeded at 1500 m above sea level by a belt with longgrass prairie, and then by a belt only 50 m wide of deciduous shrubs, with *Pinus edulis*, and *Juniperus* (pinyon belt). This is followed by the forest belts, in which *Pinus ponderosa, Pseudotsuga menziesii*, and *Picea engelmanii*, successively, achieve dominance. The timberline is reached at 3700 m above sea level, and above a narrow subalpine belt with dwarf *Picea* and *Dasiphora (Potentilla) fruticosa* bushes, the alpine belt is reached.

13. Subzonobiome of the Central Asiatic Deserts[4]

As already mentioned, the last traces of the Chinese monsoons or low pressure weather front coming from the East are still noticeable in this region, which explains why rains fall in summer and rainfall diminishes from East (Ordos, 250 mm) to West (Lop-Nor depression, 11 mm). Winter and spring are dry, and the spring ephemerals, so typical of Middle Asia, are entirely absent in central Asia. The flora is poor, with shrubby psammophytes *(Caragana, Hedysarum, Artemisia,* etc.) predominating among the east Chinese-Mongolian elements. *Stipa*, too, is represented by central Asiatic species. Buckthorn *(Hippophaë rhamnoides)* and the tallgrass chii *(Lasiagrostis splendens)* are widespread. Apart from *Populus diversifolia* and *Elaeagnus*, the floodplain forests contain *Ulmus pumila*. Among the halophytes, *Nitraria schoberi* and species of *Zygophyllum, Reaumuria, Kalidium*, and *Lycium* deserve mention.

---

4 Twenty-four photographs of the vegetation of Central Asia are to be found in Walter (1974). A detailed description is given by Walter and Box (1983)

The character of the deserts is influenced by their geological structure and the nature of the rock. They are as follows (from East to West):

1. *Ordos*. This is the region lying in the bend of the Hwang-Ho to the north of the Great Wall of China, which runs along the edge of the wandering-dune region. It joins up with the steppe region of the loess plains of the upper Hwang-Ho, nowadays cultivated and very deeply dissected by erosion gulleys. The Ordos is a *Stipa* steppe differing greatly, however, from that of eastern Europe because of the dry spring. The underlying rock of the true Ordos region is soft sandstone which has given rise to large expanses of sand and dunes with widespread *Artemisia ordosica* semidesert vegetation with *Pycnostelma* (Asclepiadaceae) (cover, 30 to 40%). In the central undrained parts, there are lakes containing Na_2CO_3 and $NaCl$.

2. *Ala-Shan*. This is a desert consisting largely of sandy wastes with barchanes. It lies to the west of the Hwang-Ho and stretches as far as the Nan-Shan Mountains in the south. To the north it borders the Gobi desert near the Gushun-Nor. Rainfall decreases from 219 mm in the east to 68 mm in the west with the potential evaporation increasing from 2400 to 3700 mm. The rainfall maximum occurs in August, the mean annual temperature is 8 °C, and the minima range from −25° to −32 °C. Ground water is present in the dune region. The encircling mountains have a higher rainfall. Above the desert and steppe altitudinal belts, mesophytic shrubs appear at an altitude of 1900 to 2500 m; these include *Lonicera, Rosa, Rhamnus,* and *Dasiphora fruticosa*. Above this, and up to 3000 m, there is coniferous forest with *Picea asperata, Pinus tabulaeformis,* and *Juniperus rigida,* succeeded by subalpine shrubs and alpine mats.

3. *Bei-Shan*. This region is west of the Ala-Shan and is an ancient elevated block rising from 1000 to 2791 m above sea level. It is bounded on the west by the Lop-Nor depression and Hami. Rainfall amounts to 39–85 mm, and potential evaporation to 3000 mm. The vegetation is low shrub-desert consisting of central Asiatic species with a few halophytes. *Picea asperuta* is found growing on the highest points.

4. *Tarim Basin and Takla-Makan*. The basin is 300 km long and 500 km wide, and is surrounded on three sides by high, snow-covered mounlains. It is the most arid part of central Asia, with hot summers and cold winters (minimum, −27.6 °C). Despite this, it is well provided with groundwater fed by the mountain rivers. The 200-km-long Tarim River has an average flow of 1200 m³/sec and forms wide floodplains. ln its lower reaches, the river continually changes its course, and its water seeps far into the central sandy desert (Fig. 137). Lop-Nor is sometimes a salt lake 100 km in diameter and is sometimes completely dried out. The sandy desert Takla-Makan is devoid of vegetation, but water can readily be obtained by sinking wells in the dune valleys.

5. *Tsaidam*. This is an elevated basin at an altitude of 2700 to 3000 m, completely surrounded by much higher mountains from which it receives its water. It is cut off from the Lop-Nor depression by the Altyn-Tag. The

Fig. 137. Tarim basin, river-seepage region with nebka landscape (dunes heaped around the *Nitraria schoberi* bushes). (Photo Petrov)

Fig. 138. Gobi desert (900 m above sea level) with sparse shrub vegetation. (Photo P. Hanelt)

Fig. 139. Oasis of Dsun-Mod (in Mongol: 100 trees) in the Transaltai Gobi, with *Populus diversifolia*. The spring is in the hollow in the background. (Photo P. Hanelt)

Fig. 140. Southern border of the Mongolian Altai (1400 m above sea level). Slopes with desert vegetation, *Populus pilosa* stands in the river valley, and tents of the nomads. (Photo P. Hanelt)

mean annual temperature is approximately 0 °C, the minimum being below −30 °C. The central part of the basin was a large lake in the Pleistocene but is today a barren salt desert. *Artemisia* semidesert is found on the sandy soil at the foot of the mountains. Tsaidam forms a transition to still higher Tibet.

6. *Gobi* (Mongol. = desert). This region is north of the above-mentioned deserts and covers the entire southern part of Outer Mongolia. It is separated from the forests and steppes to the east by the Khingan Mountains. In the west it touches on Dsungaria, which, thanks to rain originating in Atlantic cyclones, is Middle Asiatic in character. The Gobi is gradually replaced to the north by the Mongolian *Stipa* steppe with *Aneurolepidium (Agropyron)* and *Artemisia* species. Saline and gypsum soils are common in the desert. The central areas are devoid of vegetation and covered by a stony pavement, and even elsewhere the plant cover is sparse (Fig. 138), with a dry-mass production of scarcely 100–200 kg/ha, as compared with 400–500 kg/ha in the northern steppe areas. On low, brackish ground, *Nitraria sibirica, Lasiagrotis, Kalidium,* etc., are found, and on areas covered by drift sand, the saksaul *Haloxylon ammodendron* is seen. Nowhere in the entire western Gobi does groundwater come to the surface. There are some oases in the east (Fig. 139). The Mongolian Altai mountain range extends into the Gobi from the northwest, continuing as the Gobi Altai (Fig. 140). In the latter, only a steppe altitudinal belt is reached, whereas in the former, a coniferous belt is present on northern slopes, albeit completely Siberian in character, with *Larix*.

14. Subzonobiome of the Cold High-Plateau Deserts of Tibet and Pamir

Lying between the high mountain barrier of the Himalayas to the south and the Kwen-Lun and Altyn-Tag to the north is Tibet, the largest highland mass of the world, with an average height of 4200 to 4800 m above sea level. It extends 2000 km from east to west and is 1200 km wide from north to south and consists of debris-filled basins, encircled in turn by mountain ranges which are 1000 m higher still (Fig. 141). Water from melting snow forms swampy, frost-debris areas with the cyperaceous *Kobresia tibetica,* and there are occasional "salt lakes" and even sand dunes.

The monsoon still exerts an influence in the southern and eastern parts, and in the deeply incised valleys forming the upper reaches of the large southern and eastern Asiatic river systems, southeast Chinese and Himalayan forest elements appear.

The larger western and central area, the Chang Tang Desert, is characterized by the most extreme type of climate. The annual mean temperature is −5 °C, and only July has a positive mean (+8 °C). Daily temperature variations of as much as 37 °C can occur, but the rainfall seldom exceeds 100 mm. The flora, which is poor, is very young, having developed after the Ice Age. It consists of central Asiatic elements (*Ceratoides, Kochia, Reaumuria, Rheum, Ephedra, Tanacetum, Myricaria,* etc.). At the western end of the highland plateau, from which the high mountain ranges originate, is Pamir. At the Pamir Biological Station, 3864 m above sea level, Russian scientists have carried out ecophysiological investigations. The mean annual rainfall at this point, most of it falling between May and August, totals 66 mm. The air is dry, and solar radiation totals 90% of the solar constant, so that the ground warms up to 52 °C in summer, although there are only 10 to 30 nights in the entire year when there is no frost (see Fig. 133, Pamirski Post). There is no closed snow cover. The soils are so dry that they do not freeze.

Dwarf shrubs, 10 to 15 cm high, grow on the desertlike habitats: *Ceratoides papposa, Artemisia rhodantha, Tanacetum pamiricum* or *Stipa glareosa* and the cushionlike *Acantholimon diapensioides.* Along the streams in the valleys, however, there are alpine meadows (Fig. 142).

Growth in the dry habitats is extremely slow. *Ceratoides* flowers only after 25 years but lives for 100 to 300 years. The root systems are strongly developed, and their mass is 10 to 12 times that of the shoot system. Most of the roots are found in the uppermost 40 cm of the soil, that is to say, in the layers which warm up to more than 10 °C in summer. Laterally, the roots extend more than 2 m. The reserves of water in the upper meter of the skeletal soil amount to 26 mm at the most and 5 mm at the least. This is very little, but it nevertheless suffices for the scanty vegetation, even though the transpiration rate is quite high. Photosynthesis is intense only during the morning hours, and the daily production is given as 25 mg/dm^2 leaf area. Respiratory losses are slowed by the low night temperatures.

Three biogeocenes were studied: one desertlike in which *Ceratoides* predominates on talus soil; a second, more steppelike, on clay soil, with

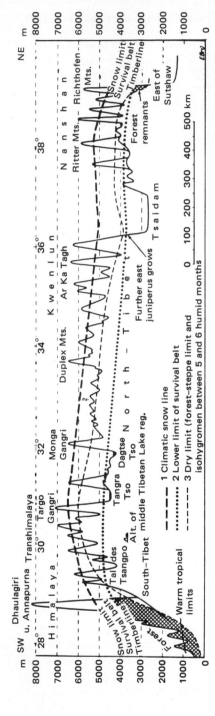

Fig. 141. Climate profile through Tibet from Southwest to Northeast and the slightly lower-lying Tsaidam desert (scale for altitude increased ×75). The temperature-dependent timberline is only theoretical in Tibet and in Tsaidam since it is lower than the forest limit set by dryness. Forest is found only on the southern slopes of the Himalayas and as a narrow altitudinal belt in the Richthofen Mountains (Nan-Shan). (From H. von Wissmann 1961)

Fig. 142. Landscape in east Pamir (4000 m above sea level), with desert steppe on talus soil. Small tussocks are *Stipa glareosa,* larger dwarf shrubs are *Ceratoides papposa.* (Photo I. A. Raikova)

Artemisia and *Stipa glareosa;* and a third with low cushionlike *Acantholimon,* etc., on stony ground with relatively good water conditions. The phytomass and water consumption are as given below.

	Biogeocene		
	Desert	Steppe	Cushion plants
Plant cover (%)	5–18	15–20	15–30
Phytomass (t/ha)	0.14–0.54	0.09–0.48	0.4–0.89
Transpiration (g/g fresh wt · h)	0.3–0.9	0.1–0.7	0.1–0.19
Water consumption during vegetational season (mm)	8–40	6–87	25–446

It is obvious that, despite the relatively high intensity of transpiration, the water requirements of the biogeocene are so small that they can be met by the precipitation. Only herbaceous, cushionlike plants in the vicinity of streams receive additional water from inflow. In general one-half to one-third of the precipitation is used to meet losses due to transpiration. The situation is complicated in the orobiomes by the fact that the sequence of altitudinal belts is highly dependent upon precipitation. In regions with less than 100 mm of precipitation, there is no true snow line, since even at more than 5500 m above sea level, the small quantities of snow evaporate owing to the strong radiation. The desert continues to the upper limits of vegetation, whereas in other regions, alpine steppe would occur at such altitudes or, where precipitation is above 500 mm, even alpine meadows could be expected. Cushionlike plants usually play a leading role in the upper alpine belt. The various altitudinal belt sequences are illustrated schematically in Walter (1974, Fig. 270, p. 321).

Zonoecotone VI/VIII – Boreonemoral Zone

Unlike the zones discussed so far, the Boreal Zonobiome VIII of the Northern Hemisphere, with coniferous-forests and cold-temperate climate, encircles the entire globe through northern Eurasia and North America. To the south, where an oceanic climate prevails, the boreal zone borders on the nemoral deciduous-forest zone, but in the regions with a continental climate, it adjoins arid steppes or semidesert (Fig. 143). There is no sharp boundary between deciduous-forest zone and coniferous-forest zone: instead, a transitional Boreonemoral Ecotone VI/VIII is intercalated between the two. This consists of either mixed stands of coniferous (mainly pine) and deciduous species or a macromosaiclike arrangement, with pure deciduous forest on favorable habitats with good soil and pure coniferous forest on less favorable habitats with poor soils. In eastern North America, different species of *Pinus* represent the conifers in the mixed stands, mainly *Pinus strobus* in the neighborhood of the Great Lakes (although *Tsuga canadensis* is also found, as well as *Juniperus virginiana* in the southeast). Pine trees are often the pioneer woody species following forest fires or occurring on abandoned arable land. Since the pines grow more rapidly than deciduous species on poor soils, they constitute the upper tree stratum, but their regeneration in such mixed stands is problematic if there is a dense deciduous undergrowth. For this reason, pine trees are only successful where fire plays a recurrent role. It has been shown that fires caused by lightning are a common occurrence in such forests, particularly on sandy soils where there is a dry litter layer in summer.

In Europe, the situation is much simpler. On the poor fluvioglacial sands extending as wide belts in front of the end-moraines in central and eastern Europe, pure pine forests can be found (Pinetum) in regions which belong, climatically speaking, to the deciduous-forest zone (psammo-peinobiomes). In eastern Europe, they are called *bor*. On rather better, loamy-sandy soils, there is an additional lower tree stratum of oak, and the forest is then called *subor,* or Querceto-Pinetum. On loamy soils, hornbeam *(Carpinus betulus)* occurs as well, and the forests, now with three strata, are termed *sugrudki* (Carpineto-Querceto-Pinetum). Finally, on loess, there are the zonal deciduous forests known as *grud,* with oak in the upper stratum and hornbeam in the lower stratum (Querceto-Carpinetum). Such forests have been drastically altered by human interference. Forest fires and felling of deciduous trees for fuel have encouraged the growth of pines, whereas the removal of pines for use as valuable building material has resulted in the

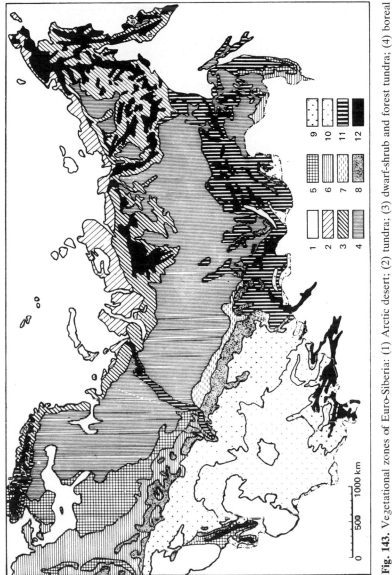

Fig. 143. Vegetational zones of Euro-Siberia: (1) Arctic desert; (2) tundra; (3) dwarf-shrub and forest tundra; (4) boreal coniferous-forest zone; (5) mixed-forest zone; (6) deciduous-forest zone; (7) small-leaved deciduous forest; (8) foreststeppe; (9) grass steppe; (10) semidesert and desert; (11) mountainous coniferous forest; (12) alpine zone. (From Walter 1960)

formation of pure deciduous forests. Yet another disrupting factor is the practice of forest grazing.

In central Europe, extensive pine forests have arisen as a result of forestry activities in what were formerly pure deciduous-forest regions, as for example in the Upper Rhine Valley. Further north (in southern Scandinavia and in central eastern Europe), spruce *(Picea abies)* and oak *(Quercus robur)* are more common. These species form a macromosaic but do not mix with each other (Klötzli 1975). Since the better soils, once the site of oak forests, are now mostly cultivated, the proportion of remaining spruce forests has risen. Furthermore, the spruce is encouraged by the forestry industry. In central Europe, the spruce forests at lower altitudes have all been planted by man, and more and more, spruce is appearing in the landscape because of its economic value. The damage to spruce, as well as to fir, however, has increased noticeably in the past several years as a result of acid rain.

The boundary between the boreonemoral and the true boreal zone coincides, in Europe, with the northern distribution limit of the oak. It runs along a latitude of 60° through southern Sweden, extends along the southern coast of Finland, thence to the middle Kama River, where the steppe borders on the boreal zone.

VIII Zonobiome of the Cold-Temperate Boreal Climate

1. Climate and Coniferous Species of the Boreal Zone

The true boreal zone (Fig. 143) commences at the point where the climate becomes too unfavorable for the hardwood deciduous species, that is to say, when summers become too short and winters too long. This is recognizable in the climate diagram as the point where the duration of the period with a daily average temperature of more than 10 °C drops below 120 days and the cold season lasts longer than 6 months (Fig. 144). The northern boundary between the boreal zone and the arctic tundra is where only approximately 30 days with a daily mean temperature above 10 °C and a cold season of 8 months are typical of the climate.

Nevertheless, in view of the large distances over which this zone extends, it would be incorrect to speak of a uniform climate. Rather, a distinction should be made between a cold oceanic climate with a relatively small temperature amplitude and a cold continental climate in which, in extreme cases, a yearly temperature span of 100 °C can be registered (from a maximum of +30 °C to a minimum of −70 °C). Temperature conditions also change from north to south, with the result that several subzonobiomes can be distinguished: northern, central, southern, and extreme continental. Extreme oceanic subzonobiomes with birch are also found in northwestern Europe and northeastern Asia (Ahti and Jalas 1968).

The floristic composition of the tree stratum also changes across this zone. The coniferous forests in North America and eastern Asia contain a large number of different species, whereas those in the Euro-Siberian region contain very few. Many species of the genera *Pinus, Picea, Abies,* and *Larix,* as well as *Tsuga, Thuja, Chamaecyparis,* and *Juniperus,* are found in North America, although the last four belong in fact to the transitional zone. The specific representatives of these genera on the Pacific Coast are different from those in the east, and only one species, *Picea glauca,* extends from Newfoundland across the Bering Strait. Apart from the species mentioned, *Picea mariana,* otherwise usually found on poor soils, is found at the timberline toward the Arctic, and *Larix laricina* is found in the continental regions. *Abies balsamea, Thuja occidentalis,* and *Pinus banksiana* also occur, *Pinus banksiana* being found on sites previously laid bare by fire. The coniferous belt in the mountain regions contains widely differing species.

In contrast to this, only two species, the spruce *(Picea abies)* and the pine *(Pinus sylvestris),* are of any importance in the boreal zone of Europe. Only in

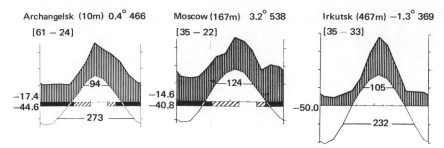

Fig. 144. Climate diagrams from the boreal zone of northern Europe, the mixed-forest zone, and the boreal zone of Siberia. The figures on the upper and lower horizontal lines indicate the number of days with a mean temperature above $+10\,°C$ and above $-10\,°C$, respectively. (From Walter and Lieth 1967)

the eastern regions is the European spruce replaced by the closely related Siberian species, *Picea obovata,* with the addition of other forest species *(Abies sibirica, Larix sibirica,* and *Pinus sibirica,* a subspecies of *Pinus cembra).* The proportion of spruce gradually decreases, until in the continental parts of eastern Siberia, spruce is entirely absent. At the same time, *Larix sibirica* is replaced by *L. dahurica.* Larch forests alone cover 2.5 million km^2 in Siberia. In northern Japan, the number of coniferous species greatly increases again.

2. Biogeocene Complexes of the European Boreal-Forest Zone

Typical of Zonobiome VIII in northern Europe is the dark spruce forest known as taiga, which occurs on podsol soils with a raw humus layer, bleached eluvial horizon, and compact B-horizon. Soils of this kind are formed from every type of parent rock in the humid boreal zone, but the fewer the bases contained in the rock, the better developed the podsol soil. Litter from spruce (förna) does not decompose readily and lies above the A_0-horizon, which consists of an organic mass of interwoven rhizomes and roots of the dwarf shrubs, as well as the mycelia of fungi. This is the raw humus layer, and it can readily be lifted off the underlying A_1-horizon (mineral soil with humus). Humic acids formed in this layer are carried down in rainwater, which completely leaches out the bases and sesquioxides (Fe_2O_3, Al_2O_3) and leaves nothing but fine, bleached quartz sand in the A_2-horizon (bleichhorizont). At the point where the underlying soil is not yet bleached, the humus and sesquioxides are precipitated because of either decreasing acidity or removal of water by the roots of the trees. This is the origin of the B-horizon, which can be either dark brown (humus podsols) or rusty red (iron podsols). Apart from the tree stratum of the spruce forests (Piceetum typicum), a herbaceous stratum and a closed mossy stratum may also be present. Predominating in the herbaceous layer are bilberries *(Vacci-*

nium myrtillus), as well as red bilberries *(Vaccinium vitis-idaea)* in the drier forests, or, very commonly in the southern zone, wood sorrel *(Oxalis aceto-sella).* The following species are also very characteristic: club mosse *(Lycopo-aium annotinum), Maianthemum bifolium, Linnaea borealis, Listera cordata, Pyrola (Moneses) uniflora,* etc. Wherever the groundwater table is high, more raw humus accumulates and peat is formed, and this in turn leads to the formation of raised bogs, where at first *Polytrichum* dominates in the mossy stratum but is later ousted by peat moss *(Sphagnum).* If running, well-oxygenated groundwater is present, the spruce forests are replaced by floodplain forests.

Apart from the spruce forests, the proportion of pine forests (Pineta) in the boreal zone is always very high, with *Pinus sylvestris* displacing spruce in dry habitats. The herbaceous stratum of these thin forests consists of heather *(Calluna vulgaris)* and red bilberries, as well as such other typical species as *Pyrola* spp., *Goodyera repens, Lycopodium complanatum,* etc. Many lichens are found in the mossy stratum *(Cladonia, Cetraria).* After forest fires, which may be caused by lightning, the burned areas are initially colonized by pine, even though the habitat is normally one favoring spruce. On such burned sites, masses of *Molinia coerulea, Calamagrostis epigeios,* or *Pteridium aquilinum* spring up, appearing in the order mentioned, according to increasing dryness of the habitat.

Although birch and aspen are the first tree species to grow on burned sites, they are later ousted by pine, beneath which spruce grows more slowly. In northern Sweden, the birch stage lasts for 150 years and the pine stage for 500 years, but the fact that fire usually recurs before the zonal vegetation has arrived at the spruce stage explains the high proportion of pine that is found. Only in habitats where there is very little danger of fire is pine entirely absent.

Corresponding types of biogeocenes are found in North America; they are, however, floristically somewhat richer in species.

3. The Coniferous Forest as a Biogeocene

Coniferous and deciduous forests differ in that the former is evergreen. The denser the stand, the less sunshine can penetrate to the forest floor. The soil beneath a spruce forest is 2 °C colder than that in the open. The snow covering, too, is thinner in the forest, so that the ground freezes to a greater depth. The frost depth beneath a dense forest stand in which the frost remained until the beginning of August was found to be 85 cm, as compared with 50 cm beneath a thinner stand from which the frost had disappeared by the beginning of June.

The roots of the spruce are very shallow, usually being confined to the upper 20 cm of the soil, or even less if the groundwater table is high. For a high productivity, spruce forests require a continuous supply of water and a medium-depth groundwater table, whereas pine, which roots more deeply, is not so sensitive to a dry soil. The total water lost annually by a typical spruce forest amounts to 250 mm in the northern taiga, 350 mm in the central taiga, and 450 mm in the southern taiga. The mean annual production of organic mass is 5.5 t/ha, and the wood production is 3 t/ha (the latter can reach 5 t/ha in the southern taiga). The largest annual increment is achieved in the north after 60 years, but in the south after only 30–40 years. The phytomass of the tree stratum of pine forests reaches a maximum of 270 t/ha, and that of the undergrowth in old stands is 20 t/ha. The quantity of litter produced by older stands on their way to maturity can exceed 1000 t/ha. This is not accumulated, however, but is continuously decomposed until a state of equilibrium between additions and losses of litter is reached at a litter mass of 50 t/ha. On very wet habitats the organic matter accumulates in the form of peat. Under such unfavorable conditions, the annual increase in dry mass of the tree stratum is often less than that of the other strata. The figures are 850 kg/ha for the tree stratum (total, 1906 kg/ha) in the herbaceous type of spruce-swamp forest and 104 kg/ha (total 1780 kg/ha) for the tree stratum in pine-raised bog.

The LAI is relatively high because of the presence on the trees of the needles of at least 2 years. In pine forests of the boreonemoral zone, the LAI is 9–10, and in the spruce forests of the taiga, it exceeds 11.

Conifers invariably possess ectotrophic mycorrhizae, the fungal hyphae greatly enlarging the range of the root system and rendering the nutrients contained in the raw humus layer more easily available to the trees. Plants in the herbaceous stratum are exposed to severe competition from the tree roots. On shallow granitic soil, all of the available water may be used up by the pine trees so that a herbaceous stratum is completely lacking and the ground is covered with lichens. Even the young pine saplings are unable to mature in the face of such root competition and, in fact, only succeed in growing in places where an old tree has died and competition from its roots is therefore lacking. Where the soil is wetter, the roots of the trees utilize the nitrogen in the soil to such an extent that only dwarf shrubs with extremely low nutrient requirements *(Vaccinium myrtillus)* can grow under the trees. However, if the roots of the trees are severed in order to exclude them from competition, conditions of illumination remaining unchanged, other more demanding species take a hold. Examples are provided by *Oxalis acetosella* or even the nitrophilic raspberry *(Rubus idaeus),* which is otherwise only found in clearings away from the competition of the tree roots. Thus it is more often the quantity of nutrients available to the plants than the amount of light which determines the composition of the herbaceous stratum.

In a spruce forest in Sweden, it has been found that a large part of the rainwater is withheld by the crowns of the trees (interception) (50% as compared to 30% in the thinner east European stands). Moss and litter layers retain a further portion of the water, and in the end, only about one-third of the total rainfall reaches the roots. In the summer months, this was found to amount to 90 mm, for the rest of the year 202 mm, which gives a total of 292 mm, a quantity which is almost completely lost by transpiration in a

40-year-old stand. In wet habitats, as much as 378 mm is lost to the atmosphere by transpiration, which means that a part of the water must be drawn from the groundwater.

Although most ecophysiological investigations have been carried out in the spruce belt of the Alps, the situation is probably analogous to that in the boreal zone. Active transpiration is paralleled by equally intense photosynthesis. The spruce possesses two kinds of needles, sun needles and shade needles. The situation is reminiscent of that in the beech, but with the difference that the active period for the evergreen spruce begins very early in spring and continues into the autumn, until the onset of occasional frost. The seasons of low nocturnal temperatures and small respiratory losses are particularly favorable for the net gain of dry substance. Nevertheless, after a night of frost, photosynthesis is temporarily inhibited, although it is not until the beginning of the cold season proper that the spruce falls into a state of dormancy in which it does not even assimilate on sunny days.

At the same time, respiration sinks to such a low level that it can hardly be measured and accounts for only negligible material losses. The needles lose their fresh green color at this time, and the chloroplasts are difficult to recognize under the microscope.

After a long period of cold, it takes a little time for photosynthesis to regain its normal level in the spring, since the photosynthetic apparatus must be reactivated. Young cembra saplings in the mountains have been shown to spend the winter beneath the snow with green needles and recommence CO_2 assimilation immediately in the higher temperatures of the spring.

The transition to winter dormancy is accompanied by a process of "hardening," or in other words, a great increase in resistance to frost (see p. 206). Conifers of the boreal zone undergo the same processes as decidous trees. Whereas in the nonhardened autumn condition, spruce needles are killed by frost at a temperature of $-7\,°C$, they are capable of enduring a temperature of $-40\,°C$ in the winter without suffering any damage. Young spruce buds are quite vulnerable to light frosts in spring, and may therefore be easily damaged by late frosts.

The resistance of the needles to frost can be artificially changed by the influence of low temperatures in late autumn and spring, and the dehardening process can be affected by the influence of normal room temperatures, especially in December and late winter. Hardening prevents the occurrence of frost damage in coniferous trees in their natural habitats even at temperatures as low as $-60\,°C$, such as occur in Siberia. Thanks to the state of winter dormancy, these trees are in a position to survive the complete darkness of the polar winters. The varying degree of adaptation achieved by the different species is reflected in their distribution. Only a few species can tolerate the extreme continental Siberian winter (the deciduous Siberian larch better than the evergreen species). Variations within a species also occur, depending upon their provenance. Spruce from the Alps behaves differently from members of the same species taken from the northern boreal zone, or again, spruce from the upper tree limit behaves differently from spruce from lower altitudes. The more extreme the conditions, the more pointed the crowns of the trees, showing that the growth of the lateral twigs is more strongly inhibited than that of the main shoot. In polar regions, the same phenomenon is also observable in pine.

Whether or not this shape results from a selection of mutants better able to withstand the weight of snow is unknown, but the same phenomenon has been observed in fir trees at the lower, dry limit in Albania, where snow is not an important factor. The most likely explanation is that whenever the general situation is unfavorable, the growth of the lateral twigs is inhibited before the main shoot suffers (the reverse is true if light conditions are poor). On dry slopes in Utah (North America), *Picea, Abies,* and *Pseudotsuga* have pointed crowns, whereas on the valley floor, where water conditions are better, the crowns of these same species are rounded. The danger due to the weight of snow is the same in both situations.

4. The Extreme Continental Larch Forests of Eastern Siberia with Thermokarst Formations

The shady coniferous forests of western Siberia, with *Picea obovata, Abies sibirica,* and *Pinus sibirica* (ssp. of *P. cembra*) are known as "dark taiga", those of eastern Siberia, with *Larix dahurica,* are known as "light taiga". The latter is a vast subzonobiome with an extreme continental boreal climate (temperature fluctuations of as much as 100°C) and can be seen on the climate diagrams in Fig. 145. In North America, a corresponding climatic region is only found around Fort Yukon in Alaska, and even this is less extreme (Fig. 155).

With annual mean temperatures as low as −10 °C, the ground in eastern Siberia is permanently frozen down to a depth of 250–400 m (permafrost soils). In the relatively warm summers, at least the upper 10–50 cm thaws out, and as much as 100–150 cm thaws in places where drainage is good. Precipitation is very low in this region (less than 250 mm), but the slow thawing of the upper soil layers compensates for this, and there is sufficient meltwater for forests to thrive. These larch forests usually have an

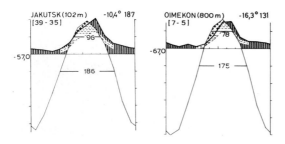

Fig. 145. Climate diagrams for the extreme cold-continental region of eastern Siberia. Oimekon is the cold pole of the Northern Hemisphere. (From Walter and Lieth 1967)

Fig. 146. Forest tundra in the Tscherski Mountain country of eastern Siberia. Larch forest *(Larix dahurica)* on the slopes, moss tundra in the cold valleys, with dwarf birch *(Betula exilis)* and *Rhododendron parviflorum.* (Photo V.N. Pavlov)

undergrowth of dwarf shrubs *(Vaccinium uliginosum, Arctous alpina),* the drier with *Vaccinium vitis-idaea* and *Dryas crenulata,* and the wetter with *Ledum palustre.* In the very dry type of forest, the ground is covered by lichens. Farther north, the thin forests change and are gradually replaced by scattered trees (redkolesye) and then by a dwarf tundra with *Betula exilis* (knee-high) and *Rhododendron parviflorum* (Fig. 146).

Especially impressive are the thermokarst conditions of this coldest part of the Northern Hemisphere. After participation in an excursion in this region, B. Frenzel wrote:

"The permafrost of Siberia (and most probably that of Alaska) has existed since the early Ice Ages. Each Ice Age contributed to its expansion, while each warm interglacial period reduced its territory and thickness. In this type of landscape, however, even the warmer climates are suitable for the new formation of permafrost, although the depth of the soil subject to annual thawing is greater than during colder periods or Ice Ages. The appearance and dissolution of permafrost are related phenomena. Ecologically and geomorphologically, these processes are especially pronounced on fine-grained sedimental formations.

During the Ice Ages, loess (an aeolic sediment) and its derivatives were deposited on large areas of the former climatic zones with extremely cold winters. In the present high continental climates of the boreal coniferous forest zone, these soils are filled up to 80% of their volume with the ice of permafrost. Local perturbations in the radiation balance and heat flux between the atmosphere and soil, such as for example, natural forest fires,

Fig. 147. Destruction of a *Larix dahurica* forest by thermokarst on the Aldan River. The upper 5–8 m of the soil on which the forest grows thaws and the mud, supersaturated with water, transports the fallen trees to the river. (Photo B. Frenzel)

which recur in a given area every 180 to 240 years, or erosion by rivers, etc., result first in an increase in the depth of thawing in summer. Since the soil was previously supersaturated with ice, the thawed layer slides downhill if situated on a slope. In Siberia, this is referred to as the "Jedom" series, meaning that the loose soil is "eaten away". The more intense summer thaw results in a decrease in soil volume. This is generally designated as thermokarst (Fig. 147). On horizontal surfaces, the soil reacts by collapsing in itself so that depressions of several kilometers in length without drainage are formed. In these so-called "alasses" the groundwater level rises, thereby drowning forests located within them (Fig. 148). These phenomena were described by travellers through Siberia as early as in the 17 th century–a clear indication that the formation of alasses is a natural process, which today, may be traced back to the end of the late glacial period (approximately 12,000–10,000 years ago). At present, alass development is being promoted by clear-cutting and building projects. Alasses are especially common in the Vilyuy basin of central and eastern Yakutia, with its highly continental climate, and in the adjacent regions. Although the centers of the alasses are at first usually filled with water, the steep slopes around the edges, which may be up to 50 m high, are better drained and insolated and are covered with a colorful steppe vegetation. Approximately ⅓ of the 900 species of higher plants in Yakutia belong to such plant communities, although their total surface area covers only a few percent of the country's total area.

If the soil of the alasses sinks slowly, and retains only a small amount of water, the dead larch or pine forests are replaced by natural meadow communities, which today are of great importance for cattle raising. The number of species is often much lower in brackish locations.

Fig. 148. An alass forming in the larch taiga of central Yakutia. Thawing of the upper 15–20 m of the permafrost causes the soil to sink to the same depth, and the absence of drainage in the basin causes drowning of the *Larix dahurica* forest. (Photo B. Frenzel)

Fig. 149. A bulgunnyacha in central Yakutia, situated in an alass with natural meadow community. The slopes of the broken, and therefore decaying, pingo are covered with a grass steppe vegetation and the top is still covered with trees. (Photo B. Frenzel)

Alass lakes may eventually again become land. This results in a new heat flux and expansion of the permafrost. Since the alass lakes contain a great amount of water, large mounds with centers of ice, known as "bulgunniachi" or "pingos", develop (Fig. 149). They increase in height until the summer insolation inhibits the existence of the ice-filled centers, or until their growth causes them to break open so that the summer heat is able to penetrate deeply

Fig. 150. Grass and herbaceous steppe on the south-exposed banks of the Aldan river in central Yakutia. The fine-grained Ice Age sediments are filled up to 80% (vol.) with the ice of permafrost, especially in the large polygonal ice wedge systems. When the ice wedges are dissected by the river, the more pure ice melts first. Starting on the higher surfaces, these are then colonized by forest as a result of the improved drainage conditions. The drier mineral soil between the ice wedges is covered with grass steppes on the upper slopes and with herbaceous steppes on the middle and lower slopes (see Fig. 151). (Photo B. Frenzel)

into the bulgunnyachi, thereby leading to their decay. The lifetime of such mounds varies from between a few decades to a maximum of several thousand years. In any case, they always contribute to the variety of biotopes, since their well-drained steep slopes with their increased insolation often offer advantageous opportunities for the settlement of colorful steppe communities. The permafrost region, in which all life appears to be extremely limited as a result of the winter cold, is full of dynamics."

The occurrence of steppe communities in Yakutia is typical on all dry steep southern slopes, which are very warm in summer, as for example on the large rivers (Figs. 150 and 151).

In general, the climate of Yakutia should at least be designated as semi-arid, as is clearly illustrated by the climate diagram in Fig. 145. Even in the euclimatotopes, the potential evaporation is higher than the low annual precipitation.

This also explains the presence of treeless areas, called "tscharany", within the larch forests, in which the high level of evaporation results in the concentration of salts. Halophytes, such as *Atriplex litoralis*, *Spergularia marina* and *Salicornia europaea*, which are also native to the seacoasts, grow on such brakish, solonized soils. On the wet salty soils, the grasses *Puccinellia tenuiflora* and *Hordeum brevisubulatum,* are also found (Walter 1974).

Since steep southern slopes in the far north are struck at noon by the low lying sun at a right angle, some steppe species are also able to grow on

Fig. 151. Herbaceous steppe on the south-exposed slope of a small side valley of the Lena within the dominant *Pinus sylvestris* taiga. The candle-shaped inflorescences of *Orostachys spinulosa,* dense fructescences of *Alyssum* sp. (lower right) and *Ephedra monosperma* (not shown) characterize these stands. This type of steppe flourishes on permafrost. (Photo B. Frenzel)

Wrangel Island at 71° N (Jurtsev 1981), such as the following typical species: *Ephedra monostachya, Stipa krylovii, Koeleria cristata, Festuca* spp. and other grasses, *Pulsatilla* spp., *Potentilla* spp. *Astragalus* and *Oxytropis* spp., *Linum perenne, Veronica incana, Galium verum, Artemisia frigida, Leontopodrum campestre, Aster alpinus* (typical of Siberain steppes) and others. These steppe islands also exist extrazonally on warm southern slopes as relics of the zonal steppes of the glacial periods, when the climate was more extremely continental. At that time, descending foehn-like winds coming off the great ice masses and warming in the process, were deflected toward the east across the ice-free periglacial surfaces and deposited the deep loess sediment.

The summers were apparently so hot that large dry cracks formed in the loess, in which freezing of permafrosts led to the accumulation of ice (Jedome series). Based on recent Russian research, it must be assumed that such periglacial steppes existed zonally across all of Eurasia and North America with an abundant steppe fauna of steppe rodents, antelopes, wild horses, as well as the woolly rhinoceros and the mammoth.

Tundra vegetation probably only grew in boggy or swampy areas around lakes (therefore, as pedobiomes) and was restricted to deeper parts of the relief. The plentiful presence of *Ephedra* and *Artemisia* pollen in the pollen spectra of peat samples from the glacial period indicates that these species grew in the surrounding steppes. In the postglacial period, the entire air circulation was altered when the ice melted; the oceans rose, the land bridge between eastern Asia and Alaska was severed, and the Gulf Stream brought warm water to the new formed polar ice. The climate of the northern latitudes was determined by the westerly currents flowing from the Aleutians on the one side, and from Iceland on the other. It became humid, and the western flanks of the continents were given an oceanic nature. The tundra vegetation expanded its range into the northern region of the former periglacial steppes and colonized the areas which had just become free of ice. The forest vegetation followed, emerging from refuges, until the tundra and forest zones assumed their present situations.

The periglacial steppe vegetation receded into the arid regions of today's continental steppes, along with its typical fauna. Animal species which were not able to keep up with the changes became extinct. This was especially true for the largest animals such as the mammoth, the woolly rhinoceros, the giant elk and others (see Walter and Breckle 1983).

These facts have been presented in order to demonstrate that today's zonal tundra vegetation and the boreal coniferous forest zone with its numerous mires are recently developed phenomena in their present form. Elevated bogs also probably did not exist earlier. Certain relicts of the periglacial steppes are found on calcareous cliffs in Central Russia. *Carex humilis,* with its scattered distribution in the recent steppe heaths of Central Europe, is also to be considered a periglacial relict. Many species occurring in the alpine mats belong genetically to typical steppe genera, such as *Astragalus, Oxytropis, Potentilla, Pulsatilla, Festuca, Avena,* and especially *Artemisia,* the edelweiss (*Leontopodium*) and *Aster alpinus.*

5. Orobiome VIII – Mountain Tundra

There are only few altitudinal belts in these northern latitudes of Zonobiome VIII , and the timberline is soon reached. The latter is formed by *Picea, Pinus sibirica,* or *Larix,* depending upon the geographical situation. However, the timberline is succeeded by mountainous tundra in the alpine belt and not by typical tundra or alpine mats like in the alps (Stanyukovich 1973). In the Alps, the first snow falls on unfrozen ground, and a thick covering of snow keeps the temperature of the soil at about 0 °C throughout the winter. The perennial herbaceous plants are thus exposed neither to severe frost nor to frost desiccation, and the vegetation consists of dense alpine mats.

The situation in the mountainous tundra is different. The ground is already frozen when the first snow falls, and the snow covering is thin and is blown away from the summits. Permafrost, which is not found in the Alps, prevails. Winter storms are violent, and the weathering due to frost action is very intense. The debris gradually moves farther and farther down the slopes (solifluction), and the fine earth is blown away. All of these factors are responsible for the bareness of the summits in the mountainous tundra and for their being termed "golzy" (Russian *goly,* bare). They are covered by lichens and a few mosses, with isolated dwarf shrubs between the rocks. The situation is very reminiscent of the windswept ridges of the Alps, with *Loiseleuria* and the same lichens (p. 211).

Conditions in the subalpine, or "podgolez", belt are rather better because drifting snow can accumulate. Mountainous tundra is encountered in the continental climate region as far as 50 °N to the south and is still found in Altai.

In the oceanic region of the boreal zone (Scandinavia, Kamchatka), there is no mountainous tundra, and the alpine belt is similar to that in the Alps. Snow in winter is plentiful, and the timberline consists of birch (*Betula* spp.).

6. Mires of the Boreal Zone (Peinohelobiomes)

The boreal zone has a humid climate, with rainfall exceeding potential evaporation, which means that the water balance is positive. If, in any way, surplus water is prevented from draining into the rivers, the groundwater table rises, and mires are formed. Since the soils of the boreal zone are poor and acid (podsols), the groundwater is acid in reaction and has a low mineral content. It is usually brown in color owing to humus sols. The situation only differs if the underlying rock is limestone. Large expanses of the boreal zone in Euro-Siberia and North America are very flat, so the groundwater table is high. As long as it remains more than 50 cm below the ground for the larger part of the year, tree growth is possible, otherwise growth of trees is inhibited, and the forests are replaced by mires. Extensive areas of the boreal zone are covered by communities on peaty soils and not by the true zonal vegetation, which is coniferous forest. In large areas of Finland, more than 40% of the total land is covered by mires, in places even 60%. The same holds true for the boreal zone of eastern Europe, and especially western Siberia, which, except in the vicinity of the rivers, is entirely covered by swamps and mires. In Kamchatka, Alaska, and Labrador, as well as the regions to the south of Hudson Bay, the situation is in places similar. For this reason, the pedobiome of the mires has to be dealt with after the coniferous forests. The dividing line between the two is often difficult to establish. In the spruce forests already mentioned, with *Polytrichum* and *Sphagnum,* peat formation is well developed.

In the geological meaning of the word, a mire ("moor" in German) must have a peat layer at least 20–30 cm deep. If the peat layer is thinner or if its content of combustible material falls below 15–30% then, in German, the term "anmoor" is applied. In the ecological sense, mires are plant communities which are dependent upon a high groundwater table but independent of the thickness of the peat layer upon which they grow.

On account of the poor aeration of the soil, the roots of the plants remain near the surface.

Three types of mires can be distinguished according to the origin of their soil water:

1. *Topogenous mires.* These are associated with a very high groundwater table and for this reason occupy the lowest portions of the relief or occur

wherever spring water is available. Many widely differing types of fens belong to this group.

2. *Ombrogenous mires, or raised bogs*. These are higher than their surroundings and are exclusively watered by the rainwater falling onto them.

3. *Soligenous mires*. These are also watered by rain but, because they are not higher than the surrounding country, also receive water draining in from surrounding slopes when the snow melts.

If the groundwater of the topogenous mires or fens contains mineral substances and is rich in nutrients, the mires are termed eutrophic or minerotrophic. Rainwater, on the other hand, is very pure and is poor in nutrient substances, so that the ombrogenous mires are said to be oligotrophic or ombrotrophic. The run-in water received by the soligenous mires, unless it comes entirely from melting snow, contains rather more nutrients. Therefore they may be oligotrophic to minerotrophic.

Since the groundwater of the boreal zone is poor in mineral salts, it is not easy to distinguish between fens and bogs, and it is more usual to speak of mesotrophic transitional bogs. If the water contains less than 1 mg of Ca per liter, the less demanding species typical of the oligotrophic mires are to be found.

Eutrophic mires or fens in which *Carex* spp. play a leading role occur in the temperate zone, regardless of climate, if the ground water contains calcium but is not saline. All of these pedobiomes are helobiomes.

We are concerned here with the low-nutrient, i.e., oligotrophic, mires, which are only found in cool to cold climates. They are all peinohelobiomes, and several distinct types can be recognized, according to their structure and topography (Fig. 152).

1. *Blanket bogs*. These have already been encountered in the extreme oceanic climate of the Atlantic heath region of the British Isles and along the west coast of Scandinavia (p. 190). They cover the entire terrain.

2. *Raised bogs*. These are typical of the rather less oceanic, northwest corner of central Europe with its heath regions, the entire boreonemoral zone, and the southern part of the boreal zone. In their typical form, they are devoid of trees, but as the climate becomes drier and more continental, pine grows out into the bogs to form the so-called forest-raised bogs. The entire southern margin of the boreal zone consists of such bog forests (Fig. 152).

3. *Aapa mires, or string bogs*. These are found north of the raised-bog zone, but above all in Fennoscandia. They are gently sloping soligenous mires with slightly raised ombrotrophic ridges, or strings, running at right angles to the slope. Inbetween the ridges are elongated hollows filled with minerotrophic water (Finnish *rimpis,* Swedish *flarke*). The entire bog presents a descending, terraced aspect reminiscent of terraced rice fields. The ridges are in part the result of the lateral pushing effect of the ice covering the rimpis in winter (see Fig. 154).

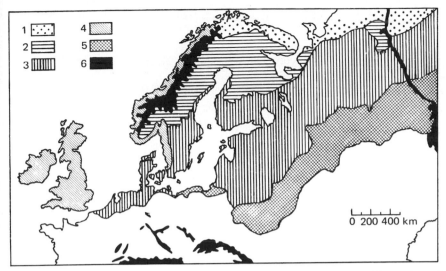

Fig. 152. Distribution of the various types of mires (bogs) in northern Europe: (1) palsa bogs; (2) aapa mires; (3) typical raised bogs; (4) "blanket" bogs; (5) forest-raised bogs; (6) mountain bogs. The white patches in the southern regions have mainly topogenous fens. (Modified from Katz and Eurola; Walter 1964/68)

4. *Palsa bogs of the peat-hummock tundra.* These begin beyond the boreal zone in the forest tundra, in regions where the annual mean temperature is below −1 °C. Ice in the ground itself is partly responsible for the formation of the peat hummocks, which may be as much as 20 to 35 m long, 10 to 15 m wide, and 2−3 m (max. 7 m) high. On slightly elevated ground, the snow cover is thinner, and the frost can penetrate more rapidly into the peaty soil. Layers of ice are formed which attract water from the unfrozen surrounding peat. As the ice thickens, it pushes up the peat, and since not all of the ice melts in summer, at least part of the hummock remains. As a result, the snow covering in the following year is still thinner, and the soil freezes even more rapidly. From year to year, the ice becomes thicker, and the peat hummocks with their core of ice grow higher and higher. In summer, the structure as a whole sinks into the ground somewhat, giving rise to a ditchlike depression filled with water, in which dwarf birch *(Betula nana)* and cotton grasses *(Eriophorum)* grow (Fig. 153).

The tops of the peat hummocks (palsen) dry out in summer, crack, and undergo wind erosion. It can be assumed that the majority of palsen are subfossil structures from a period with a colder climate and are now in a state of disintegration. These may be considered thermokarst phenomena of a smaller scale.

Whereas the raised bogs of Eurasia and North America are confined to the oceanic regions, the aapa and palsa bogs are circumpolar in distribution.

Fig. 153. Palsa, or peat hummock bog, in northern Finland. (Photo E. Walter)

7. The Ecology of Raised Bogs

Peat mosses (*Sphagnum* spp.) play the largest part in the formation of raised bogs. They contain large numbers of dead cells which easily fill up with water by capillary action, so that the cushionlike plants are of a spongy nature and contain many times their own dry weight in water. As the plants grow upward, the lower ends die off and are converted into peat. As the cushions grow larger and larger, they merge with one another, until finally the entire area presents the watch-crystal aspect of a typical raised bog. Because peat mosses cannot tolerate drying-out, they require uniformly damp, cool summers. Since they colonize only poor, acid soils, the podsols are well-suited to their requirements. For this reason, raised bogs usually originate in boreal coniferous forests which have gradually become wetter.

In a large, growing, raised bog it is possible to distinguish a very wet, only slightly convex *high area,* a better drained and relatively steep *marginal slope,* and a surrounding *minerotrophic fen,* termed a "lagg." The high area is not absolutely flat, but consists of small hummocks, the "bults," which extend above the mossy areas, and "schlenken," or small hollows in the mossy carpet, which are filled with water and in which such hygrophilic peat plants as *Carex limosa* or *Scheuchzeria* grow. Several schlenken may unite to form the bog-pools known as "blänken", or "kolke". They are usually 1.5 to 2 m deep and are filled with soft detritus. Surplus water drains off the high areas in small gullys called "rüllen".

Only a few flowering plants are able to survive in the raised bogs, and they have to be species that are undemanding as regards nutrients. Examples are provided by *Eriophorum vaginatum* and *Trichophorum caespitosum,* in addition to the dwarf shrubs *Andromeda polifolia, Vaccinium oxycoccus, V. vitis-idaea, V. uliginosum, Calluna vulgaris,* and *Empetrum.* In Atlantic regions, *Narthecium* is found; in the east, *Ledum palustre* and *Chamaedaphne calyculata,* and in the north, *Rubus chamaemorus, Betula nana,* and *Scheuchzeria palustris.*

The second ecological factor affecting the survival of other plants, apart from the scarcity of nutrient materials, is the danger of their being overgrown by the peat mosses. The viable tips of the latter are the substrate upon which the flowering plants have to germinate. Depending upon the amount of water available, the peat mosses grow 3.5 to 10 cm annually, and if the flowering plants are not to be smothered by the mosses, they are obliged to raise their shoot bases by this amount, either by elongating the rhizome or by putting out adventitious roots. In places where the peat mosses grow more slowly, such as on relatively dry hummocks or well-drained slopes, it is easier for other plants to survive, and here dwarf shrubs are usually seen. Definite zonation can be distinguished on the individual hummocks: *Eriophorum vaginatum* and *Andromeda* at the base, *Vaccinium oxycoccus* further up, and other dwarf shrubs at the top. The top of the hummock is often so dry that *Sphagnum* is replaced by other mosses *(Polytrichum strictum, Entodon schreberi)*, and even by lichens *(Cladonia* spp., *Cetraria)*.

Trees (pine, spruce) confronted with *sphagnum* moss are at a great disadvantage since not only is the base of the tree trunk fixed, but growth is also extremely slow on such a poor substrate. Often not more than the topmost twigs are higher than the hummocks. Bog forest occurs only in places where the growth of the peat mosses is limited by a drier climate. Once a bog is drained, peat moss ceases to grow, and the area is rapidly converted into heath in which dwarf shrubs assume a leading role, soon to be joined by trees such as birch, pine, or spruce. This is the state of the majority of contemporary bogs in central Europe.

Even though the water is just as poor in nutrients as the rest of the bog, species typical of minerotrophic soil often grow along the runnels or on the edges of the pools. It appears that running water or water agitated by waves provides the plants with more nutrients than stagnant water, in which a mere diffusion of nutrients takes place. Because of their high water content, bog soils warm up very slowly and are therefore cold habitats, which explains the presence of northern arctic floristic elements, including relicts of the glacial period. Furthermore, on a raised bog, these relicts do not have to compete with more rapidly growing and demanding species.

With the exception of *Drosera* spp., which supplement their nitrogen supplies considerably by digesting the insects caught on their leaves, all of the other plants are xeromorphic, despite there being water to excess at their disposal. This is ascribed to a lack of nitrogen. It has generally been observed that *xeromorphism occurs whenever the growth of the plant is inhibited by a deficiency,* for example, lack or excess of water (leading to poorly oxygenated soil), low soil temperatures which hinder the uptake of nitrogen, or direct nitrogen deficiency. This xeromorphosis is a symptom of deficiency, and it is therefore more accurate to use the term *peinomorphosis* (Greek *peine*, hunger).

A survey of the mires of northwestern Europe has was published by Overbeck (1975).

8. The Western Siberian Lowlands – the Largest Bog Region of the Earth

This region, comprising the Ob-Irtysh Basin, constitutes a peinohelobiome of almost inconceivable extent. It stretches over 800 km from the forest tundra in the north to the steppes in the south and as much as 1800 km from the Urals in the west to the Yenisey River in the east. Forty percent of the entire peat deposits of the Earth are located in this region, and the bogs, together with more than 100,000 bog lakes, store a volume of water said to be the equivalent of twice the annual drainage of the immense Ob-Irtysh system.

Settlements are found only along the rivers that serve for communication. A closer study of the bog region proper began only a few decades ago (Popov 1971–1975).

Topography, climate, and the hydrological situation can be considered as the factors responsible for the formation of the bogs, which have replaced the "dark taiga".

This vast basin has a foundation of Mesocenozoic layers. The ice ages of the Pleistocene left no marks, except that alluvial and, in places, water-impermeable sediments were deposited, which promoted the retention of water. Peat mosses *(Sphagnum* spp.), which initiate peat formation, readily took hold in the wet and poor-quality podsol soils. With an annual precipitation of 500 mm, the climate is humid; evaporation accounts for only 240–300 mm, and drainage for a further 127–270 mm. As far as temperature is concerned, the climate is very continental, with 174 days free of frost and 100 days on which the daily mean is above 10 °C. The summers are thus relatively warm, and as a result, plant production and peat growth are considerable. There are even brief periods of drought, so that the danger of forest fires cannot be excluded. Burned areas rapidly turn into bog.

The hydrological conditions of the region are of very special importance. The rivers have not cut deep into the ground and are extremely sinuous, so that drainage is poor. The spring high water begins in the upper reaches of the Ob and Irtysh 1.5 months before the snow melts in the lower reaches, i.e., when the rivers are still covered with ice in the north. When the ice begins to break, it forms high walls which dam the water on the upstream side. And since the sources of the Ob are fed by the glaciers of the Altai range, summer high water soon follows. Thus the rivers run high (12 m above low water) for almost the whole of the short Siberian summer, and the low watersheds are flooded and unite with the bog lakes to produce one enormous expanse of water.

The rivers of western Siberia do not, therefore, drain the region. On the contrary, their waters congest the area and promote bog formation. The latter process began as far back as the subarctic period of the postglacial era. The starting point was provided by the wide, shallow depressions containing water poor in mineral salts, in which *Scheuchzeria* bogs could develop, with *Eriophorum vaginatum* and *Sphagnum* species. Mesotrophic *Scheuchzeria* peats corresponding to this period are found at the base of the oldest peat profiles at a depth of 4–7 m. The oligotrophic phase is indicated by the appearance of the most important peat moss species, *Sphagnum fuscum*, beginning in the mid-postglacial. At this time, the bogs upvaulted and the groundwater table rose, and as a result, the adjacent forests became water-logged. *Sphagnum* species took hold beneath the dying trees, and the bogs rapidly spread in all directions. In all younger bog profiles (the majority of bogs are young), the peat is 3–4 m thick, and the lowest horizon invariably includes pinewood and the remains of bark. The oligotrophic phase, with *Sphagnum fuscum* peat, commences immediately above this.

The bogs of western Siberia are mostly string bogs with a mean gradient of 0.0008–0.004. On the strings grow the dwarf form of *Pinus sylvestris (P. willkommii), Ledum palustre*, and the dwarf shrubs *Chamaedaphne calyculata, Andromeda polifolia*, and *Oxycoccus microcarpus*, as well as scattered individuals of *Rubus chamaemorus* and *Drosera rotundifolia*. The moss layer consists of *Sphagnum fuscum*. Patches of lichens *(Cladonia* spp., *Cetraria)* are rare.

Growing in the hollows (wet depressions between the strings or ridges) is *Eriophorum vaginatum* with *Sphagnum balticum* or *Scheuchzeria* (or *Carex limosa)* with *Sphagnum majus*, although *Rhynchospora alba* with *Sphagnum cuspidatum* can also be encountered.

The string bogs tend to a regression on the wet watersheds, which results in the formation of bog lakes, particularly where recent tectonic movements involve subsidence. Aerial geological surveys have revealed a subsidence of about 0.07–0.25 mm per year in some regions. This is sufficient to disrupt the very labile equilibrium between strings and hollows and bring about the accumulation of increasing amounts of water. This surplus of water initiates the phenomena connected with regression. Oxygen deficiency ensues, even in the uppermost peat layers, and methane gas is formed. If drillings are made in such places, fountains of liquid peat are forced out by the escaping methane gas. Naturally escaping gas kills off the vegetation, and pools form on the dead patches. These pools gradually unite to form larger ones, which again grow in size as wave action causes the banks to collapse. Bog lakes of all sizes, together with the wet hollows, form a hydrological system which represents an ecological unit known as a peinohydrobiome (Fig. 154) on account of its low nutrient content [ash content of only 2%–4% (Walter 1977)].

In places where an independent drainage system develops, the string bogs may become dry. Where the bog runnels cut into the peat, the banks are better drained, and a narrow strip of forest consisting of *Pinus sylvestris, Betula,* and *Pinus cembra* ssp. *sibirica* may develop.

The above description of raised bogs applies to the taiga zone. Further north, the thickness of the peat layers decreases due to the shorter vegetational period and smaller production. Different types of bogs are found to the south of the taiga.

In the forest-steppe region, which is Zonoecotone VII/VIII in Siberia, with birch-aspen forests, the calcium content of the groundwater is already high. Slightly domed eutrophic *Hypnaceae* mires with *Carex* (sedge) species

Fig. 154. Endless expanses of the western Siberian bog region. String bogs, with bog lakes on the horizon. Aerial photograph taken from a Zeppelin on August 17, 1929. (Photo by Archives of the Zeppelin Museum, Friedrichshafen)

predominate, and peat growth is inhibited by the greater dryness of the climate. On such mires islands of oligotrophic forest bogs ("ryami") can develop.

In the southern areas, mires are only found in the lowest-lying places, chiefly in the wide river valleys. The ash content of the peat may be very high (19%). "Hummock and hollow" fens are common, the hummocks, "bults", consisting of old clumps of *Carex caespitosa* and *C. omskiana*. Such fens are helobiomes.

Still further south, in the northern steppe zone, the climate is semiarid. Instead of a river system, innumerable small undrained lakelets are present in the Baraba depression, a situation similar to that in the Pampa (p. 239). Some of these lakes become brackish. They are surrounded by eutrophic mires or even halophytic swamps with halophytes and thus represent a transition to a halohelobiome.

Zonoecotone VIII/XI – Forest Tundra

Just as the forest-steppe forms a Zonoecotone VI/VII between forest and steppe proper, a macromosaic arrangement of forest and tundra, Zonoecotone VIII/IX, provides a transition from the boreal forest zone to the treeless tundra. The first sign of a transition is the occurrence of scattered treeless patches, usually on raised ground, within the forest region. These become more frequent toward the north until only scattered islands of forest remain, finally consisting of low, stunted, malformed specimens. Whereas this zone of stunted trees is quite narrow in the mountains, it may extend for hundreds of kilometers on flat land.

The tree most typical of oceanic regions is birch; that of the extreme continental regions, larch; and that elsewhere, spruce.

The polar timberline can be assumed to be determined by factors similar to those governing the timberline in the Alps. Frost desiccation is accentuated by winter storms, and the forest extends farther north on slopes of the river valleys in places with snow and sheltered from wind, where well-drained soils thaw out to a greater depth in summer, or where the rivers flowing from the south carry warmer water. But failure to regenerate is also held to be one of the factors determining the timberline in polar regions. At their northermost limits of distribution, trees rarely produce seeds capable of germination, and the seeds that are produced are often eaten by animals or swept by storms far up to the north over the smooth, snowy surfaces to regions where they can no longer develop. Only few of them germinate in warm niches. The thick blankets of moss and lichens found in forest-tundra provide a very unsuitable substrate for the establishment of tree seedlings.

Man and his accompanying herds of reindeer play an important role in this part of the world, both on account of the damage done by the animals and, more importantly, as a result of wood utilization since the natural growth rate of the trees is extremely slow. As a rule, a tree seedling succeeds in establishing itself only if the temperature has been especially favorable for 2 years in succession. Even so, its further growth is very slow, and after 20 to 25 years, the tree may scarcely be taller than the plants in the herbaceous layer, the annual increase in height amounting to only 1 to 2 cm. The size of annual rings of the stem is closely correlated with the July temperatures.

Open areas in the forest-tundra are usually occupied by dwarf-shrub tundra, which also constitutes the southernmost true tundra (Fig. 143).

During the warm period of the postglacial era, the forests extended considerably farther to the north, as evidenced by tree stumps found in the peat of the present tundra.

Zonobiome of the Arctic Tundra Climate

1. Climate and Vegetation of the Tundra[5]

The largest tundra region completely devoid of forest is an area of 3 million km^2 in northern Siberia. At most, there are 188 days in the year with a mean temperature above 0 °C, and sometimes as few as 55. The low summer temperatures are partially due to the large amount of heat required to melt the snow and thaw out the ground. Winters are rather mild in the oceanic regions but extremely cold in the continental regions (Fig. 155). However, the cold pole still falls within the forest region, near Verchoyansk and Oimekon, although the mean annual temperature at this point is −16.1 °C and the permafrost extends far down into the ground (Fig. 145). The depth to which the ground freezes in winter has no influence upon the vegetation, since the growth of plants depends only on the thickness of the upper soil layer thawing out in summer.

In the southern tundra, the growing season commences in June and lasts until September. The wind is of great importance here because it causes irregular drifting of snow, which, in turn, is responsible for the mosaiclike arrangement of the vegetation. In winter, storms can reach a wind velocity of 15–30 m/sec.

Precipitation is slight, often being less than 200 mm, but since potential evaporation is also very low, the climate is humid. Surplus water is unable to seep into the ground because of the permafrost and thus extensive swamps are formed. The amount of peat formed, is negligible, however, because plant productivity is low. Snowfall amounts to 19–50 cm annually, the raised ground being blown free of snow so that the abrasive action of snow and ice are decisive mechanical factors influencing the vegetation.

The steep, stony southern slopes that warm up relatively well in summer, when the sun is low in the sky, often look like flower gardens. Together with the banks of streams and rivers, these slopes constitute the most favorable habitats in this zone. Flat, raised ground with stone nets (polygonal soil) and gentle slopes subject to solifluction are, on the other hand, sparsely colonized. Interminable stretches of land are covered with dwarf birch and dwarf willows and with *Eriophorum* and *Carex* spp. The drier soils support a pure lichen tundra, whereas mosses predominate on wet ground (*Sphagnum*

---

5 Twenty-three photographs of the vegetation of the Eurasian tundra can be found in Walter (1974)

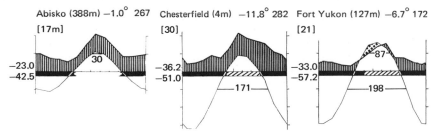

Fig. 155. Climate diagrams from the forest tundra of Sweden (oceanic), the tundra of northern Canada, and the extreme continental boreal region of Alaska. (From Walter and Lieth 1967)

spp. are not found). As far as air temperature is concerned, meteorological measurements carried out at a height of 2 m are no indication of the temperature of the low plant cover. By the time the air temperature has risen to 0 °C, the ground has usually already thawed out to a depth of 0.5 m, and the development of the vegetation is in progress. In fact, the daytime temperature of the plants is often 10 °C higher than that of the air, but in spite of this, the summer is often too short for the seeds to ripen. In an attempt to overcome this problem, half of the plant species in Greenland, for example, produce their flower buds in the preceding season, so that no delay is involved in their coming into flower at the beginning of the following warmer season. Buds and leaves usually spend the winter safely beneath a covering of snow, despite temperatures as low as −30 °C, conditions under which open flowers would die. Reports that *Cochlearia arctica* survives the winter in flower beneath the snow are based on a misunderstanding.

Of particular interest are the aperiodic species, such as the tiny crucifer *Braya humilis*. Their development is protracted over several years and can be broken off temporarily for the winter at any stage. These species, therefore, are unaffected by the shortness of the summer and may flower either at the beginning of the vegetational season or at a later date, the buds having been formed as much as 2 years previously (aperiodic species). Dispersal of seeds and fruits is carried out by the wind skating them over the snow in 84% of the species and by water in 10%; berries are found only in the forest tundra. The size of the seeds reflects the low productivity of the tundra, 75% of the species producing seeds weighing less than 1 mg. The majority of seeds in the tundra require the influence of the low winter temperatures before they are capable of germination, but in spring they are then in a position to germinate rapidly and have time to accumulate some reserves before the autumn. Various grasses, as well as *Polygonum, Stellaria,* and *Cerastium* spp., accounting for 1.5% of the total species, are viviparous (pseudoviviparous). Open spaces, such as occur on the lower Lena, can be colonized rapidly because of the extremely prolific seed production. The majority of tundra species are hemicryptophytes or chamaephytes. The brief period of vegetation at low temperatures is unfavorable for annuals or therophytes (in contrast to conditions in the desert), which are therefore represented only by *Koenigia islandica,* three species of *Gentiana, Montia lamprosperma,* two species of *Pedicularis,* and a few others. Most of the plants develop thick roots which serve as storage organs. Individual plants, including herbaceous specimens, may live more than 100 years, the dwarf shrubs reaching an age of 40 to 200 years.

Nitrogen supply presents a problem of great importance in this region. Leguminosae *(Oxytropics, Hedysarum, Astragalus)* possess root nodules lying immediately beneath the surface, where the soil warms up in summer. If there is no nitrogen in the soil, then only mosses and lichens are found. Animal dung is of great value as a source of nitrogen. It has recently been shown that *Dryas drummondii,* a pioneer species growing in Alaska, has nodules similar to those of *Alnus,* and that the nitrogen content of the soil rises from 33 kg/ha to 400 kg/ha during the pioneer stage of colonization.

In some of the trough valleys in the interior of Peary Land (northern Greenland), at a latitude of 80 °, the climate is completely different from that of the rest of the Arctic. As a result of the descending winds blowing from the interior in summer, there is no rain, and desertlike conditions prevail, with salt efflorescence and an alkaline soil supporting a few halophytes. Apart from these plants, vegetation is not completely lacking because the snow drifting down from the mountains in winter provides a source of water when it melts in the spring. Because the soil thaws out to a depth of about 1 m, the water can seep down and is then available for plants such as *Braya purpurescens,* which can develop taproots more than 1 m in length. Fifty-nine days of the year are frost-free, and the mean July temperature is 6 °C.

2. Ecophysiological Investigations

Arctic plants have a well-balanced water budget, and their cell-sap concentration lies between 7 and 20 atm. The fact that some arctic plants exhibit xeromorphic features can be attributed to an inherited peinomorphosis, caused by nitrogen deficiency, similar to that seen in plants of raised bogs. A low soil temperature renders the uptake of nitrogen more difficult. Photosynthesis and the resulting production of organic matter are of vital importance. The maximum CO_2 assimilation does not exceed 12 mg/ $dm^2 \cdot hr$, and on cloudy days, the CO_2 uptake curve temporarily sinks below zero. But since photosynthesis can usually be carried on around the clock, with a minimum at midnight when illumination is poor, uptake of CO_2 on a summer's day amounts to 100 mg/dm^2/day, which corresponds to 60 mg of starch.

Such quantities suffice for the accumulation of adequate reserves in the summer. The primary production of the plant cover in one year in the subarctic regions of Swedish Lapland near Abisko (growth season, 111 days) amounts to 2500 kg/ha; in Alaska (growth season, 70 days), 830 kg/ ha; and in the extreme Arctic (growth season, 60 days), only 30 kg/ha. In an arctic willow scrub in Greenland, the phytomass was 5.5 t/ha.

The "tundra biome" (Zonobiome IX) is being intensively studied within the framework of the U.S. IBP Tundra Biome Program (Barrow, Alaska), as well as in Canada, Scandinavia and on the Taymyr peninsula in northern Siberia.

3. Animal Life in the Arctic Tundra

The extensive tundra of Siberia provides us with one of the few remaining opportunities of encountering a fauna in its original state and studying its influence upon the vegetation. The majority of larger vertebrates leave the tundra in winter, the birds migrating to the south, and only the lemmings and ground squirrels being left behind. The polar fox and the snowy owl leave the northernmost regions where prey has become scarce.

Lemmings neither hibernate nor lay up stores for the winter. Rather they remain active beneath the hard covering of snow and exist mainly upon the young buds of the Cyperaceae. Although weighing only 50 g, a lemming requires 40–50 kg of fresh plant material annually. As a rule, it colonizes well-drained southern slopes and builds a nest of Cyperaceae shoots in the vicinity of its feeding area, which, for one family, usually covers 100–200 m^2. An entire colony occupies about 1–1.5 ha and destroys 90 to 94% of the entire vegetation. In such places, *Eriophorum angustifolium* does not succeed in flowering. The number of lemmings reaches a maximum, on an average, every 3 years. The dried-out vegetation is not eaten but forms a kind of hay in spring (1–1.2 t/ha), which is then washed together and forms peaty hummocks. When they abandon their winter quarters, the lemmings proceed to build on higher ground, throwing up as much as 250 kg of earth per hectare.

On disrupted habitats such as these, a characteristic plant community is found which is the beginning of a secondary succession. The same situation is produced by the burrowing of ground squirrels, and in this manner, the plant cover is held in a continuously dynamic state. Flocks of water birds, mainly geese, arriving in spring, destroy 50 to 80% of the plant cover by pecking off the young shoots of *Oxytropis* and pulling up the starchy rhizomes of *Eriophorum*. Either solifluction sets in on the bare ground or a thick covering of moss develops.

The nesting and flocking places of these birds are well manured, and support nitrophilous species such as *Rhodiola, Stellaria, Polemonium, Myosotis, Draba, Papaver,* etc.

Although the reindeer only remain in the tundra in winter if large snowfree areas are available for grazing, they too must be considered as belonging to these regions. In summer, the animals graze singly and have little effect on the vegetation, but in the autumn, when they gather together in large herds, their trampling is very obvious. Lichens and dwarf shrubs are destroyed, so that grassy communities consisting of *Deschampsia* and *Poa* gain ground. The number of wild reindeer is decreasing nowadays in favor of the domesticated reindeer. Animals of prey, such as the polar fox, have only a small direct effect upon the plant cover.

See Holding et al. (1974) concerning the decomposers.

4. The Cold Arctic Desert – Solifluction

From south to north, the three following subzonobiomes can be distinguished in the arctic tundra: (1) dwarf shrub tundra in the region covered by forest in postglacial times, (2) true moss and lichen tundra, and (3) cold desert, commencing where plant growth becomes sparse. Oceanic and continental regions can also be distinguished (Aleksandrova 1971). Days on which the temperature rises above and sinks below zero again are very frequent in the cold desert and are responsible for the phenomenon known as solifluction. In the tundra itself, local ice formation leads to an increase in volume of the wet soil and to the formation of peat hummocks below the plant cover (frost hillocks, or "bult" tundra).

Even on slopes with a very slight gradient, the soil is pushed downward stepwise, giving the impression of cattle paths (shallow terraces running parallel to the isohypses); Fig. 156 shows a cross section through one of these steps. Earth movements of this kind become increasingly pronounced toward the north.

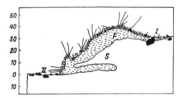

Fig. 156. Streaming movements of the soil on a gentle slope in the Arctic (Alaska). The fibrous peat layer (F) with the living plant cover, has moved about 30 cm, from I to II. A fold has formed in the process, which is partly trapped in the soil (S). (After Hanson 1950)

In places where, in autumn, a wet unfrozen layer is sandwiched between the permafrost layer below and a freezing layer above, the ground may burst through the upper frozen crust in places and cover the vegetation with a layer of liquid clay. A bare patch several centimeters higher than the surroundings is thus formed (Figs. 157 and 158), and this landscape is called "patchy tundra".

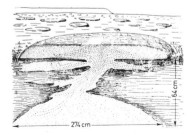

Fig. 157. Patchy tundra with a vertical section through one of the patches. Explanation in text. (According to Govorukhin, from Walter 1974)

Fig. 158. Patchy soil tundra in the Tsherski Mountains of eastern Siberia at 1100 m above sea level. *Betula exilis* and *Rhododendron parviflorum* predominate. Frost patches are clearly recognizable in the foreground. (Photo V. N. Pavlov)

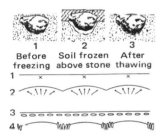

Fig. 159. Schematic representation of the processes involved in freezing and thawing of the ground: (1–2) stones working themselves out of the ground (hatched area is frozen ground); (3) stone after it has reached the surface. Below, stone-network formation: (1) × shows freezing loci; (2) arrows indicating direction in which the stones travel; (3) original position of stones in the ground; (4) final position, after formation of the frost-stone network, or polygonal soil (cross section). (From Walter 1960)

Another effect of frost is that stones gradually work up to the soil surface. Figure 159 illustrates this process. As the upper layer of the soil freezes, it draws up water from below and increases in volume, at the same time carrying the stones of the freezing layer upward. Beneath each stone, a cavity is left, into which fine sand collects, with the result that when the ground thaws, the stone is higher than before. If the process is repeated time and again on the days when the temperature rises above zero and then sinks again, the stone finally reaches the surface. The soil usually freezes from a number of loci separated by distances of one or more meters. The stones are not only lifted out of the soil but are also displaced laterally. The result is a network of stones between the freezing loci, known as polygonal or stone-net soil (Fig. 159). Solitary plants take refuge between the stones where movement is at a minimum, but the roots reach down below the center of the polygon (the plants are pushed to one side). If this takes place on a slope, the stones are also pushed down the gradient, which gives rise to a "stone stream", or striped soils.

Earth movements of this nature mean that the arctic vegetation is constantly disturbed and suffers accordingly, as can be observed in Iceland (Lötschert 1974) and even more clearly on Spitzbergen. Solifluction plays a similar role in mountainous regions, particularly in the upper alpine and nival belts, although the areas involved are not as large as in the Arctic.

As regards the composition of the arctic vegetation, floristic variety around the north pole is relatively small.

5. Antarctic and Subantarctic Islands

Terrestrial conditions in the Antarctic have been dealt with by Holdgate (1970, vol. 2, parts XII-XIII).

Only two flowering plants have been found on the edges of the icecovered Antarctic continent: *Colobanthus crassifolius* (Caryophyllaceae) and a grass, *Deschampsia antarctica*. In recent times, *Poa 'pratensis* has been imported, but otherwise, only mosses, terrestrial algae, and lichens are to be found. They are confined to places on the coast that are sometimes free of snow, steep cliffs, and talus, but their phytomass is negligible. Bacteria and fungi have been found in soil samples.

The sea surrounding the Antarctic, with its continuous westerly storms, is scattered with many tiny islands, most of them south of the 50th parallel. They are bare of trees since the summers are cool, although the winters are not cold. Conditions are almost isothermic on the islands, e.g., the hourly temperature fluctuates between 2.8 ° and 7.7 °C on the Macquerie Islands (at 54° 3′ S) over the entire year. Drizzling rain and fog are typical of the weather here. The Antarctic islands are sometimes referred to as wind deserts, since only in sheltered places is there a somewhat more luxuriant vegetation.

The most common plant on the Kerguelen Islands is the dense, cushion-shaped *Azorella selago* (Umbelliferae). In earlier times, the seafarers were aware of the antiscorbutic action of the large-leaved Kerguelan cabbage, *Pringlea antiscorbutica,* and used it as a fresh vegetable. *Acaena* species are common on all of the islands, and besides many mosses, ferns, and lichens, tussock grassland *(Festuca* and *Poa* spp.) also occurs. As in all very windy habitats, a variety of cushion-plants are characteristic of the sub-antarctic.

Summary

Phytomass and Primary Production of the Various Vegetational Zones and of the Entire Biosphere

The *geobiosphere* is that thin layer at the earth's surface in which living organisms exist and biological cycling takes place. It includes the upper horizons of the soil in which plants root, the atmosphere near the ground, (insofar as organisms penetrate this space), and all the surface waters.

More than 99% of the earth's biomass is phytomass, to which we shall limit our discussion. Amounts of phytomass are distinctly related to vegetational zones.

Because accurate determination of phytomass and primary production is difficult, only gross estimates have been available until recently. However, in 1970, Bazilevich, Rodin and Rozov published (in Russian) more accurate calculations, based on the rapidly accumulating literature, for the various thermal zones and bioclimatic regions of the earth. These authors calculated mean phytomass and mean annual primary production for the various regions as dry mass (in metric tons) per hectare. On the basis of measurements of the areas covered by the individual regions, excluding rivers, lakes, glaciers, and permanent snow, total phytomass and total annual primary production for the various regions were obtained (see following table). The sum of these figures is the phytomass and annual production of the land surface of the earth. In addition, the table gives corresponding data for the waters of the earth. The values involved are potential i.e., they are based on natural vegetation uninfluenced by man.

The authors distinguished five thermal zones: (1) Arctic, (2) Boreal, (3) Temperate, (4) Subtropical, and (5) Tropical. The first two zones have humid climates; the three others are subdivided into (a) humid, (b) semiarid, and (c) arid (cf. the map in Fig. 160). The correspondence between this classification and that of this author's is as follows:

Thermal zones and climatic regions of Bazilevich et al. (1970)	Zonobiome
Zone 1	ZB IX
Zone 2	ZB VIII
Zone 3 (humid, semiarid, and arid)	ZBs VI and VII
Zone 4 (humid, semiarid, and arid)	ZBs III, IV, and V (poleward of $23\frac{1}{2}°$ latitude, the Tropics of Cancer and Capricorn)
Zone 5 (humid, semiarid, and arid)	ZBs I, II, and III (equatorward of $23\frac{1}{2}°$ latitude)

Distribution of Potential Production of the Earth. (After Bazilevich et al. 1970)

Climate zones	Area ($km^2 \times 10^6$)	Phytomass		Primary production	
		Total ($t \times 10^9$)	Average (t/ha)	Total ($t/hr \times 10^9$)	Average (t/ha · yr)
Polar	8.05	13.8	17.1	1.33	1.6
Boreal	23.2	439	189	15.2	6.5
Temperate	7.39	254	342	9.34	12.6
Humid					
Semiarid	8.10	16.8	20.8	6.64	8.2
Arid	7.04	8.24	11.7	1.99	2.8
Subtropical					
Humid	6.24	228	366	15.9	25.5
Semiarid	8.29	81.9	98.7	11.5	13.8
Arid	9.73	13.6	14.9	7.14	7.3
Tropical					
Humid	26.5	1166	440	77.3	29.2
Semiarid	16.0	172	107	22.6	14.1
Arid	12.8	9.01	7.0	2.62	2.0
Geo-biosphere					
Land area	133.0	2400	180	172	12.8
Glaciers	13.9	0	0	0	0
Biohydrosphere					
Lakes and rivers	2.0	0.04	0.2	1.0	5.0
Oceans	361.0	0.17	0.005	60.0	1.7

A comparison of terrestrial production with that in the oceans (see table) shows that the latter, with 60×10^9 t, equals only about one-third that of the land although the surface area of the oceans is almost three times larger. It is remarkable that the phytomass in the oceans is minute in comparison with the 300-times-greater primary production. This is understandable since plankton plants are single-celled and divide and multiply continuously. In contrast, terrestrial primary production is 7% of the corresponding phytomass.

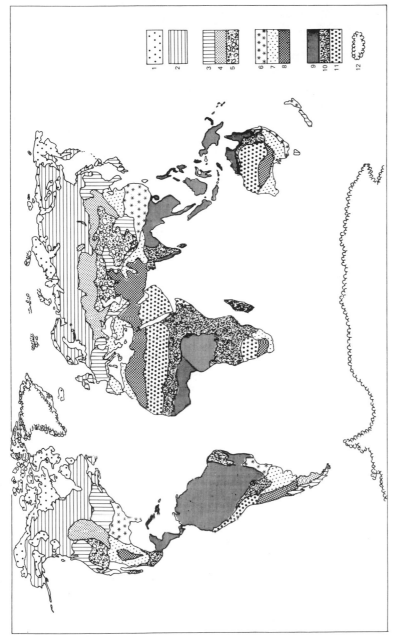

Fig. 160. Thermal zones and bioclimatic regions of Bazilevich et al. (1) Arctic zone. (2) Boreal zone. (3–5) Temperate zone: 4 semiarid region; 5, arid regions. (6–8) Subtropical zone: 6, humid region; 7, semiarid region; 8, arid region. (9–11) Tropical zone. 9, humid region; 10, semiarid region; 11, arid region. (12) glaciers and firn areas (permanent snow). (According to Bazilevic et al. 1970)

For the mass of consumers and decomposers on all the continents, a figure of only 20×10^9 t dry weight is given. This is less than 1% of the phytomass. In the oceans, organisms of these trophic levels amount to 3×10^9 t, or more than 15 times the phytomass. In contrast to the single-celled plants, the consumers in the oceans include large animals which are exploited for human consumption. Various example have already been given to illustrate the comparatively low zoomass of the large terrestrial consumers.

The phytomass on land is mostly wood in the forests. This is 82% of the total phytomass on all the continents, although forests cover only 39% of the land area. The principal part of the forest phytomass, about 50%, is found in tropical forests, about 20% is found in the boreal zone, and about 15% is found in the subtropical and temperate zones.

The phytomass in deserts is only 0.8% of the land total, a very small amount considering that deserts occupy 22% of the land area.

The average phytomass in forests of the humid regions rises continuously with increasingly favorable temperature conditions, from 189 t/ha in the boreal zone to 440 t/ha in the tropics. On the other hand, the average phytomass in the tropical arid regions is, at 7 t/ha, the smallest. Drought combined with continual high temperatures is particularly unfavorable for plant growth.

As for average yearly primary production, on land it is 12.8 t/ha, i.e. 7 times that of the oceans and about 2.5 times that of the lakes and rivers with their aquatic and swamp vegetation.

Approaching the equator, the primary production of the humid land regions increases. It doubles from the boreal to the temperate zones and doubles again from the temperate zones to the subtropics, but increases little from the subtropics to tropics. The differences between the humid and semiarid regions are not so great as those for the respective phytomasses, since the wood in the forest does not contribute to production and it is the leaf area that is important. The relatively high production of 13.8 and 7.3 t/ ha in the subtropical semiarid and arid regions, respectively, is noteworthy. Production depends on the often very luxuriant and productive ephemeral vegetation which can develop during the favorable, cooler rainy part of the year.

The total yearly potential primary production of the biosphere on land, in the oceans, and in lakes and rivers is 233×10^9 t/ha · yr. The land masses contribute 172×10^9 t, the lakes and rivers 1×10^9 t, and the oceans 60×10^9 t.

Lieth and Whittaker (1975) give rather diffferent figures in their latest publication. Instead of calculating the potential production, they worked out the actual production, also considering the areas under cultivation. Their values for terrestrial production are thus lower. The most exact figures available for primary production are given by these authors, who arrive at the figure of 121.7×10^9 t dry mass for an area of 149×10^6 km^2 of land.

At that time, a population of 3 billion human beings, with a biomass of 0.2 $\times 10^9$ t can be assumed to have consumed approximately the total agricultural product, which is put at 0.7% of the primary production of the biosphere. Energy consumption is estimated to be 2.8×10^{18} cal since only a part of the energy taken up as food is utilized. These figures do not seem particularly high, but it can be assumed that, since then, the rapid population explosion has resulted in a significant increase in consumption.

Conclusion from an Ecological Point of View

This book is a summary of the great *natural, ecological relationships* of the entire geobiosphere, the comprehension of which is essential for an accurate prediction of the dangers arising from mankind's increasing interference in natural systems. The effects of this intervention are so numerous and so profound that they cannot be discussed within the framework of this review. Thanks to his mental capacities, mankind has created—next to the natural world—his own apparently independent world—one of a technically oriented world economy. As a result of increasing urbanization, he has become more and more estranged from nature. He is losing the ground beneath his feet, considers everything technically possible and believes in unlimited economic growth.

Based on detailed objective investigations, the Club of Rome warned as early as 1972 of the Utopian nature of this attitude, and predicted an economic crisis unless counter-measures were taken immediately. Nothing was done, however, and in the meantime the crisis has arrived. Again, we are being cautioned not to be overly pessimistic and are told to watch for rosy hopes of economic growth somewhere out on the horizon. Although everyone speaks of "ecology", a basic change in our way of thinking has not taken place. *The economic forces still have priority over all else. The destruction of the environment, on which the existence of mankind is dependent, continues across the entire globe.* Local damage is simply hidden by cosmetic measures, although the problems are on a *global level*. The two greatest dangers to be mentioned here are (1) the population explosion and (2) excessive technological developments. These are discussed in detail by Walter (1984).

1. The Population Explosion in the Developing Countries

Aurelio Peccio, president of the Club of Rome, in the German edition of his publication „Die Zukunft in unserer Hand" ("The Future in our Hand"), again warned emphatically that the earth's population is increasing at such an alarming rate that something must be done about it immediately. According to Peccio, 223 children are born on the earth each minute, or 321,000 each day and 120 million each year. Presently, the population is increasing at a rate of 1 billion every 10 years. By the year 2000, it will be increasing by this same amount every 3 years. If it were possible, in the next 10 years, to keep all

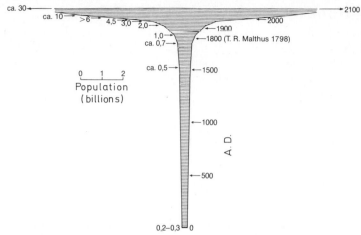

Fig. 161. Growth of the Earth's population from 0 A. D. to the present in the mushroom shape of an atomic bomb explosion. It took millions of years from the first appearance of man on earth for the human population to reach one billion in the middle of the nineteenth century. One century later the population had reached two billion; in 37 more years, three billion; and in only 13 more years, four billion. It has now passed the four and a half billion mark, and should exceed six billion around the turn of the next millennium and ten billion in about 2030 A. D. It is not possible to depict at this scale the population level in 2100 A.D. because by this time the top of mushroom cloud will have doubled in width to over thirty billion

newborn children alive, which is obviously our goal, there would be 1.2 billion children on earth under 10 years of age to be provided for and sent to school. In another 10 years jobs would have to be created for them and they would, in turn, begin to have their own children.

The population explosion may be compared with the mushroom shape of an atomic bomb, in which the size of the population is represented by the width of the curve, as is illustrated above (Fig. 161).

In the year 1 A.D., the earth's population may be estimated to have been approximately 200—300 million, and until around 1800, the stem of the "atomic bomb" increased only insignificantly in width, meaning that the population had only grown slowly. But at about that time, the head of the mushroom began to form. The warning of the approaching danger in 1798 by T. R. Malthus would have been in time, but it was not heeded. The industrial age was just beginning in western Europe and workers were needed. It was especially these countries, with their medical advances, in which the population in the cities rose significantly. Later, as a result of a higher standard of living and the increased desire for a more comfortable life, the number of children per family decreased so rapidly that the population of the industrial countries has presently almost ceased to grow.

The catastrophic situation in all of the developing countries, in which millions of people are undernourished or starving, is due to *the population explosion, which must be met with counter-measures immediately*. Hunger is only a symptom. It is the natural result of such uncontrolled growth, a phenomenon which holds true for all living beings in ecological systems, including mankind. The unlimited exploitation by one species at the expense of all others cannot be upheld.

No foreign aid programs are capable of annulifying this law of nature. Humans, with their earthly bodies, which must be nourished, are a part of nature. Although well-intentioned, food-aid programs are especially dangerous, since they only tend to provoke the population to increase more, as if one were trying to extinguish a fire with oil.

A foreign graduate of the agricultural school in Hohenheim in West Germany, who later became a professor of agricultural sciences in his own country warned in a radio show: "Keep your hands off the developing countries! They must solve their problems on their own. Foreign aid programs only inhibit this." Notably honest in this respect is also the statement of the foreign aid consultant of the World Church Council, Johann Freyer, which is based on his experiences, although it caused outrage among a wide range of uninformed individuals. According to a newspaper article in March last year, he believes that the food delivery programs are causing terrible damage. Since the food is distributed among the poorest, the farming population no longer makes an effort, thereby also becoming welfare recipients. The domestic production then collapses and the number of welfare recipients increases even more in an endless cycle. More recent mottos, such as "Help them to help themselves" consider neither the life styles nor the ways of thinking of the native populations.

Their ways of life have been determined by strict cultural laws over thousands of years. Their cultural levels were optimally adapted to their environments, otherwise they could not have survived over the centuries.

Even colonial rule, which still existed when I began my research in East Africa (discussed in Walter 1984), did not change much in this respect. The power struggles between the individual tribes were stopped, and unsettled land was converted to farmland or plantations on which workers were able to earn money. This began a gradual inclusion in the European economic system. Sudden independence and the attempt to develop unified countries with democratic rule within the borders of the former colonies resulted in chaos and tribal warfare everywhere. The uneducated masses were unable to compensate and to fill the void for which Europeans required several thousand years of development. The Europeans of 2000 years ago would certainly also not have been able to accomplish this successfully. The strict cultural rules restricting sexual relationship were also no longer effective, and the door was opened to an uninhibited reproductive drive, followed by an enormous increase in the birth rate. Private possessions, in the European sense, had been unknown, as was therefore the drive for private initiative. Most foreign aid workers return home deeply disappointed, as my own experiences confirm.

As long as a project is directed with the necessary leadership, the work is accomplished willingly and diligently, even in the missions. As soon as the guidance ceases, nothing else is undertaken, and only the outer façade is kept intact, which however, is insufficient. Those who never worked practically in the developing countries themselves often voice the argument, "But something must be done; we have to help the developing countries". It should not be forgotten, however, that these are sovereign countries, which are very distrustful and consider any fundamental advice as an attempt at neocolonialism. This is especially true for advice regarding the control of the population explosion. As long as this is not achieved, however, any form of help is senseless or even detrimental. Of course, it is true that the development of commercial industries, SOS childrens' villages, care for the blind etc. is of a great help for the people it serves and must be praised, but it does nothing to change the basic catastrophic situation, which continues to grow worse. Even the more or less secondary argument that foreign aid is also helping to open up new markets for western industrial products, for which the developing countries represent an unlimited demand, is not based on reality.

The products can only be delivered on credit, which cannot be repaid and on which interest cannot be collected. Such countries as Brasil and Mexico, although rich in natural resources, exemplify this fact.

In addition, it must be recognized that an economical system is being forced on the developing countries, for which cannot be said today whether it is capable of guaranteeing mankind a lasting existence, or whether it will itself someday pop like a shiny soap bubble. In the past, every civilization which had become estranged from nature, eventually collapsed and was replaced by "barbarians", whose way of life was more in unison with nature.

2. Excessive Technological Developments in the Industrial Countries

Technological advances have made it possible to continually raise the standard of living in the industrial countries. This progress is measured by the gross national product or by the average per capita income. A further desired goal is to shorten the number of working hours as much as possible, and thereby assure the longest possible amount of leisure time so that everyone is able to enjoy a high level of self-determination. This ideal has already been achieved—not in one of the industriliazed countries, but in the smallest island nation on earth—in Nauru in the Pacific Ocean (approximately 2° S and 164° E). There, one should expect to find the happiest people on earth. The average per capita income is above that of the wealthiest industrialized countries. The weekly working hours number zero and the amount of free time is unlimited (report in IWZ, 8–14 January, 1983). Children are born as pensioners. The island is 21.4 km^2 in size and rises to 60 m above sea level. The population numbers 4000 Nauruanians.

A deposit of fossil guano, several feet deep, is located on the island—the purest phosphate deposits known. These were discovered by the German Colonial Office in 1900, which soon began exploitation. After World War I, Great Britain, Australia and New Zealand subsequently continued the exploitation with various degrees of intensity. In 1968, the Nauruanian chief, Hammer du Robert, was able to negotiate the island's independence within the British Commonwealth, and in 1979, the phosphate reserves became the property of the Nauruanians. Since then, Nauruanians no longer have to work. This is done by foreign "guest workers" from Australia, New Zealand, Hong Kong, Taiwan, etc., who are not allowed to obtain Nauruanian citizenship. An annual yield of 2 million tons of phosphate is being sold on the world market. The main pastimes of Nauru's inhabitants is sleeping, eating (corpulency is considered beautiful) and watching television (favorites are Mickey Mouse and wild-west films as well as Australian advertisements). Their obesity makes sports too strenuous. They have the most modern automobiles for driving along the 18-km-long road around the island. Empty beer cans litter the countryside. One hobby is fishing in high-horsepower motor boats. They have provided themselves with the luxury of "Air Nauru" airlines, which runs on a deficit with six jets flown by Australian pilots to Melbourne, Hong Kong, Manila and Samoa, as well as a luxury ship line.

The future is being secured by saving two-thirds of the income in the Nauru Royalties Trust, and securely investing it in real estate, hotels and businesses in foreign countries. The "Nauru house" in Melbourne, with its 50 floors, is the highest commercial building in Australia.

There is a "but" in all this, however. It is estimated that the phophate reserves are only enough to last another 4—6 years. All that will remain is a sterile coral landscape with 10—20 m-high tooth-shaped cliffs. If asked why they do not use their resource more frugally, Nauruanians reply that they are no different from the rest of the world, and they love money just as much as the Europeans and Americans do. They just live from day to day as long as they have enough.

The industrialized countries are behaving similarly. All warnings that the natural resources will soon be depleted have not changed anything. More important are the next elections, and unpleasant decisions are postponed for later, even the increasingly urgent ecological problems. It cannot be denied that most people in the industrialized nations do not know how to utilize their free time; it has to be organized and commercialized at the same time. The holiday industry has become a lucrative business, as is exemplified by the numerous travel agencies and mass holiday quarters in the rapidly growing domestic and foreign holiday resorts with their amusement centers. The holiday guest does not have to worry about anything; he can passively let everything happen for him and only has to pay the appropriate price. In foreign countries, he lives in a ghetto, just as if he were at home, even though the desperation of the developing countries cannot be overseen. What is the profit of this mass tourism besides the consumption of photographic material? Other than that it is only a passive assimilation similar to the flood of information from the mass media. It comes up on the observer so fast that there no chance of it being subject to serious thought. This also holds true for education in the schools, as well as in the universities. The amount of information to be conveyed increases constantly and there is no time left for the critical discussion of problems. Independent thinking is not promoted.

Many believe that the thinking can be left to the computers. Science has been divided into innumerable specialized disciplines and is approaching the danger of becoming a Tower of Babel. The masses have rendered fruitful discussions in small circles impossible. A mass lecture is not much different from a television program. The students passively absorb the information and only begin to study a few weeks before the examinations. This type of knowledge does not last very long.

It has been warned that people are developing an increasingly hostile attitude towards technology. The real problem, however, is an increasingly inhuman technology. Technology, which is supposed to help mankind by making his life more comfortable and pleasant, *has developed its own independent dynamics and is forcing the human masses more and more into its own mold and into dependency.*

It should also not be forgotten that the main purpose of technology has always been the development of weapons. Wars have always given technology its greatest impulses towards further advances. New discoveries were always used immediately for weapons technology. If it had not been for the two World Wars, technology and mass production would not yet have attained their present levels. Although the weapons arsenal is enough to extinguish mankind ten times over, the armament race goes on with no end in sight. Unfortunately, experience has taught that new weapons are always put into use.

The inhumanity of technology is also expressed by the destruction of the environment. While in Central Europe large areas of forest are sacrificed to technology every year, the environmentally conscious corporations in Japan are trying to increase the amount of forest. All steel works of the Nippon Steel Coop., all plants and research institutes of Honda Motors Co. and of Topay Industries, the power plants of the Tokyo Electric Co. and of the Kansai Electric Co. etc. reforest the areas around their industrial complexes to have them serve as air filters and recreational areas. The indiginous tree species have already attained a height of 10 m (Miyawaki 1983). In Europe, however, the buildings are merely cement blocks surrounded by parking lots, or naked lawns at best. The rest of the environment is poisoned. Although the permissible values for the individual poisons may not be exceeded, no one knows whether these levels are effective for the accumulation of a number of different poisonous substances. Prime examples are the dying forests of the recreational areas of Central Europa and the lead-cadmium poisonings of agricultural soils. Although today hazardous substances are concentrated in human milk, breast-feeding has not been forbidden only because the substitutes contain just as much of these dangerous chemicals. People who are now older obtained healthy nutrition at least in their youth, but the generation growing up today is exposed to hazardous substances in even their mothers' milk.

The increasing use of technology in agriculture is an especially serious matter. Formerly, the large, mostly autonomous farms, which ran without external sources of energy, were the only industries existing in a kind of harmonius equilibrium with the environment. These are now being replaced by agricultural factories with huge amounts of organic refuse, which has become difficult to dispose of. These farms have been absorbed into the mainstream of the world economy, and have lost their ability to withstand economic crises. Technology, therefore, is undermining mankind's natural basis of life!

For this reason, ecologists cannot be expected to accept this excessive technology with open arms. They must feel obligated to warn repeatedly of the approaching danger. People can do without very much and they can live on very little if they have to, but they need clean air to breath, clean water to drink, food free of poisons, and the healthy bodily strength provided by nature. To be poor but healthy should be preferred to the alternative of being wealthy and half dead.

The products of technology are not objects of necessity. Instead, they only serve for greater comfort or prestige. World-wide propaganda is used to raise the demand artificially. Everyone should be able to have everything. The main purpose of technology is not to serve human interests; but rather for profit and economic interests (especially those of large corporations). Rationalization in industry (robots, microelectronics) is increasingly forcing human workers out of the production process, and is degrading them to mere consumers of the mass-produced wares. But how can the masses buy industrial products if their earnings are not assured because they have become unemployed? The economic force of competition is brought as an argument —an endless spiral! Technology has made people neither happier nor healthier. Civilization illnesses of physical and psychological nature are constantly on the rise. The increase in the average life span can only be attributed to the greater amount of medicine, the costs of which are rising astronomically.

Looking back over the past eight decades of my own life, it is difficult to find criteria according to which the merits of technology should be judged. Any judgement is certainly subjective. Stress was definitely not a problem in earlier times. Even the ocean crossings for research excursions were a pleasant recovery before and after the work was done and allowed for a gradual adaptation. This does not hold true for modern trips by airplane; even the transition to the new climate, the new environment and new time zone is much too abrupt.

Mass processing would not be possible without technology. It has led to the increasing number of unemployed, who have become a heavy burden on the future. Even the most optimistic economists no longer reckon with economic growth as it was a decade ago. Even the population of Central

Europe is too large. Humaneness and close human contact are becoming more and more rare, being reduced to short and uncomiting telephone calls. Each person has become a number: an identification number, a tax number, a social security number, various customer numbers, etc. A name is only used on the envelope—but for how long? What most people once regarded as a relationship to nature is unkwnon to today's youth in the technologically advanced countries. They have no idea what has been denied them.

It is the *quality* of life and not the material standard of living which is most important; the inner being and not the outward appearance. The quality of life and the standard of living are not necessarily opposites. Experience has proven, however, that the more one values outward appearance, the more one loses of one's inner life, which is therefore, never spoken of.

The quality of life is also expressed outwardly in a healthy and natural way of living with the sensible use of one's vitality, rejection of all addictive substances and the preference for a simple life. He who has a true relationship with nature and who is familiar with nature's enormous variety and greatness, does not only come to know the external environment, which has been discussed in this book, to which we physically belong and which we can investigate by using our intelligence, but he also comes to know the other aspect of mankind—his inner consiousness, which is not subject to logic, and for which philosophers have developed many different complicated designations, and which is generally referred to as the "soul". It is part of the transcendental and absolute and the almightiness of love, therefore of God. These thoughts cannot be expressed in words, nor can they be proven. It is the *free decision* of each individual to acknowledge this and to regard himself as part of the whole. Without this recognition, there is no true freedom for mankind. It makes one independent of the judgement of others, providing an inner security, peace and composure, as well as an inner joy.

This has nothing to do with mortality or eternity. The absolute knows no limits. It is within us and around us. This is the most important conclusion for the youth searching for the purpose of life, the result of a long life dedicated to the research of life on the entire earth. It was a life full of miraculous wonder in a time in which the wonder of miracles is not believed in, and which has lost touch with the center of all things.

It is always necessary to swim against the dirty current before arriving at the pure source ascending from the deep.

This is not the right place to go into these thoughts in detail. They are discussed in „Bekenntnisse eines Ökologen" ("Testimony of an Ecologist"), which was actually written more for previous collaborators, students and friends. Here, another miracle occurred: a scientific publisher published this work in 1980; the resonance was so great that a new revised edition has appeared every year since (Walter 1984, 4th ed.).

References

Abd el Rahman, A.A., El Hadidy, M.N. (1958) Observations on the water output of the desert vegetation along Suez Road. Egypt. J. Bot. **1**, 19–38

Ahti, T., Ahti, L., Jalas, I. (1968) Vegetation zones and their sections in Northwestern Europe. Ann. Bot. Fenn. **5**, 169–211

Ahti, L., Konen, T. (1974) A scheme of vegetation zones for Japan and adjacent regions. Ann. Bot. Fenn. **11**, 59–88

Aleksandrova, V.D. (1971) On the principles of zonal subdivision of arctic vegetation. Bot. Z. **56**, 3–21 (in Russian)

Aleksandrova, V.D. (1977) Geobotanical classification of the arctic and antarctic. Komarovskye Chteniya XXIX, Leningrad, 186 pp. (in Russian)

Anderson, G.D., Herlocker, D. J. (1973), Soil factors affecting the distribution of the vegetation types and their utilisation by wild animals in Ngorongoro crater, Tanzania. J. Ecol. **61**, 627–651

Axelrod, D. L. (1973) History of the Mediterranean ecosystems in California. Ecol. Studies, Vol. 11, pp. 83–112. Springer, Berlin-Heidelberg-New York

Barbour, M.G., Major, J. (1977) Terrestrial vegetation of California, 1002 pp. (with colored vegetation map).Wiley-Interscience Publ., New York

Bazilevic, N.I., Rodin, L.E., Rozov, N.N. (1970) Untersuchungen der biologischen Produktivität in geographischer Sicht. V. Tag. Geogr. Ges. USSR, Leningrad 1970 (in Russian)

Beadle, N.C.W. (1981) The Vegatation of Australia. Vegetationsmonographien der einzeln. Großräume, Bd. IV. Stuttgart, 690 pp

Beard, J.S. (1953) The savanna vegetation of northern tropical America. Ecol. Monogr. **23**, 149–215

Blasco, F. (1977) Outlines of ecology, botany and forestry of the mangals of Indien subcontinent. In: Ecosystems of the World, ed. V.J. Chapman, Vol. I, S. 241–260

Bliss, L.C., Wielgolaski, F. E., eds. (1973) Primary production and production processes, Tundra Biome. 250 pp., Proc. Conf. Dublin 1973. Swedish IBP Comm., Stockholm

Böcher, T.W., Hjerting, J.P., Rahn, K. (1972) Botanical studies in the Atuel Valley, Mendoza Province, Argentina. Dansk Botan. Ark. **22**, 195–358

Bourlière, F., ed. (1983) Tropical savannas. Ecosystems of the World **13**, 730 pp

Brande, A. (1973) Untersuchungen zur postglazialen Vegetationsgeschichte der Neretwa-Niederung (Dalmatien, Herzegowina) Flora **162**, 1–44

Braun-Blanquet, J. (1925) Die *Brachypodium ramosum-Phlomis bychnites* Assoziation der Roterdeböden Südfrankreichs. C. Schrötel-Festschr., Veröff. Geobot. Inst. Rübel (Zürich), H.3

Breckle, S.-W. (1976) Zur Ökologie und zu den Mineralstoffverhältnissen absalzender und nichtabsalzender xerohalophyten. Habil.-Schr. Bonn, 170 S., Cramer

Brünig, E.F. (1973) Species richness and stand diversity in relation to site and succession. Amazonia **4**, 293–320

Bucher, E.F. (1982) Chaco und Caatinga — South American arid savannas, woodlands and thickets, pp. 48–79. In: Huntley, B.J. and Walker, B.H. (eds.), s. diese

Cabrera, A.L., Willink, A. (1973) Biogeografia de America Latina, 120 pp. and 1 map. Washington, D.C.

Cannel, M.G.R. (1982) World forest biomass and primary production dates, 391 pp. Academic Press, London-New York

Castri, F. di (1973) Climatographical comparison between Chile and the western coast of North America. Ecol. Studies, Vol. 7, pp. 21–36. Springer, Berlin-Heidelberg-New York

Castri, F. di, Mooney, H.A., eds. (1973) Mediterranean type ecosystems. Ecol. Studies, Vol. 7. Springer, New York-Heidelberg-Berlin

Cernusca, A. (1976) Bestandsstruktur, Bioklima und Energiehaushalt von alpinen Zwergstrauchbeständen. Oecologia Plantarum **11**, 71–102

Chapman, V.J. (1976) Mangrove Vegetation, 477 pp., Vaduz

Clements, F.E., Weaver, Y.E., Hanson, H.C. (1929) Plant competition. Carnegie Inst. of Washington. Publ. 398

Countinho, L.M. (1964) Untersuchungen über die Lage des Lichtkompensationspunktes einiger Pflanzen zu verschiedenen Tageszeiten mit besonderer Berücksichtigung des "de-Saussure-Effektes" bei Sukkulenten. Beiträge zur Phytologie (Walter-Festschrift) in Arbeiten d. Landw. Hochsch. Hohenheim, Vol. 30, Stuttgart

Countinho, L.M. (1969) Novas observações sôbre a occorrência do "Effeito de de Saussure" e suas relações com a succulência, a temperatura folhear e os movimentos estomaticos. Bol. 331, Fac. Fil., Ciênc. e Letr. da Univ. São Paulo, Botan. **24**, 77–102

Countinho L.M.: Ecological effect of fire in Brasilien Cerrado, pp. 273–291. In: Huntley, B.J. and Walker, B.H. (eds.), s. diese

Cumming, D.H.M.: The influence of large herbivores on savanna structure in Africa pp. 217–245. In: Huntley, B.J. and Walker, B.H. (eds.), s. diese

Dafis, SP., Landolt, E. (1975) Zur Vegetation und Flora Griechenlands. Veröff. Geobot. Inst. Zürich, H 50, 237 S.

Dinger, B.E., Patten, D. T. (1974) Carbon dioxyde exchange and transpiration in species of Echinocereus (Cactaceae) as related to their distribution within the Pinaleno Mountains, Arizona. Oecologia **14**, 389–411

Duvigneaud, P. (1974) La synthèse écologique. Population, communautés, écosystèmes, biosphére, noosphére. Paris, 296 pp

Edmonton, R.L., ed. (1982) Analysis of coniferous forest ecosystems in the Western United States. US/IBP Synthesis Series **14**, 419 pp

Edwards, C.A., Reichle, D.E., Crossley, D.A., Jr. (1970) The role of soil invertebrates in turnover of organic matter and nutrients. Ecol. Studies, Vol. 1, pp. 147–172. Springer, Berlin-Heidelberg-New York

Eiten, G. (1982) Brazilian „savannas", pp. 25–79. In: Huntley, B.J. and Walker B.H. (eds.), s. diese

Ellenberg, H. (1975) Vegetationsstufen in perhumiden bis perariden Bereichen der tropischen Anden. Phytocoenologia **2**, 368–387. Göttingen

Ellenberg, H. (1978) Vegetation Mitteleuropas mit den Alpen, 981 pp. Ulmer, Stuttgart

Ernst, W., Walker, G.H. (1973) Studies on hydrature of trees in miombo woodland in South Central Africa. J. Ecol. **61**, 667–686

Evenari, M., Shanan, L., Tadmor, N. (1971) The Negev: The Challenge of a Desert. 345 pp. Harvard, Cambridge, Mass

Frangi, J. (1975) Sinopsis de la comunidades vegetales y el medio de las sierras de Tandil (Provincia Buenos Aires) Bol. Soc. Argentina de Botan. **16**, 293–319

Freitag, H. (1971a) Die natürliche Vegetation des südostspanischen Trockengebiets. Bot. Jb. (Stuttgart) **91**, 147–308

Freitag, H. (1971b) Die natürliche Vegetation Afghanistans. Vegetatio **22**, 285–344

French, N.R., ed. (1979) Perspectives in grassland ecology. Ecol. Studies, Vol. 32, 204 pp. Springer, Berlin-Heidelberg-New York

Gaussen, H. (1954) Théorie et classification des climats et microclimats. 8 me Congr. Internat Bot. Paris, Sect. 7 et 3, pp. 125–130

Gigon, A. (1974) Ökosysteme. Gleichgewichte und Störungen. In: Leibundgut, H., Hg., Landschaftsschutz und Umweltpflege. Huber, Frauenfeld, S. 16–39

Golley, F.B., Medina, E., eds. (1975) Tropical ecological systems. Ecol. Studies, Vol. 11. Springer, New York-Heidelberg-Berlin

Goryschina, T. K., ed. (1974) Biological production and the factors in an oak-forest of the forest-steppe. Trudy Lesn. Opytn. St. Lenigradsk. Univ., Les na Vorskle **6**, 7–213 (in Russian)

Grebentshchikov, O.S. (1972) Ecologic-geographical conformity in the plant cover of the Balkan. Peninsula. Akad. Nayk Ser. Geogr. Nr. 4, Moskva (in Russian)

Gruhl, H. (1975) Ein Planet wird geplündert. Die Schreckensbilanz unserer Politik. 376 S. S. Fischer, Frankfurt

Haines, B. (1975) Impact of leaf-cutting ants on vegetation development at Barro Colorado Island. Ecol. Studies, Vol. 11, pp. 99–111. Springer, Berlin-Heidelberg-New York

Hallé, F., Oldeman, R.A.A., Tomlinson, P.B. (1978) Tropical trees and forests. An architectural analysis, 441 pp. Springer, Berlin-Heidelberg-New York

Halvorson, W.L., Patten, D.T. (1974) Seasonal water potential changes in Sonoran Desert shrubs in relation to topography. Ecology 55, 173–177

Hamilton III, W.J., Seely, M.K. (1976) Fog basking by the Namib Desert beetle. *Onymacris unguicularis.* Natur 262, 284–285

Hanson, H.C. (1950) Vegetation and soil profiles in some solifluction and mound areas in Alaska. Ecology 31, 606–630

Hennig, I. (1975) Die La Sal Mountains, Utah. Akad. Wiss., Mainz, Math. Naturw. Klasse Nr. 2, Wiesbaden

Holdgate, M.W., ed. (1970) Antarctic Ecology, Vol. 2, 394 pp. New York

Holding, A.J., Heal, O.W., Maclean, S. F., Flanagan, P.W., eds. (1974) Soil organisms and decomposition in tundra. Proc. Microbiol. Meet., Fairbanks 1973. Swedish IBP Comm., Stockholm

Hopkins, B. (1974) Forest and Savanna (West Africa), 2nd ed., 154 pp., Ibadan-London

Horvat, I., Glavač, V., Ellenberg, H. (1974) Vegetation Südosteuropas. 768 S. mit 2 Vegetationskarten. Gustav Fischer, Stuttgart

Hueck, K., Seibert, P. (1972) Vegetationskarte von Südamerika. Veget. Monogr. d. einz. Großräume, Bd. II a. Gustav Fischer, Stuttgart

Hulten, E. (1932) Süd-Kamtschatka. Vegetationsbilder (Jena), 23. Reihe, Heft 1/2

Huntley, B.J., Morris, J.W. (1978) Savanna ccosystem project. Phase I summary and phase II progress. South Africa Nat. Sc. Progr., Rep. No 29, 52

Huntley, B.J., Walker, B.H., eds. (1982) Ecology of tropical savannas. Ecol Studies, Vol. 42, 669 pp. Springer, Berlin-Heidelberg-New York

Hüttel, Cl. (1975) Root distribution and biomass in three Ivory Coast rain forests plots. Ecol. Studies, Vol. 11, pp. 123–130. Springer, Berlin-Heidelberg-New York

Iwaki, H., Monsi, M., Midorikawa, B. (1966) Dry matter production of some herb communities in Japan. The Eleventh Pacific Science Congress, Tokyo, August-September

Johansson, D. (1974) Ecology of vascular epiphytes in West Africa rain forest. Acta Phytogeogr. Suecica 59, 1–129, Uppsala

Kämmer, F. (1974) Klima und Vegetation von Teneriffa besonders im Hinblick auf den Nebelniederschlag. Scripta Geobot. (Göttingen), Nr. 7, 78 S.

Kämmer, F. (1982) Flora und Fauna von Makaronesien, 179 S. Selbstverlag, Freiburg i. B.

Kira, T., Ono, Y., Hosokawa, T., eds. (1978) Biological production in a warm temperate evergreen oak forest of Japan. JIBP Synthesis (Tokyo), Vol. 18

Klötzli, F. (1975) Edellaubwälder im Bereich der südlichen Nadelwälder Schwedens. Ber. Geobot. Inst. Rübel (Zürich) 43, 23–53

Knapp, R. (1973) Die Vegetation von Afrika. Veget. Monogr. d. einz. Großräume, Bd. III, 626 S., Stuttgart

Knodel, T., Kull, U. (1974) Ökologie und Umweltschutz, 162 pp. Metzler, Stuttgart

Kreeb, K. H. (1979) Ökologie und menschliche Umwelt. Geschichte-Bedeutung-Zukunftsaspekte, 204 pp. UTB 808, Gustav Fischer, Stuttgart

Küchler, A.W. (1974) A new vegetation map of Kansas. Ecology 55, 586–604

Kühnelt, W. (1975) Beiträge zur Kenntnis der Nahrungsketten in der Namib (Südwestafrika). Verh. Ges. f. Ökologen, Wien

Kummerow, J. (1981) Structure of roots and root systems, Vol. 11, pp. 269–288. Ecosystems of the World, Amsterdam

Künkel, G., ed. (1976) Biogeography and ecology in the Canary Islands. Dr. W. Junk, The Hague

Lamotte, M. (1975) The structure and function of a tropical savanna ecosystem. Ecol. Studies, Vol. 11, pp. 179–222. Springer, Berlin-Heidelberg-New York

Lamotte, M. (1982) Consumption and decomposition in tropical grassland ecosystem at Lamto, Ivory Coast, pp. 414–429. In: Huntley, B.J. and Walker, B.H. (eds.), s. diese

Lauer, W., Klaus, D. (1975) Geoecological investigation on the timberline of Pico de Orizaba, Mexico. Arctic and Alpine Res. 7, 315–330

Lavagne, A. (1972) La végétation de l'Ile de Port Cros, 30 pp., Marseille

Lavagne, A., Moutte, P. (1974) Bull. Carte Végétation de la Provence et des Alpes du Sud, 129 pp., Marseille

Levina, F. Ja. (1964) The vegetation of the semideserts in the northern Caspian plain, 344 pp., Moskva-Leningrad. (in Russian)

Lewis, J.P. Collantes, M.B. (1975) La vegetacion de la Provincia de Santa Fe. Bol. Soc. Argentina de Bot. 16, 151–179

Lieth, H., Whittaker, R.H., eds. (1975) Primary productivity of the biosphere. Ecol. Studies, Vol. 14. Springer New York-Heidelberg-Berlin

Logan, R. F. (1960) The Central Namib Desert, South West Africa. Publication 758, 162 pp. Nat. Ac. Sc., Washington D.C.

Longman, K.A., Jenik, J. (1974) Tropical forest and its environment (Ghana). 196 pp., Thetford, Norfolk.

Lötschert, W. (1974) Über die Vegetation frostgeformter Böden auf Island. Ber. Forschungsst. Neori As (Island), Nr. 16, 1–15

Mani, M.S., ed. (1974) Ecology and Biography in India, 773 pp. Junck, The Hague

Mann, H.S., ed. (1977) The spectre of desertification. Ann. Arid Zone (Jodphur) 16, 279–394

Mann, H.S., Lahiri, A.N., Pareek, O.P. (1976) A study on the moisture availability and other conditions of unstabilized dunes etc. Ann. Arid Zone (Jodphur) 15, 270–286

Medina, E. (1968) Bodenatmung und Streuproduktion verschiedener tropischer Pflanzengemeinschaften. Ber. Dtsch. Bot. Ger. 81, 159–168

Medina, E. (1974) Dark CO_2-fixation, habitat preference and evolution within the Bromeliaceae. Evolution 28, 677–686

Menault, J.C., Cesar, J. (1982) The structure and dynamics of a West African Savanna, pp. 80–100. In: Huntley, B.J. and Walker, B.H. (eds.), s. diese

Meusel, H., et al. (1971) Beiträge zur Pflanzengeographie des Westhimalajas. Flora 160, 137–194, 370–432, 573–606

Miller, P.C., ed. (1981) Resources by chaparral and matoral. Ecol. Studies, Vol. 39, 455 pp. Springer, Berlin-Heidelberg-New York

Monod, Th. (1954) Modes, contractés "et diffus" de la Végétation saharienne. Biology of deserts, pp. 35–44, London

Montgomery, G.G., Sunquist, M.E. (1975) Impact of sloths on neotropical forest. Energy flow and nutrient cycling. Ecol. Studies, Vol. 11, pp. 69–98. Springer, Berlin-Heidelberg-New York

Möller, C.M., Müller, D., Nielsen (1954) Ein Diagramm der Stoffproduktion im Buchenwald. Ber. Schweiz. Bot. Ges. 64, 487–494

Mooney, H.A., Parsons, D.J. (1973) Structure and function of the Californian chapparral–an example from San Dimas. Ecol. Studies, Vol. 11, pp. 83–112. Springer, Berlin-Heidelberg-New York

Moser, W., Brzoska, W., Zachhuber, K., Larcher, W. (1977) Ergebnisse des IBP-Projekts „Hoher Nebelkogel 3184 m". Sitz. ber. Österr. Akad. d. Wiss., Mathemat.-naturw. Kl., Abt. I, 186, 387–419, Wien

Müller, D., Nielson, J. (1965) Production brute, pertes par respiration et production nette dans le forêt ombrophile tropicale. Forstl. Forsgsv. in Denmark 29, 69–160

Netshayeva, N.T., ed. (1975) Biogeocens of the Eastern Karakun. Akad. Nauk SSSR, Repetek. St., 74 pp. Ashkhabad. (in Russian)

Nobel, P.S. (1976) Water relation and photosynthesis of a desert plant, *Agave deserti*. Plant Phyiol. 58, 576–582

Nobel, P.S. (1977) Water relation of flowering of *Agave deserti*. Bot. Gaz. 138, 1–6

Nobel, P.S. (1977) Water relations and photosynthesis of Barrel Cactus *(Ferocactus acanthoides)* in the Colorado desert. Oecologia (Berl.) **27,** 117–133

Numata, M., Miyawaki, A., Itow. D., (1972) Natural and seminatural vegetation in Japan. Blumea **20,** 435–481

Oberdorfer, E. (1965) Pflanzensoziologische Studien auf Teneriffa und Gomera. Beitr. Naturk. Forsch. SW-Deutschl. **24,** 47–104

Odum, E.P. (1971) Fundamentals of ecology. 3. ed., 574 pp. Saunders Co., Philadelphia-London-Toronto

Ogawa, H., Yoda, K., Kirot. (1961) A preliminary survey of the vegetation of Thailand. Nature life SE Asia **1,** 21–157

Overbeck, F. (1975) Botanisch-Geologische Moorkunde. 719 S. Neumünster

Ozenda, P. (1975) Sur les étages de végétation dans les montagnes du bassin méditerranéen. Doc. Cartogr. Ecol. (Grenoble) **16,** 1–32

Peccio, A. (1981) Die Zukunft in unserer Hand, 244 S. Verlag Fritz Molden (deutsche Ausgabe)

Popov, A.I., ed. (1971–75) Die natürlichen Verhältnisse Westsibiriens. Lief. I–V. Verlag d. Moskauer Univ. (in Russian)

Quintanilla, V. (1974) Les formations végétales du Chili temperé. Doc. Cartogr. Ecol. (Grenoble) **14,** 33–80

Rauh, W. (1973) Über Zonierung und Differenzierung der Vegetation Madagaskars. Akad. Wiss. Mainz, Math.-Naturwiss. Klasse 1. Wiesbaden

Rawitscher, F. (1948) The water economy of the „Campos cerrados" in southern Brazil. J. Ecol. **36,** 237–268

Rodin, L.E., Bazilevich, N.I., Gradusov, B.P., Yarilova, E.A. (1977) Trockensavanne von Rajputan (Wüste Thar). Aridnye pochvy, ikh genesis, geokhimia, ispol'novaniye. S. 195–225, Moskva 1977 (in Russian)

Rutherford, M.C. (1982) Woody plant biomass distribution in *Burkea africana* savannas, pp. 120–141. In: Huntley, B.J. and Walker, B.H. (eds.), s. diese

Ruthsatz, B. (1977) Pflanzengesellschaften und ihre Lebensbedingungen in den Andinen Halbwüsten Nordwest-Argentiniens, 168 pp. Dissertationes Botanicae, Bd. 39. J. Cramer, Vaduz

Sakai, A. (1970) Freezing resistance in willows from different climates. Ecology **51,** 485–491

Sakai, A. (1970a) Mechanism of desiccation damage of conifers wintering in soil frozen areas. Ecology **51,** 657 664

Sakai, A. (1971) Freezing resistance of relicts from the arcto-tertiary flora. New Phytol. **70,** 1199–1205

Sakai, A. (1971a) Freezing resistance of conifers. Silvae Genetica (Frankfurt a. M.) **20,** 53–100

Sakai, A. (1973) Characteristic of winter hardiness in extremely hardy twigs of woody plants. Plant and Cell Physiol. **14,** 1–9

Sakai, A. (1973a) Freezing resistance of trees in North America with reference to tree regions. Ecology **54,** 118–126

San José, J.J., Medina, E. (1975) Effect of fire, organic matter production and water balance in a tropical savanna. Ecol. Studies, Vol. 11, pp. 251–264. Springer, Berlin-Heidelberg-New York

Scholander, P. F. (1968) How mangroves desalinate seawater. Plant Physiol. **21,** 251–261

Scholander, P.F., Van Dam, L., Scholander, S.I. (1955) Gas exchange in the roots of mangroves. Am. J. Bot. **42,** 92–98

Schulze, E.-D. (1970) Der CO_2-Gaswechsel der Buche *(Fagus silvatica* L.) in Abhängigkeit von den Klimafaktoren im Freiland. Flora **159,** 177–232

Schulze, E.-D., Ziegler, H., Stichler, W. (1976) Environmental control of crassulacean metabolism in *Welwitschia mirabilis* Hook Fil. Oecologia **24,** 323–334

Seely, M. K. (1978) Grasland productivity: The desert end of the curve. South Afric. J. Sci. **74,** 295–297

Seely, M.K. (in print) Plant productivity and rainfall under the extreme desert conditions of the Central Namib

Seely, M.K., Hamilton III, W.J. (1976) Fog catchment sand trenches by Tenebrionid beetles, *Lepidochora*, from the Namib Desert. Science **193**, No. 4252

Specht, R.L. (1973) Structure and functional response of ecosystems in the mediterranean climate of Australia. Ecol. Studies, Vol. 7, pp. 113–120. Springer, Berlin-Heidelberg-New York

Stanjukovich, K.V. (1973) The mountains of USSR, 412 pp., Duschanbe (in Russian)

Sunding, P. (1972) The vegetation of Gran Canaria, 186 pp., with 35 tables and 2 vegetation maps. Norske Vid.-Akad. Oslo, I Mat.-Naturv. Klasse Ny Serie No. 29

Tranquillini, W. (1964) Photosynthesis and dry matter production of trees at high altitudes. Formation of Wood (in For. Trees, Acad. Press New York)

Taylor, A.R. (1973) Ecological aspects of lightning in forests. Ann. Proc. Tall Timber Fire Ecol. (Tallahassee) **13**, 455–482

Tinley, K.L. (1982) The influence of soil moisture balance on ecosystem patterns in Southern Africa, pp. 175–192. In: Huntley, B.J. and Walker, B.H. (eds.), s. diese

Troll, C. (1967) Die klimatische und vegetationsgeographische Gliederung des Himalaya-Systems. Ergeb. Forsch.-Untern. Nepal Himalaya, Bd. 1, pp. 353–388

Vareschi, V. (1980) Vegetationsökologie der Tropen. 253 S. Stuttgart

Walter, H. (1939) Grasland, Savanne und Busch der ariden Teile Afrikas in ihrer ökologischen Bedingtheit. Jb. Wiss. Bot. **87**, 750–860

Walter, H. (1960) Standortslehre. Phytologie, Vol. 3, part 1, 2nd ed. Ulmer, Stuttgart

Walter H. (1964/68) Die Vegetation der Erde, Vol. 1, 1964 (3. ed. 1973); Vol. 2, 1968. Gustav Fischer, Jena

Walter, H. (1967) Die physiologischen Voraussetzungen für den Übergang der autotrophen Pflanzen vom Leben im Wasser zum Landleben. Z. f. Pflanzenphysiologie **56**, 170–185

Walter, H. (1973) Ökologische Betrachtungen der Vegetationsverhältnisse im Ebrobecken (Nordost-Spanien). Acta Bot. Acad. Sc. Hungaricae **19**, 193–402

Walter, H. (1974) Die Vegetation Osteuropas, Nord- und Zentralasiens, 452 pp., Gustav Fischer, Stuttgart

Walter, H. (1975a) Über ökologische Beziehungen zwischen Steppenpflanzen und alpinen Elementen. Flora **164**, 339–346

Walter, H. (1975b) Betrachtungen zur Höhenstufenfolge im Mediterrangebiet (insbesondere in Griechenland) in Verbindung mit dem Wettbewerbsfaktor. Veröff. Geobot. Inst. Zürich, **55**, 72–83

Walter, H. (1976) Die ökologischen Systeme der Kontinente (Biogeosphäre). Prinzipien ihrer Gliederung mit Beispielen, 131 S. Gustav Fischer, Stuttgart

Walter, H. (1977) The oligotrophic peatlands of Western Siberia – the largest peino-helobiome in the world. Vegetatio (The Hague) **34**, 167–178

Walter, H. (1979) Allgemeine Geobotanik (UTB 284), 2. Aufl., 260 S. Ulmer, Stuttgart

Walter, H. (1984) Bekenntnisse eines Ökologen. Erlebtes in acht Jahrzehnten und auf Forschungsreisen in allen Erdteilen, 4. Aufl., Stuttgart

Walter, H., Box, E.O. (1983) Overview of Eurasien continental deserts and semideserts, pp. 3–269. In: Ecosystems of the World, Vol V, Amsterdam

Walter, H., Kreeb, K. (1970) Die Hydratation and Hydratur des Protoplasmas der Pflanzen und ihre öko-physiologische Bedeutung. Protoplasmatologia, Vol. II C/6, 306 pp. Springer, Wien-New York

Walter, H., Lieth, H. (1967) Klimadiagramm-Weltatlas. VEB Gustav Fischer, Jena

Walter, H., Medina, E. (1969) Die Bodentemperatur als ausschlaggebender Faktor für die Gliederung der subalpinen und alpinen Stufe in den Anden Venezuelas. Ber. Dtsch. Bot. Ges. **82**, 272–281

Walter, H., Stadelmann, E. (1968) The physiological prerequisites for the transition of autotrophic plants from water to terrestrial life. Bio Science **18**, 694–701

Walter, H., Straka, H. (1970) Arealkunde, Floristisch-Historische Geobotanik, Vol. 3, Part 2, ed. 2, 478 pp. Stuttgart

Walter, H., Volk, O.H. (1954) Grundlagen der Weidewirtschaft in Südwestafrika. Ulmer, Stuttgart

Walter, H., Harnickell, E., Mueller-Dombois, D. (1975) Klimadiagramm-Karten der einzelnen Kontinente und ökologische Klimagliederung der Erde. Gustav Fischer, Stuttgart. (English ed, 1975, "Climate-Diagram Maps of the Individual Continents and the Ecological Climate Regions of the Earth," Springer, New York-Berlin-Heidelberg.)

Weaver, J.E. (1954) North American prairie. Lincoln, 348 pp

Went, F.W., Stark, N. (1968) Mycorrhiza. Bio Science **18,** 1035–1039

Whyte, R.O. (1974) Tropical Grazing Lands. 222 pp. The Hague

Willert, J. von, Eller, B.M., Brinckmann, E., Baasch, R. (1982) CO_2 gas exchange and transpiration of *Welwitschia mirabilis* Hook fil. in the Central Namib Desert, Oecologia (Berl.) **55,** 21–29

Wissmann, v.T. (1960/61) Stufen und Gürtel der Vegetation und des Klimas in Hochasien und seinen Randgebieten. Erdkunde (Bonn) **14,** 249–272

Yurtsev, B.A. (1981) Relikte von Steppenkomplexen in Nordostasien, 168 pp. „Nauka", Novosibirsk (in Russian)

Zinke, P.J. (1973) Analogies between the soil and vegetation types of Italy, Greece and California. Ecol. Studies, Vol. 7, pp. 61–82

Subject Index

Numbers in *italics* denote pages with illustrations